Lecture Notes in Mathematics 1793

Editors:
J.-M. Morel, Cachan
F. Takens, Groningen
B. Teissier, Paris

Springer
Berlin
Heidelberg
New York
Hong Kong
London
Milan
Paris
Tokyo

Jorge Cortés Monforte

Geometric, Control and Numerical Aspects of Nonholonomic Systems

Springer

Author

Jorge Cortés Monforte

Systems, Signals and Control Department
Faculty of Mathematical Sciences
University of Twente
P.O. Box 217
7500 AE Enschede
Netherlands

e-mail: j.cortesmonforte@math.utwente.nl
http://www.math.utwente.nl/ssb/cortesmonforte.htm

Cataloging-in-Publication Data applied for.

Die Deutsche Bibliothek - CIP-Einheitsaufnahme

Cortés Monforte, Jorge:
Geometric, control and numeric aspects of nonholonomic systems / Jorge
Cortés Monforte. - Berlin ; Heidelberg ; New York ; Barcelona ; Hong Kong ;
London ; Milan ; Paris ; Tokyo : Springer, 2002
  (Lecture notes in mathematics ; 1793)
  ISBN 3-540-44154-9

Cover illustration by María Cortés Monforte

Mathematics Subject Classification (2000): 70F25, 70G45, 37J15, 70Q05, 93B05, 93B29

ISSN 0075-8434
ISBN 3-540-44154-9 Springer-Verlag Berlin Heidelberg New York

Springer-Verlag Berlin Heidelberg New York a member of BertelsmannSpringer
Science + Business Media GmbH

http://www.springer.de

© Springer-Verlag Berlin Heidelberg 2002
Printed in Germany

Typesetting: Camera-ready TeX output by the author

SPIN: 10884692   41/3142/ du - 543210 - Printed on acid-free paper

A mi padre, mi madre, Ima y la Kuka

Preface

Nonholonomic systems are a widespread topic in several scientific and commercial domains, including robotics, locomotion and space exploration. This book sheds new light on this interdisciplinary character through the investigation of a variety of aspects coming from different disciplines.

Nonholonomic systems are a special family of the broader class of mechanical systems. Traditionally, the study of mechanical systems has been carried out from two points of view. On the one hand, the area of Classical Mechanics focuses on more theoretically oriented problems such as the role of dynamics, the analysis of symmetry and related subjects (reduction, phases, relative equilibria), integrability, etc. On the other hand, the discipline of Nonlinear Control Theory tries to answer more practically oriented questions such as which points can be reached by the system (accessibility and controllability), how to reach them (motion and trajectory planning), how to find motions that spend the least amount of time or energy (optimal control), how to pursue a desired trajectory (trajectory tracking), how to enforce stable behaviors (point and set stabilization),... Of course, both viewpoints are complementary and mutually interact. For instance, a deeper knowledge of the role of the dynamics can lead to an improvement of the motion capabilities of a given mechanism; or the study of forces and actuators can very well help in the design of less costly devices.

It is the main aim of this book to illustrate the idea that a better understanding of the geometric structures of mechanical systems (specifically to our interests, nonholonomic systems) unveils new and unknown aspects of them, and helps both analysis and design to solve standing problems and identify new challenges. In this way, separate areas of research such as Mechanics, Differential Geometry, Numerical Analysis or Control Theory are brought together in this (intended to be) interdisciplinary study of nonholonomic systems.

Chapter 1 presents an introduction to the book. In Chapter 2 we review the necessary background material from Differential Geometry, with a special emphasis on Lie groups, principal connections, Riemannian geometry and symplectic geometry. Chapter 3 gives a brief account of variational principles in Mechanics, paying special attention to the derivation of the nonholonomic equations of motion through the Lagrange-d'Alembert principle. It also presents various geometric intrinsic formulations of the equations as well as several examples of nonholonomic systems.

The following three chapters focus on the geometric aspects of nonholonomic systems. Chapter 4 presents the geometric theory of the reduction and reconstruction of nonholonomic systems with symmetry. At this point, we pay special attention to the so-called nonholonomic bracket, which plays a parallel role to that of the Poisson bracket for Hamiltonian systems. The results stated in this chapter are the building block for the discussion in Chapter 5, where the integrability issue is examined for the class of nonholonomic Chaplygin systems. Chapter 6 deals with nonholonomic systems whose constraints may vary from point to point. This turns out in the coexistence of two types of dynamics, the (already known) continuous one, plus a (new) discrete dynamics. The domain of actuation and the behavior of the latter one are carefully analyzed.

Based on recent developments on the geometric integration of Lagrangian and Hamiltonian systems, Chapter 7 deals with the numerical study of nonholonomic systems. We introduce a whole new family of numerical integrators called nonholonomic integrators. Their geometric properties are thoroughly explored and their performance is shown on several examples. Finally, Chapter 8 is devoted to the control of nonholonomic systems. After exposing concepts such as configuration accessibility, configuration controllability and kinematic controllability, we present known and new results on these and other topics such as series expansion and dissipation.

I am most grateful to many people from whom I have learnt not only Geometric Mechanics, but also perseverance and commitment with quality research. I am honored by having had them as my fellow travelers in the development of the research contained in this book. Among all of them, I particularly would like to thank Manuel de León, Frans Cantrijn, Jim Ostrowski, Francesco Bullo, Alberto Ibort, Andrew Lewis and David Martín for many fruitful and amusing conversations. I am also indebted to my family for their encouragement and continued faith in me. Finally, and most of all, I would like to thank Sonia Martínez for the combination of enriching discussions, support and care which have been the ground on which to build this work.

Enschede, July 2002   *Jorge Cortés Monforte*

Table of Contents

Preface VII

1 Introduction 1
   1.1 Literature review 3
   1.2 Contents 5

2 Basic geometric tools 13
   2.1 Manifolds and tensor calculus 13
   2.2 Generalized distributions and codistributions............. 17
   2.3 Lie groups and group actions 18
   2.4 Principal connections 23
   2.5 Riemannian geometry 24
      2.5.1 Metric connections 26
   2.6 Symplectic manifolds 29
   2.7 Symplectic and Hamiltonian actions 30
   2.8 Almost-Poisson manifolds 32
      2.8.1 Almost-Poisson reduction 33
   2.9 The geometry of the tangent bundle 34

3 Nonholonomic systems 39
   3.1 Variational principles in Mechanics 39
      3.1.1 Hamilton's principle 39
      3.1.2 Symplectic formulation.......................... 42
   3.2 Introducing constraints................................ 43
      3.2.1 The rolling disk 45
      3.2.2 A homogeneous ball on a rotating table 47
      3.2.3 The Snakeboard................................ 49
      3.2.4 A variation of Benenti's example 50
   3.3 The Lagrange-d'Alembert principle 51

3.4   Geometric formalizations 55
      3.4.1   Symplectic approach 55
      3.4.2   Affine connection approach 58

4   **Symmetries of nonholonomic systems** 63
    4.1   Nonholonomic systems with symmetry 63
    4.2   The purely kinematic case 67
          4.2.1   Reduction 68
          4.2.2   Reconstruction 77
    4.3   The case of horizontal symmetries 80
          4.3.1   Reduction 80
          4.3.2   Reconstruction 81
    4.4   The general case 84
          4.4.1   Reduction 87
    4.5   A special subcase: kinematic plus horizontal 98
          4.5.1   The nonholonomic free particle modified 100

5   **Chaplygin systems** 103
    5.1   Generalized Chaplygin systems 103
          5.1.1   Reduction in the affine connection formalism 104
          5.1.2   Reconstruction 107
    5.2   Two motivating examples 107
          5.2.1   Mobile robot with fixed orientation 107
          5.2.2   Two-wheeled planar mobile robot 109
    5.3   Relation between both approaches 112
    5.4   Invariant measure 114
          5.4.1   Koiller's question 114
          5.4.2   A counter example 119

6   **A class of hybrid nonholonomic systems** 121
    6.1   Mechanical systems subject to constraints of variable rank ... 121
    6.2   Impulsive forces 123
    6.3   Generalized constraints 126
          6.3.1   Momentum jumps 129
          6.3.2   The holonomic case 134
    6.4   Examples 134
          6.4.1   The rolling sphere 135
          6.4.2   Particle with constraint 138

7   Nonholonomic integrators 141

7.1   Symplectic integration 142

7.2   Variational integrators 143

7.3   Discrete Lagrange-d'Alembert principle 145

7.4   Construction of integrators 148

7.5   Geometric invariance properties 153

    7.5.1   The symplectic form 154

    7.5.2   The momentum 154

    7.5.3   Chaplygin systems.............................. 156

7.6   Numerical examples 166

    7.6.1   Nonholonomic particle 166

    7.6.2   Mobile robot with fixed orientation with a potential .. 168

8   Control of mechanical systems 171

8.1   Simple mechanical control systems...................... 171

    8.1.1   Homogeneity and Lie algebraic structure 173

    8.1.2   Controllability notions 174

8.2   Existing results 175

    8.2.1   On controllability 176

    8.2.2   Series expansions............................... 176

8.3   The one-input case 178

8.4   Systems underactuated by one control................... 179

8.5   Examples.. 190

    8.5.1   The planar rigid body 190

    8.5.2   A simple example 191

8.6   Mechanical systems with isotropic damping 193

    8.6.1   Local accessibility and controllability 194

    8.6.2   Kinematic controllability 198

    8.6.3   Series expansion................................ 199

    8.6.4   Systems underactuated by one control 202

References... 203

Index.. 217

List of Figures

3.1 Illustration of a variation c_s and an infinitesimal variation X of a curve c with endpoints q_0 and q_1. 40

3.2 The rolling disk 46

3.3 A ball on a rotating table 47

3.4 The Snakeboard model. Figure courtesy of Jim Ostrowski. 49

3.5 A prototype robotic Snakeboard. Figure courtesy of Jim Ostrowski. 50

3.6 A variation of Benenti's system. 51

4.1 Plate with a knife edge on an inclined plane 78

4.2 Illustration of the result in Theorem 4.3.2 83

4.3 G-equivariance of the nonholonomic momentum mapping. 89

5.1 A mobile robot with fixed orientation 108

5.2 A two-wheeled planar mobile robot 109

6.1 Possible trajectories in Example 6.3.1 128

6.2 The rolling sphere on a 'special' surface 135

7.1 Energy behavior of integrators for the nonholonomic particle with a quadratic potential. Note the long-time stable behavior of the nonholonomic integrator, as opposed to classical methods such as Runge Kutta. 167

7.2 Illustration of the extent to which the tested algorithms respect the constraint. The Runge Kutta technique does not take into account the special nature of nonholonomic systems which explains its bad behavior in this regard. 168

7.3 Energy behavior of integrators for a mobile robot with fixed orientation. 170

7.4   Illustration of the extent to which the tested algorithms respect the constraints $\omega_1 = 0$ and $\omega_2 = 0$. The behavior of the nonholonomic integrator and the Benchmark algorithm are indistinguishable. 170

8.1   Table of Lie brackets between the drift vector field Z and the input vector field Y^{lift}. The (i,j)th position contains Lie brackets with i copies of Y^{lift} and j copies of Z. The corresponding homogeneous degree is $j - i$. All Lie brackets to the right of $\mathcal{P}_{-1}$ exactly vanish. All Lie brackets to the left of $\mathcal{P}_{-1}$ vanish when evaluated at $v_q = 0_q$. Figure courtesy of Francesco Bullo. 173

8.2   Illustration of the proof of Theorem 8.4.2. $R^{(p-1)}$ denotes $(a^{(p-1)}_{s_{p-1}s_p})^2 - a^{(p-1)}_{s_{p-1}s_{p-1}}a^{(p-1)}_{s_p s_p}$. The dashed lines mean that one cannot fall repeatedly in cases $A3$ or B without contradicting STLCC. 187

8.3   The planar rigid body. 190

8.4   The level surface $\phi(x, y, z) = 0$. 192

List of Tables

6.1 Possible cases. The rank of $\mathcal{D}$ is denoted by ρ. 127

6.2 The two cases that may arise in studying the jump of momenta. 130

1 Introduction

NONHOLONOMIC systems are present in a great variety of environments: ranging from Engineering to Robotics, wheeled vehicle and satellite dynamics, manipulation devices and locomotion systems. But, what is a nonholonomic system? First of all, it is a mechanical system. But among these, some are nonholonomic and others are not. Which is the distinction? What makes a mechanical system nonholonomic is the presence of *nonholonomic constraints*. A constraint is a condition imposed on the possible motions of a system. For instance, when a penny is rolling without slipping over the floor, it is satisfying the condition that the linear velocity of the point of contact with the surface is zero, otherwise the penny would slip. Another example is given by a robotic manipulator with various links: we can think of each link as a rigid body that can move arbitrarily as long as it maintains the contact with the other links imposed by the joints. Holonomic constraints are those which can be expressed in terms of configuration variables only. This is the case of the robotic manipulator mentioned above. Nonholonomic constraints are those which necessarily involve the velocities of the system, i.e. it is not possible to express them in terms of configuration variables only. This is the case of the rolling penny.

Numerous typical problems from Mechanics and Control Theory appear in a natural way while investigating the behavior of this class of systems. One of the important questions concerns the role played by the dynamics of the system: in some nonholonomic problems, as we shall see, dynamics is crucial – these are the so-called *dynamic systems*, as, for instance, the Snakeboard [29, 148] or the rattleback [36, 77, 250]; in others, however, it is the kinematics of the system which plays the key role – the *kinematic systems* [117]. Another interesting issue concerns the presence of symmetry, in connection with the reduction of the number of degrees of freedom of the problem, the reconstruction problem of the dynamics and the role of geometric and dynamic phases, which are long studied subjects in the Mechanics literature (see [159]). Other topics include the study of (relative) equilibria and stability, the notion of complete integrability of nonholonomic systems, etc. On the control side, relevant problems arising when studying nonholonomic systems are, among others, the development of motion and trajectory planning strategies, the design of point and trajectory stabilization algorithms, the accessibility and controllability analysis,...

This wealth of questions associated with nonholonomic systems explains the fact that, along history, nonholonomic mechanics has been the meeting point for many scientists coming from different disciplines. The origin of the study of nonholonomic systems is nicely explained in the introduction of the book by Neimark and Fufaev [188],

> "The birth of the theory of dynamics of nonholonomic systems oc-
> curred at the time when the universal and brilliant analytical for-
> malism created by Euler and Lagrange was found, to general amaze-
> ment, to be inapplicable to the very simple mechanical problems of
> rigid bodies rolling without slipping on a plane. Lindelöf's error, de-
> tected by Chaplygin, became famous and rolling systems attracted
> the attention of many eminent scientists of the time..."

The stage of what we might call "classical" development of the subject can be placed between the end of the 19th century and the beginning of the 20th century. At this point, the development of the analytical mechanics of nonholonomic systems was intimately linked with the problems encountered in the study of mechanical systems with holonomic constraints and the developments in the theory of differential equations and tensor calculus. It was the time of the contributions by Appell, Chaplygin, Chetaev, Delassus, Hamel, Hertz, Hölder, Levi-Civita, Maggi, Routh, Vierkandt, Voronec, etc.

The work by Vershik and Faddeev [244] marked the introduction of Differential Geometry in the study of nonholonomic mechanics. Since then, many authors have studied these systems from a geometric perspective. The emphasis on geometry is motivated by the aim of understanding the structure of the equations of motion of the system in a way that helps both analysis and design. This is not restricted to nonholonomic mechanics, but forms part of a wider body of research called *Geometric Mechanics*, which deals with the geometrical treatment of Classical Mechanics and has ramifications into Field Theory, Continuum and Structural Mechanics, Partial Differential Equations, etc. Geometric Mechanics is a fertile area of research with fruitful interactions with other disciplines such as Nonlinear Control Theory (starting with the introduction of differential-geometric and topological methods in control in the 1970s by Agračhev, Brockett, Gamkrelidze, Hermann, Hermes, Jurdjevic, Krener, Lobry, Sussmann and others; see the books [105, 189, 211, 224]) or Numerical Analysis (with the development of the so-called *geometric integration*; see the recent books [160, 210]). Many ideas and developments from Geometric Mechanics have been employed in connection with other disciplines to tackle practical problems in several application areas. Examples are ubiquitous and we only mention a few here: for instance, the use of the affine connection formalism and the symmetric product in the design of motion planning algorithms for point to point reconfiguration and point stabilization [42], and in the development of decoupled trajectory planning algorithms for robotic manipulators [45]; the use of the theory of reduction (principal connections, geometric phases, relative equilibria) and series expansions on Lie groups to study motion control and stability issues in underwater vehicle exploration (see [139, 140] and references therein), and optimal gaits in dynamic robotic locomotion [75]; the use of the technique of the augmented potential in the analysis and design of oscillatory controls for micromechanical systems [13, 14]; the interaction with dynamical systems theory in computing homoclinic and heteroclinic orbits for the NASA's Genesis Mission to collect solar wind samples [121]; the use of Dirac structures, Casimir functions and passivity techniques in robotic and industrial applications [228]; and more.

The present book aims to be part of the effort to better understand non-holonomic systems from the point of view of Geometric Mechanics. Our interest is in the identification and analysis of the geometric objects that govern the motion of the problem. Exciting modern developments include the nonholonomic momentum equation, that plays a key role in explaining the generation of momentum, even though the external forces of constraint do no work on the system and the energy remains constant; geometric phases that account for displacements in position and orientation through periodic motions or *gaits*; the use of the nonholonomic affine connection in the modeling of several control problems with applications to controllability analysis, series expansions, motion planning and optimal control; the stabilization of unstable relative equilibria evolving on semidirect products; and much more.

1.1 Literature review

IN the following, we provide the reader with a brief review of the literature on nonholonomic systems. There are many works on the subject, so the exposition here should not be taken as exhaustive. Complementary discussions can be found in [25, 59, 103, 188].

There are many classical examples of nonholonomic systems that have been studied (see the books [188, 207]). Routh [208] showed that a uniform sphere rolling on a surface of revolution is an integrable system in the classical sense; Vierkandt [249] treated the rolling disk and showed that the solutions of the reduced equations are all periodic; Chaplygin [62, 63] studied the case of a rolling sphere on a horizontal plane, allowing for the possibility of an nonuniform mass distribution. Another classical example which has attracted much interest (due to its preferred direction of rotation and the multiple reversals it can execute) is the wobblestone or rattleback [36, 77, 250]. Other examples include the plate on an inclined plane and the two-wheeled carriage [188], the nonholonomic free particle [207], etc.

In the modern literature, there are several approaches to the dynamics of nonholonomic systems. Many of them originated in the course of the study of symmetries and the theory of reduction. Koiller [120] describes the reduction of the dynamics of Chaplygin systems on a general manifold. He also considers the case when the configuration manifold is itself a Lie group, studying the so-called Euler-Poincaré-Suslov equations [125]. The Hamiltonian formalism is exploited by Bates, Śniatycki and co-workers [17, 18, 80, 220] to develop a reduction procedure in which one obtains a reduced system with the same structure as the original one. Lagrangian reduction methods following the exposition in [164, 165] are employed in [29]. In this latter work, the nonholonomic momentum map is introduced and its evolution is described in terms of the *nonholonomic momentum equation*. Both approaches, the Hamiltonian and the one via Lagrangian reduction, are compared in [122] (see also [222]). The geometry of the tangent bundle is employed in [50, 57, 137] to obtain the dynamics of the systems through the use of projection mappings. Several authors have investigated what has been called almost-Poisson brackets ("almost" because they fail to satisfy the Jacobi identity) in connection with stability issues [53, 123, 156, 241]. Interestingly, it has been shown in [241] that the almost Poisson bracket is integrable if and only if the constraints are holonomic. Nonholonomic mechanical systems with symmetry are also treated in [24, 239] within the framework of Dirac structures and implicit Hamiltonian systems. Stability aspects adapting the energy-momentum method for unconstrained systems [155] are studied in [261] (see also [214]).

The language of affine connections has also been explored within the context of nonholonomic mechanical systems. Synge [235] originally obtained the *nonholonomic affine connection*, whose geodesics are precisely the solutions of the Lagrange-d'Alembert equations. His work was further developed in [243, 244] and, recently, it has been successfully applied to the modeling of nonholonomic control systems [27, 47, 143, 144]. This has enabled the incorporation of nonholonomic dynamics into several lines of research within the framework of affine connection control systems, such as controllability analysis, series expansions, motion planning, kinematic reductions and optimal control.

Other relevant contributions to nonholonomic mechanics include [52, 88, 126, 135, 175, 180, 213] on various approaches to the geometric formulation of time-dependent nonholonomic systems; [174] on the geometrical meaning of Chetaev's conditions; [157, 202] on the validity of these conditions and various alternative constructions; [81, 254] on the Hamiltonian formulation of nonholonomic systems; [127] on systems subject to higher-order nonholonomic constraints; [97] on the existence of general connections associated with nonholonomic problems and [22, 33, 118, 173, 214, 262] on the stabilization of equilibrium points and relative equilibria of nonholonomic systems.

Another line of research has been the comparison between nonholonomic mechanics and vakonomic mechanics. The latter was proposed by Kozlov [10, 124] and consists of imposing the constraints on the admissible variations before extremizing the action functional. This variational nature has been intensively explored from the mathematical point of view [55, 96, 129, 171, 245]. It is known that both dynamics coincide when the constraints are holonomic, a result slightly extended by Lewis and Murray [145] to integrable affine constraints. Cortés, de León, Martín de Diego and Martínez [70, 168] developed an algorithm to compare the solutions of both dynamics, recovering the result of Lewis and Murray, and others of Bloch and Crouch [26], and Favretti [84]. Lewis and Murray [145] also performed an experiment with a ball moving on a rotating table and concluded that it is nonholonomic mechanics that leads to the correct equations of motion. Other authors have reached the same conclusion through different routes [260]. Nevertheless, it should be mentioned that the vakonomic model has interesting applications to constrained optimization problems in Economic Growth Theory and Engineering problems, see for example [71, 154, 172, 212].

In the control and robotics community, the study of driftless systems is a major subject of interest. These control systems are of the form $\dot{x} = \sum_i u_i g_i(x)$, i.e no drift is present. The control vector fields g_i generate a distribution $\mathcal{D}$ and then the velocity state $\dot{x}$ necessarily verifies $\dot{x} \in \mathcal{D}$. These problems are often called nonholonomic systems, though second-order dynamics do not appear into the picture. As shown for instance in [117], when studying the control problem of motion generation by internal shape changes, kinematic nonholonomic systems can be interpreted as driftless systems. Some intensively studied issues regarding driftless systems include the design of stabilizing laws [37], either discontinuous [11, 28, 54] or time-varying [67, 177, 182, 203], the search for conditions to transform the equations into various normal forms [200, 236], and the development of oscillatory controls for trajectory planning and constructive controllability [38, 152, 187, 234].

1.2 Contents

TO assist the reader, this section presents a detailed description of the mathematical context in which the various aspects of nonholonomic systems dealt with in this book have been developed. We put a special emphasis on the interrelation of nonholonomic mechanics with applications such as undulatory locomotion, mobile robots, hybrid control systems or numerical methods.

Nonholonomic reduction and reconstruction of the dynamics Nonholonomic systems with symmetry have been a field of intensive research in the last years [18, 29, 50, 51, 68, 120, 156, 241]. In Geometric Mechanics, this study is part of a well-established (and still growing) body of research known as the theory of reduction of systems with symmetry, which started in the 1970s with the seminal works by Smale [218, 219], Marsden and Weinstein [166] and Meyer [178], and since then has been devoted to the study of the role of symmetries in the dynamics of mechanical systems (see [163, 190]). An important objective driving the progress in this area has been the identification of relevant geometric structures in the description of the behavior of the systems. This has led to nice geometric formulations of the reduction and reconstruction of the dynamics, which unveil crucial notions such as geometric and dynamic phases, relative equilibria, the energy-momentum technique in the stability analysis, etc.

These developments have had a considerable impact on applications to robotic locomotion [75, 117, 195, 197] and control of mechanical systems [47, 76, 196], especially to undulatory locomotion. *Undulatory robotic locomotion* is the process of generating net displacements of a robotic mechanism via periodic internal mechanism deformations that are coupled to continuous constraints between the mechanism and its environment. Actuable wheels, tracks, or legs are not necessary. In general, undulatory locomotion is "snake-like" or "worm-like," and includes the study of hyper-redundant robotic systems [66]. However, there are examples, such as the Snakeboard, which do not have biological counterparts. The modeling of the locomotion process by means of principal connections has led to a more complete understanding of the behavior of these systems in a variety of contexts. Issues such as controllability, choice of gait or motion planning strategies are considerably simplified when addressed using the language of phases, holonomy groups and relative equilibria directions.

In Chapter 4, we develop a geometric formulation of the reduction and reconstruction of the dynamics for nonholonomic systems with symmetry. We start by introducing a classification of systems with symmetry, depending on the relative position of the symmetry directions with respect to the constraints. We treat first the purely kinematic or principal case, in which none of the symmetries are compatible with the constraints. We obtain that the reduction gives rise to an unconstrained system, with an external nonconservative force that is in fact of gyroscopic type. These results are instrumental in the following chapter, where we specialize our discussion to Chaplygin systems. We also discuss the reconstruction procedure and prove that the total phase in this case is uniquely geometric, i.e. there is no dynamic phase. Then, we deal with the horizontal case, which is the only case in which the reduction procedure respects the category of systems under consideration. The reconstruction of the dynamics is also explored, showing the parallelisms with the unconstrained case [159].

Finally, we discuss the reduction in the general case. The momentum equation is derived within our geometric setting, and this is the starting point to develop a full discussion of the almost-Poisson reduction. Special attention is paid to the almost-Poisson bracket. As a particular case of these results, we establish the appropriate relation in the horizontal case between the original almost Poisson bracket and the reduced one. The chapter ends with a detailed study of a special case where the reduction can be decomposed in a two-step procedure, a horizontal and a kinematic one.

Integrability of Chaplygin systems An important topic which is receiving growing attention in the literature concerns the identification and characterization of a suitable notion of complete integrability of nonholonomic systems (see e.g. [10, 16, 27, 83, 107, 125, 248]). As is well known, an (unconstrained) Hamiltonian system on a $2n$-dimensional phase space is called completely integrable if it admits n independent integrals of motion in involution. It then follows from the Arnold-Liouville theorem that, when assuming compactness of the common level sets of these first integrals, the motion in the $2n$-dimensional phase space is quasi-periodic and consists of a winding on n-dimensional invariant tori (see e.g. [10], Chapter 4). For the integrability of a nonholonomic system with k constraints one needs, in general, $2n - k - 1$ independent first integrals. It turns out, however, that for a nonholonomic system which admits an invariant measure, "only" $2n - k - 2$ first integrals are needed in order to reduce its integration to quadratures, and in such a case – again assuming compactness of the common level sets of the first integrals – the phase space trajectories of the system live on 2-dimensional invariant tori [10]. Several authors have studied the problem of the existence of invariant measures for some special classes of nonholonomic systems. For instance, Veselov and Veselova [248] have studied nonholonomic geodesic flows on Lie groups with a left-invariant metric and a right-invariant nonholonomic distribution (the so-called LR systems). Kozlov [125] has treated the analogous problem for left-invariant constraints. Their results have been very useful for finding new examples of completely integrable nonholonomic dynamical systems [83, 107, 248].

In Chapter 5, we focus our attention on generalized Chaplygin systems. Systems of this type are present in Mechanics [188], robotic locomotion [117] and motions of micro-organisms at low Reynolds number [216]. The special feature about Chaplygin systems is that, after reduction, they give rise to an unconstrained system subject to an external force of gyroscopic type. We present a coordinate-free proof of this fact, together with a characterization of the case where the external force vanishes. In his pioneering paper on the reduction of nonholonomic systems with symmetry, Koiller has made a conjecture concerning the existence of an invariant measure for the reduced dynamics of generalized Chaplygin systems (see [120], Section 9). Based on several known examples of such systems which do admit an invariant measure, Koiller suggests that this property may perhaps hold in general. One of the main results of Chapter 5 is the derivation of a necessary and sufficient condition for the existence of an invariant measure for the reduced dynamics of a generalized Chaplygin system whose Lagrangian is of pure kinetic energy type. This condition then enables us to disprove Koiller's conjecture by means of a simple counter example.

Dynamics of nonholonomic systems with generalized constraints
Chapter 6 deals with nonholonomic systems subject to generalized constraints, that is, linear constraints that may vary from point to point. One could think of simple examples that exhibit this kind of behavior. For instance, imagine a rolling ball on a surface which is rough on some parts but smooth on the rest. On the rough parts, the ball will roll without slipping and, hence, nonholonomic linear constraints will be present. However, when the sphere reaches a smooth part, these constraints will disappear. Geometrically, we model this situation through the notion of a generalized differentiable codistribution, in which the dimension of the subspaces may vary depending on the point under consideration. This type of systems is receiving increasing attention in Engineering and Robotics within the context of the so-called *hybrid mechanical control systems* [46, 91, 92], and more generally, *hybrid systems* [39, 242]. Within this context, the engineering objective is to analyze and design systems that accomplish various tasks thanks to their hybrid nature. This motivation leads to problems in which discontinuities, locomotion and stability interact. Examples include hopping and (biped and multi-legged) walking robots, robots that progress by swinging arms, and devices that switch between clamped, sliding and rolling regimes. A nice work in this direction, which also contains many useful references, is provided by [150].

This study fits in with the traditional interest in systems subject to impulsive forces from Theoretical Physics and Applied Mathematics (see [39] for an excellent overview on the subject and the December 2001 special issue of *Philosophical Transactions: Mathematical, Physical and Engineering Sciences* on "Non-smooth mechanics"). Starting with the classical treatment, the Newtonian and Poisson approaches [6, 108, 188, 198, 207], the subject has continued to attract attention in the literature and has been approached by a rich variety of (analytical, numerical and experimental) methods, see for instance [116, 227, 229, 230]. Recently, the study of such systems has been put into the context of Geometric Mechanics [100, 101, 102, 130].

In Chapter 6 we establish a classification of the points in the configuration space in regular and singular points. At the regular points, the dynamics is described by the geometric formalism discussed in Chapter 3. The singular points precisely correspond to the points where the discrete dynamics drives the system. For these points, we define two subspaces related to the constraint codistribution, whose relative position determines the possibility of a jump in the system's momentum. We derive an explicit formula to compute the "post-impact" momentum in terms of the "pre-impact" momentum and the constraints. Applications to switched and hybrid dynamical systems are treated in several examples to illustrate the theory.

Nonholonomic integrators In the last years there has been a huge interest in the development of numerical methods that preserve relevant geometric structures of Lagrangian and Hamiltonian systems (see [160, 210] and references therein). Several reasons explain this effervescence. Among them, we should mention the fact that standard methods often introduce spurious effects such as nonexistent chaos or incorrect dissipation. This is especially dramatic in long time integrations, which are common in several areas of application such as molecular dynamics, particle accelerators, multibody systems and solar system simulations. In addition, in the presence of symmetry, the system may exhibit, via Noether's theorem, additional conserved quantities we would like to preserve. Again, standard methods do not take this into account[1].

Mechanical integrators are algorithms that preserve some of the invariants of the mechanical system, such as energy, momentum or the symplectic form. It is known (see [86]) that if the energy and the momentum map include all integrals belonging to a certain class, then one cannot create constant time step integrators that are simultaneously symplectic, energy preserving and momentum preserving, unless they integrate the equations *exactly* up to a time reparameterization. (Recently, it has been shown that the construction of energy-symplectic-momentum integrators is indeed possible if one allows varying time steps [109], see also [167]). This justifies the focus on mechanical integrators that are either symplectic-momentum or energy-momentum preserving (although other types may also be considered, such as volume preserving integrators, methods respecting Lie symmetries, integrators preserving contact structures, methods preserving reversing symmetries, etc[2]).

---

[1] A quote from R.W. Hamming [98] taken from [60] gives an additional explanation of a more philosophical nature:

> "...an algorithm which transforms properly with respect to a class of transformations is more basic than one that does not. In a sense the invariant algorithm attacks the problem and not the particular representation used..."

In fact, many people have employed this kind of integrators, such as the implicit Euler rule, the mid-point rule or leap-frog method, some Newmark algorithms in nonlinear structural mechanics, etc., although they were often unaware of their geometric properties.

[2] A list with different types of integrators may be found in the web page of the Geometric Integration Interest Group (http://www.focm.net/gi/). We thank Miguel Angel López for this remark.

Based on certain applications, such as molecular dynamics simulation or multibody systems, the necessity of treating holonomic constraints in discrete mechanics has also been discussed in the literature. Examples include the Shake algorithm [209] and the Rattle algorithm [5] (see [134] for a discussion of the symplectic character of these methods), general Hamiltonian systems (i.e. not necessarily mechanical) subject to holonomic constraints [106, 205, 206], the use of Dirac's theory of constraints to find unconstrained formulations in which the constraints appear as invariants [133], energy-momentum integrators [89, 90], etc.

Variational integrators are symplectic-momentum mechanical integrators derived from a discretization of Hamilton's principle [12, 183, 246, 247]. Different discrete Lagrangians result in different variational integrators, including the Verlet algorithm and the family of Newmark algorithms (with $\gamma = 1/2$) used in structural mechanics [110, 217]. Variational integrators handle constraints in a simple and efficient manner by using Lagrange multipliers [256]. It is worth mentioning that, when treated variationally, holonomic constraints do not affect the symplectic or conservative nature of the algorithms, while other techniques can run into trouble in this regard [133].

In Chapter 7, we address the problem of constructing integrators for nonholonomic systems. This problem has been stated in a number of recent papers [59, 256], including the presentation of open problems in symplectic integration given in [176]. Our starting point to develop integrators in the presence of nonholonomic constraints is the introduction of a discrete version of the Lagrange-d'Alembert principle. This follows the idea that, by respecting the geometric structure of nonholonomic systems, one can create integrators capturing the essential features of this kind of systems. Indeed, we prove that the *nonholonomic integrators* derived from this discrete principle enjoy the same geometric properties as its continuous counterpart: on the one hand, they preserve the structure of the evolution of the symplectic form along the trajectories of the system; on the other hand, they give rise to a discrete version of the nonholonomic momentum equation. Moreover, in the presence of horizontal symmetries, the discrete flow exactly preserves the associated momenta. We also treat the purely kinematic case, where no nonholonomic momentum map exists: we show that the nonholonomic integrator passes to the discrete reduced space yielding a generalized variational integrator in the sense of [110, 194]. In case the continuous gyroscopic force vanishes, we prove that the reduced nonholonomic integrator is indeed a variational integrator.

Control of nonholonomic systems Mechanical control systems provide a challenging research area that falls between Classical Mechanics and Nonlinear Control, pervading modern applications in science and industry. This has motivated many researchers to address the development of a rigorous control theory applicable to this large class of systems: much work has been devoted to the study of their rich geometrical structure, both in the Hamiltonian framework (see [23, 189, 193, 240] and references therein) and on the Lagrangian side, which is receiving increasing attention during the last years (see [31, 40, 117, 141, 142, 168, 196] and the plenary presentations [139, 185]). In particular, the affine connection formalism has turned out to be very useful for modeling different types of mechanical systems, such as natural ones (with Lagrangian equal to kinetic energy minus potential energy) [146, 147], with symmetries [43, 47, 76], with nonholonomic constraints [143], etc. and, on the other hand, it has led to the development of some new techniques and control algorithms for approximate trajectory generation in controller design [42, 45, 169].

Chapter 8 provides the reader with an introduction to affine connection control systems. We expound basic notions and review known results concerning the controllability properties of underactuated mechanical systems such as (configuration) accessibility and controllability, kinematic controllability, etc. Underactuated mechanical control systems are interesting to study both from a theoretical and a practical point of view. From a theoretical perspective, they offer a control challenge as they have non-zero drift, their linearization at zero velocity is not controllable in the absence of potential forces, they are not static feedback linearizable and it is not known if they are dynamic feedback linearizable. That is, they are not amenable to standard techniques in control theory [105, 189]. From the practical point of view, they appear in numerous applications as a result of design choices motivated by the search for less costly devices. Even more, fully actuated mechanical systems may temporarily suffer actuator failures, turning them into underactuated systems.

One of the most basic and interesting aspects of underactuated mechanical systems is the characterization of its controllability properties. The work by Lewis and Murray [146, 147] has rendered strong conditions for configuration accessibility and sufficient conditions for configuration controllability. The conditions for the latter are based on the sufficient conditions that Sussmann obtained for general affine control systems [232]. It is worth noting the fact that these conditions are not invariant under input transformations. As controllability is the more interesting property in practice, more research is needed in order to sharpen the configuration controllability conditions. Whatever these conditions might be, they will turn out to be harder to check than the ones for accessibility, since controllability is inherently a more difficult property to establish [111, 223]. Lewis [142] investigated and fully solved the single-input case, building on previous results by Sussmann for general scalar-input systems [231]. The recent work by Bullo [41] on series expansions for the evolution of a mechanical control system starting from rest has given the necessary tools to tackle this problem in the much more involved multi-input case. In Chapter 8, we characterize local configuration controllability for systems whose number of inputs and degrees of freedom differs by one.

Examples include autonomous vehicles, robotic manipulators and locomotion devices. Interestingly, the differential flatness properties of this type of under-actuated mechanical control systems have also been characterized precisely in intrinsic geometric terms [204]. It is remarkable to note that local controllability has not been characterized yet for general control systems, even for the single input case (in this respect see [99, 231, 232]).

The other important topic treated in Chapter 8 is the extension of previous controllability analyses and series expansion results [41, 45, 146] to systems with isotropic dissipation. The motivation for this work is a standing limitation in the known results on controllability and series expansions. The analysis in [41, 45, 76, 146] applies only to systems subject to no external dissipation, i.e., the system's dynamics is fully determined by the Lagrangian function. With the aim of developing more accurate mathematical models for controlled mechanical systems, we address the setting of dissipative or damping forces. It is worth adding that dissipation is a classic topic in Geometric Mechanics (see for example the work on dissipation induced instabilities [30], the extensive literature on dissipation-based control [7, 191, 240], and recent works including [31, 32, 184, 192]).

Remarkably, the same conditions guaranteeing a variety of local accessibility and controllability properties for systems without damping remain valid for the class of systems under consideration. This applies to small-time local controllability, local configuration controllability, and kinematic controllability. Furthermore, we develop a series expansion describing the evolution of the controlled trajectories starting from rest, thus generalizing the work in [41]. The technical approach exploits the homogeneity property of the affine connection model for mechanical control systems.

As the reader will have already observed, geometry plays a key role in the various problems raised along this introduction. Indeed, our primary concern throughout the present book will be the understanding of the geometric structure of nonholonomic systems, and the use of this knowledge in the approach to the above mentioned topics.

2 Basic geometric tools

THIS chapter gives a brief review of several differential geometric tools used throughout the book. For a more thorough introduction we refer to [1, 2, 138, 149, 163].

The basic concept on which the notions presented in this chapter are built is that of *differentiable manifold*. It was Poincaré who, at the end of the 19th century, found that the prevailing mathematical model of his time was inadequate, for its underlying space was Euclidean, whereas for a mechanical system with angular variables or constraints the phase space might be nonlinear. In this way, he was led by his global geometric point of view to the notion of differentiable manifold as the phase space in Mechanics. This was the starting point of the developments that culminated in what we nowadays know as Modern Differential Geometry.

The chapter is organized as follows. Section 2.1 presents some basic notions on tensor analysis and exterior calculus on manifolds. In Section 2.2 we define the important concepts of generalized distributions and codistributions. Section 2.3 contains a basic account of Lie group theory and Section 2.4 reviews the notion of principal connection. The following two sections are devoted to Riemannian and symplectic manifolds, respectively. Section 2.7 deals with symplectic and Hamiltonian actions. Section 2.8 presents the concept of Poisson manifold and in Section 2.9 we have collected several facts concerning the geometry of the tangent bundle and of a Lagrangian system. References for further study are provided at each section.

2.1 Manifolds and tensor calculus

A BASIC understanding of Differential Geometry is assumed. In this chapter, we quickly review some notation and notions we will need later. The manifolds we deal with will be assumed to belong to the C^∞-category. We shall further suppose that all manifolds are finite-dimensional, paracompact and Hausdorff, unless otherwise stated. The notation we use is common to many standard reference books such as [1, 2, 119, 253].

The tangent bundle of a manifold Q is the collection of all the tangent vectors to Q at each point. We will denote it by TQ. The tangent bundle projection, which assigns to each tangent vector its base point, is denoted by $\tau_Q : TQ \longrightarrow Q$. Given a tangent space T_qQ, we denote the dual space, i.e. the space of linear functions from T_qQ to $\mathbb{R}$, by T_q^*Q. The cotangent bundle T^*Q of a manifold Q is the vector bundle over Q formed by the collection of all the dual spaces T_q^*Q. Elements $\omega \in T_q^*Q$ are called dual vectors or covectors. The cotangent bundle projection, which assigns to each covector its base point, is denoted by $\pi_Q : T^*Q \longrightarrow Q$.

Let $f : Q \longrightarrow N$ be a smooth mapping between manifolds Q and N. We write $Tf : TQ \longrightarrow TN$ to denote the tangent map or differential of f. There are other notations such as f_* and Df. The set of all smooth mappings from Q to N will be denoted by $C^\infty(Q, N)$. When $N = \mathbb{R}$, we shall denote the set of smooth real-valued functions on Q by $C^\infty(Q)$.

A *vector field* X on Q is a smooth mapping $X : Q \longrightarrow TQ$ which assigns to each point $q \in Q$ a tangent vector $X(q) \in T_qQ$ or, stated otherwise, $\tau_Q \circ X = \mathrm{Id}_Q$. The set of all vector fields over Q is denoted by $\mathfrak{X}(Q)$. An integral curve of a vector field X is a curve satisfying $\dot{c}(t) = X(c(t))$. Given $q \in Q$, let $\phi_t(q)$ denote the maximal integral curve of X, $c(t) = \phi_t(q)$ starting at q, i.e. $c(0) = q$. Here "maximal" means that the interval of definition of $c(t)$ is maximal. It is easy to verify that $\phi_0 = \mathrm{Id}$ and $\phi_t \circ \phi_{t'} = \phi_{t+t'}$, whenever the composition is defined. The flow of a vector field X is then determined by the collection of mappings $\phi_t : Q \longrightarrow Q$. From the definition, they satisfy

$$\frac{d}{dt}\left(\phi_t(q)\right) = X(\phi_t(q)), \quad t \in (-\epsilon_1(q), \epsilon_2(q)), \ \forall q \in Q.$$

Similarly, a *one-form* α on Q is a smooth mapping $\alpha : Q \longrightarrow T^*Q$ which associates to each point $q \in Q$ a covector $\alpha(q) \in T_q^*Q$, i.e. $\pi_Q \circ \alpha = \mathrm{Id}_Q$. The set of all one-forms over Q is denoted by $\Omega^1(Q)$.

Both notions, vector fields and one-forms, are special cases of a more general geometric object, called *tensor field*. A tensor field t of contravariant order r and covariant order s is a C^∞-section of T_s^rQ, that is, it associates to each $q \in Q$ a multilinear map

$$t(q) : \underbrace{T_q^*Q \times \cdots \times T_q^*Q}_{r \text{ times}} \times \underbrace{T_qQ \times \cdots \times T_qQ}_{s \text{ times}} \longrightarrow \mathbb{R}.$$

It is common to say that t is a (r, s)-tensor field. The *tensor product* of a (r, s)-tensor field, t, and a (r', s')-tensor field, t', is the $(r + r', s + s')$-tensor field $t \otimes t'$ defined by

$$t \otimes t'(q)(\omega_1, \ldots, \omega_r, \mu_1, \ldots, \mu_{r'}, v_1, \ldots, v_s, w_1, \ldots, w_{s'})$$
$$= t(q)(\omega_1, \ldots, \omega_r, v_1, \ldots, v_s) \cdot t'(q)(\mu_1, \ldots, \mu_{r'}, w_1, \ldots, w_{s'}),$$

where $q \in Q$, v_i, $w_i \in T_q Q$ and ω_j, $\mu_j \in T_q^* Q$.

A special subset of tensor fields is $\Omega^k(Q) \subset T_k^0 Q$, the set of all $(0, k)$ skew-symmetric tensor fields. The elements of $\Omega^k(Q)$ are called k-forms.

The *alternation map* $A : T_k^0 Q \longrightarrow \Omega^k(Q)$ is defined by

$$A(t)(v_1, \ldots, v_k) = \frac{1}{k!} \sum_{\sigma \in \Sigma_k} \text{sign}(\sigma) t(v_{\sigma(1)}, \ldots, v_{\sigma(k)}),$$

where Σ_k is the set of k-permutations. It is easy to see that A is linear, $A_{|\Omega^k(Q)} = \text{Id}$ and $A \circ A = A$.

The *wedge* or *exterior product* between $\alpha \in \Omega^k(Q)$ and $\beta \in \Omega^l(Q)$ is the form $\alpha \wedge \beta \in \Omega^{k+l}(Q)$ defined by

$$\alpha \wedge \beta = \frac{(k+l)!}{k! \, l!} A(\alpha \otimes \beta).$$

There are several possible conventions for defining the constant appearing in the wedge product. The one here conforms to [1, 2, 225], but not to [119]. Some important properties of the wedge product are the following:

1. $\wedge$ is bilinear and associative.

2. $\alpha \wedge \beta = (-1)^{kl} \beta \wedge \alpha$, where $\alpha \in \Omega^k(Q)$ and $\beta \in \Omega^l(Q)$.

The *algebra of exterior differential forms*, $\Omega(Q)$, is the direct sum of $\Omega^k(Q)$, $k = 0, 1, \ldots$, together with its structure as an infinite-dimensional real vector space and with the multiplication $\wedge$.

When dealing with exterior differential forms, another important geometric object is the *exterior derivative*, d. It is defined as the unique family of mappings $d^k(U) : \Omega^k(U) \longrightarrow \Omega^{k+1}(U)$ ($k = 0, 1, \ldots$ and $U \subset Q$ open) such that [1, 253],

1. d is a $\wedge$ antiderivation, i.e. d is $\mathbb{R}$-linear and $d(\alpha \wedge \beta) = d\alpha \wedge \beta + (-1)^k \alpha \wedge d\beta$, where $\alpha \in \Omega^k(U)$ and $\beta \in \Omega^l(U)$.

2. $df = p_2 \circ Tf$, for $f \in C^\infty(U)$, with p_2 the canonical projection of $T\mathbb{R} \cong \mathbb{R} \times \mathbb{R}$ onto the second factor.

3. $d \circ d = 0$.

4. d is natural with respect to inclusions, i.e. if $U \subset V \subset Q$ are open, then $d(\alpha_{|U}) = (d\alpha)_{|U}$, where $\alpha \in \Omega^k(V)$.

Let $f : Q \longrightarrow N$ be a smooth mapping and $\omega \in \Omega^k(N)$. Define the *pullback* $f^* \omega$ of ω by f as $f^* \omega(q)(v_1, \ldots, v_k) = \omega(f(q))(T_q f(v_1), \ldots, T_q f(v_k))$, where $v_i \in T_q Q$. Note that the pullback defines a mapping $f^* : \Omega^k(N) \longrightarrow \Omega^k(Q)$. The main properties related with the pullback are the following,

1. $(g \circ f)^* = f^* \circ g^*$, where $f \in C^\infty(Q, N)$ and $g \in C^\infty(N, W)$.

2. $(\mathrm{Id}_Q^*)_{|\Omega^k(Q)} = \mathrm{Id}_{\Omega^k(Q)}$.

3. If $f \in C^\infty(Q, N)$ is a diffeomorphism, then f^* is a vector bundle isomorphism and $(f^*)^{-1} = (f^{-1})^*$.

4. $f^*(\alpha \wedge \beta) = f^*\alpha \wedge f^*\beta$, where $f \in C^\infty(Q, N)$, $\alpha \in \Omega^k(N)$ and $\beta \in \Omega^l(N)$.

5. d is natural with respect to mappings, that is, for $f \in C^\infty(Q, N)$, $f^*d\omega = df^*\omega$.

Given a vector field $X \in \mathfrak{X}(Q)$ and a function $f \in C^\infty(Q)$, the *Lie derivative of f with respect to X*, $\mathcal{L}_X f \in C^\infty(Q)$, is defined as $\mathcal{L}_X f(q) = df(q)[X(q)]$. The operation $\mathcal{L}_X : C^\infty(Q) \longrightarrow C^\infty(Q)$ is a derivation, i.e. it is $\mathbb{R}$-linear and $\mathcal{L}_X(fg) = \mathcal{L}_X(f)g + f\mathcal{L}_X(g)$, for any $f, g \in C^\infty(Q)$.

The collection of all ($\mathbb{R}$-linear) derivations θ on $C^\infty(Q)$ forms a C^∞-module, with the external law $(f\theta)(g) = f(\theta g)$. This module is indeed isomorphic to $\mathfrak{X}(Q)$. In particular, for each derivation θ, there is a unique $X \in \mathfrak{X}(Q)$ such that $\theta = \mathcal{L}_X$. This is often taken as an alternative definition of vector field (see, for instance, [3]).

Given two vector fields, $X, Y \in \mathfrak{X}(Q)$, we may define the $\mathbb{R}$-linear derivation $[\mathcal{L}_X, \mathcal{L}_Y] = \mathcal{L}_X \circ \mathcal{L}_Y - \mathcal{L}_Y \circ \mathcal{L}_X$. This enables us to define the *Lie derivative of Y with respect to X*, $\mathcal{L}_X Y = [X, Y]$ as the unique vector field such that $\mathcal{L}_{[X,Y]} = [\mathcal{L}_X, \mathcal{L}_Y]$. Some important properties are,

1. If $\phi \in C^\infty(Q, N)$ is a diffeomorphism, $[\phi_* X, \phi_* Y] = \phi_*[X, Y]$.

2. $\mathcal{L}_X$ is natural with respect to restrictions, i.e. for $U \subset Q$ open, $[X_{|U}, Y_{|U}] = [X, Y]_{|U}$ and $(\mathcal{L}f)_{|U} = \mathcal{L}_{X_{|U}}(f_{|U})$, for $f \in C^\infty(Q)$.

3. $\mathcal{L}_X(f \cdot Y) = \mathcal{L}_X f \cdot Y + f \cdot \mathcal{L}_X Y$, for $f \in C^\infty(Q)$.

Indeed, the operator $\mathcal{L}_X$ can be defined on the full tensor algebra of the manifold Q (see [1, 2, 253]).

There is also another natural operator associated with a vector field X. Let $\omega \in \Omega^k(Q)$. The *inner product or contraction of X and ω*, $i_X\omega \in \Omega^{k-1}(Q)$, is defined by $i_X\omega(q)(v_1, \ldots, v_{k-1}) = \omega(q)(X(q), v_1, \ldots, v_{k-1})$, where $v_i \in T_qQ$. The operator i_X is a $\wedge$ antiderivation, namely, it is $\mathbb{R}$-linear and $i_X(\alpha \wedge \beta) = (i_X\alpha) \wedge \beta + (-1)^k \alpha \wedge (i_X\beta)$, where $\alpha \in \Omega^k(Q)$. Also, for $f \in C^\infty(Q)$, we have that $i_{fX}\alpha = f i_X\alpha$.

Finally, we conclude this section by stating some relevant properties involving d, i_X and $\mathcal{L}_X$. For arbitrary $X, Y \in \mathfrak{X}(Q)$, $f \in C^\infty(Q)$ and $\alpha \in \Omega^k(Q)$, we have

1. $d\mathcal{L}_X\alpha = \mathcal{L}_X d\alpha$.

2. $i_X df = \mathcal{L}_X f$.

3. $\mathcal{L}_X \alpha = i_X d\alpha + d i_X \alpha$.

4. $\mathcal{L}_{fX} \alpha = f \mathcal{L}_X \alpha + df \wedge i_X \alpha$.

5. $i_{[X,Y]} \alpha = \mathcal{L}_X i_Y \alpha - i_Y \mathcal{L}_X \alpha$.

2.2 Generalized distributions and codistributions

WE introduce here the notion of generalized distributions and codistributions. These notions will be key in the geometrical modeling of nonholonomic dynamical systems. The exposition here is taken from [237].

Definition 2.2.1. *A generalized distribution (respectively codistribution) $\mathcal{D}$ on a manifold Q is a family of linear subspaces $\{\mathcal{D}_q\}$ of the tangent spaces $T_q Q$ (resp. $T_q^* Q$). A generalized distribution (resp. codistribution) is called differentiable if $\forall q \in Dom\,\mathcal{D}$, there is a finite number of differentiable local vector fields $X_1, \ldots, X_l$ (resp. 1-forms $\omega_1, \ldots, \omega_l$) defined on some open neighborhood U of q in such a way that $\mathcal{D}_{q'} = span\{X_1(q'), \ldots, X_l(q')\}$ (resp. $\mathcal{D}_{q'} = span\{\omega_1(q'), \ldots, \omega_l(q')\}$) for all $q' \in U$.*

We define the rank of $\mathcal{D}$ at q as the dimension of the linear space $\mathcal{D}_q$, i.e. $\rho : Q \longrightarrow \mathbb{R}$, $\rho(q) = \dim \mathcal{D}_q$. For any $q_0 \in Q$, if $\mathcal{D}$ is differentiable, it is clear that $\rho(q) \geq \rho(q_0)$ in a neighborhood of q_0. Therefore, ρ is a lower semicontinuous function. If ρ is a constant function, then $\mathcal{D}$ is called a regular distribution (resp. codistribution). For most part of the book, we shall consider regular (co)distributions. However, in Chapter 6 we shall treat the special case of nonholonomic systems with constraints given by a generalized codistribution.

For a generalized differentiable (co)distribution $\mathcal{D}$, a point $q \in Q$ will be called *regular* if q is a local maximum of ρ, that is, ρ is constant on an open neighborhood of q. Otherwise, q will be called a *singular* point of $\mathcal{D}$. The set R of regular points of $\mathcal{D}$ is obviously open. But, in addition, it is dense, since if $q_0 \in S = Q \setminus R$, and U is a neighborhood of q_0, U necessarily contains regular points of $\mathcal{D}$ ($\rho_{|U}$ must have a maximum because it is integer valued and bounded). Consequently, $q_0 \in \bar{R}$.

Note that in general R will not be connected, as the following example shows:

Example 2.2.2. Let us consider $Q = \mathbb{R}^2$ and the generalized differentiable codistribution $\mathcal{D}_{(x,y)} = span\{\phi(x)(dx - dy)\}$, where $\phi(x)$ is defined by

$$\phi(x) = \begin{cases} 0 & x \leq 0 \\ e^{-\frac{1}{x^2}} & x > 0 \end{cases}$$

The singular points are those of the y-axis, and the connected components of R are the half-planes $x > 0$ (where the rank is 1) and $x < 0$ (where the rank is 0).

In the following we specialize our discussion to codistributions. The definition of the same concepts for distributions is straightforward.

Given a generalized codistribution $\mathcal{D}$, we define its annihilator $\mathcal{D}^o$ as the generalized distribution given by

$$\mathcal{D}^o : \mathrm{Dom}\,\mathcal{D} \subset Q \longrightarrow TQ$$
$$q \longmapsto \mathcal{D}_q^o = (\mathcal{D}_q)^o = \{v \in T_qQ \,|\, \alpha(v) = 0, \forall \alpha \in \mathcal{D}_q\}\,.$$

Notice that if $\mathcal{D}$ is differentiable, $\mathcal{D}^o$ is not differentiable, or even continuous, in general (the corresponding rank function of $\mathcal{D}^o$ will not be lower semicontinuous). In fact, $\mathcal{D}^o$ is differentiable if and only if $\mathcal{D}$ is a regular codistribution.

An immersed submanifold N of Q will be called an *integral submanifold* of $\mathcal{D}$ if T_nN is annihilated by $\mathcal{D}_n$ at each point $n \in N$. N will be an integral submanifold of maximal dimension if

$$T_nN^o = \mathcal{D}_n\,, \text{ for all } n \in N\,.$$

In particular, this implies that the rank of $\mathcal{D}$ is constant along N. A *leaf* L of $\mathcal{D}$ is a connected integral submanifold of maximal dimension such that every connected integral manifold of maximal dimension of $\mathcal{D}$ which intersects L is an open submanifold of L. $\mathcal{D}$ will be a *partially integrable* codistribution if for every regular point $q \in R$, there exists one leaf passing through q. $\mathcal{D}$ will be a *completely integrable* codistribution if there exists a leaf passing through q, for every $q \in Q$. In the latter case, the set of leaves defines a general foliation of Q. Obviously, any completely integrable codistribution is partially integrable.

N being an integral submanifold of $\mathcal{D}$ is exactly the same as being an integral submanifold of its annihilator $\mathcal{D}^o$, and so on.

In Example 2.2.2, the leaves of $\mathcal{D}$ are the half-plane $\{x < 0\}$ and the half-lines of slope 1 in the half-plane $\{x > 0\}$. Given any singular point, there is no leaf passing through it. Consequently, $\mathcal{D}$ is not a completely integrable codistribution, but it is partially integrable.

2.3 Lie groups and group actions

AN important and ubiquitous structure appearing in Mechanics is that of a Lie group. We refer the reader to [163, 253] for details and examples related to the discussion of this section.

Let G be a group, that is, a set with an additional internal operation
$\cdot : G \times G \longrightarrow G$, usually called multiplication, satisfying the following defining
properties

1. Associativity: $g \cdot (h \cdot k) = (g \cdot h) \cdot k$, for all g, h and $k \in G$.

2. Identity element: there is a distinguished element e of G, called the identity, such that $e \cdot g = g = g \cdot e$, for all $g \in G$.

3. Inverses: for each $g \in G$, there exists an element g^{-1} with the property $g^{-1}g = e = gg^{-1}$.

The special feature about Lie groups is that, in addition to the multiplication, they also carry a structure of smooth manifold, in such a way that both structures are compatible. More precisely,

Definition 2.3.1. *A group G equipped with a manifold structure is said to be a Lie group if the product mapping $\cdot$ and the inverse mapping $g \longrightarrow g^{-1}$ are both C^∞-mappings.*

A Lie group H is said to be a Lie subgroup of a Lie group G if it is a submanifold of G and the inclusion mapping $i : H \hookrightarrow G$ is a group homomorphism.

For $g \in G$, we denote by $L_g : G \longrightarrow G$ and $R_g : G \longrightarrow G$ the left and right multiplications by g, respectively, i.e., $L_g(h) = gh$ and $R_g(h) = hg$. This allows us to consider the adjoint action of G on G defined by

$$Ad : G \times G \longrightarrow G$$
$$(g, h) \longmapsto Ad_g(h) = L_g R_{g^{-1}} h = ghg^{-1}.$$

Roughly speaking, the adjoint action measures the non-commutativity of the multiplication of the Lie group: if G is Abelian, then the adjoint action Ad_g is simply the identity mapping on G. In addition, when considering motion along non-Abelian Lie groups, a choice must be made as to whether to represent translation by left or right multiplication. The adjoint action provides the transition between these two possibilities.

Example 2.3.2. Basic examples of Lie groups which will appear in this book include the non-zero complex numbers $\mathbb{C}^*$, the unit circle $\mathbb{S}^1$, the group of $n \times n$ invertible matrices $GL(n, \mathbb{R})$ with the matrix multiplication, and several of its Lie subgroups: the group of rigid motions in 3-dimensional Euclidean space, $SE(3)$; the group of rigid motions in the plane, $SE(2)$; and the group of rotations in $\mathbb{R}^3$, $SO(3)$. More examples can be found, for instance, in [186, 253].

Definition 2.3.3. *A real Lie algebra is a vector space $\mathfrak{L}$ over $\mathbb{R}$ with an operation $[\cdot,\cdot] : \mathfrak{L} \times \mathfrak{L} \longrightarrow \mathfrak{L}$, called Lie bracket, satisfying*

1. *Bilinearity over $\mathbb{R}$: $[\sum \alpha^i X_i, \sum \beta^j Y_j] = \sum \alpha^i \beta^j [X_i, Y_j]$, for α^i, $\beta^j \in \mathbb{R}$ and X_i, $Y_j \in \mathfrak{L}$,*

2. *Skew-symmetry: $[X, Y] = -[Y, X]$, for X, $Y \in \mathfrak{L}$,*

3. *The Jacobi identity: $[X, [Y, Z]] + [Y, [Z, X]] + [Z, [X, Y]] = 0$, for X, Y, $Z \in \mathfrak{L}$.*

If $e_1, \ldots, e_m$ is a basis of $\mathfrak{L}$ (as vector space), then the structure constants c^d_{ab} of $\mathfrak{L}$ relative to this basis are uniquely determined by

$$[e_a, e_b] = c^d_{ab} e_d.$$

Example 2.3.4. The set of vector fields on a general manifold Q carries a natural Lie algebra structure. For any $X, Y \in \mathfrak{X}(Q)$, define $[X, Y] = \mathcal{L}_X Y$. It is easy to verify that this operation is a Lie bracket.

For a Lie group G, we consider the set of left-invariant vector fields on G, $\mathfrak{X}_l(G)$. This means that $X \in \mathfrak{X}_l(G)$ if and only if $TL_g(X) = X$ for all $g \in G$. The set $\mathfrak{X}_l(G)$ is a Lie subalgebra of $\mathfrak{X}(G)$, meaning that the Lie bracket of two left-invariant vector fields is also a left-invariant vector field. The Lie algebra $\mathfrak{X}_l(G)$ is called the Lie algebra associated with G and is commonly denoted by $\mathfrak{g}$. Note that $\mathfrak{g}$ can be identified with $T_e G$, since for each $\xi \in T_e G$, $X_\xi(g) = T_e L_g \xi$ is a left-invariant vector field.

Let ξ be an element of the Lie algebra $\mathfrak{g}$. Consider the associated left invariant vector field, X_ξ. Let $\phi_\xi : \mathbb{R} \longrightarrow G$ be the integral curve of X_ξ passing through e at $t = 0$. By definition, we have that $\frac{d}{dt}\big|_{t=0}(\phi_\xi(t)) = \xi$. The *exponential mapping* of the Lie group, $\exp : \mathfrak{g} \longrightarrow G$, is defined by $\exp(\xi) = \phi_\xi(1)$.

For non-Abelian Lie groups, the non-commutativity of the Lie group multiplication implies that we can also consider the above notions replacing "left" by "right". In Geometric Mechanics, this exactly corresponds to the body and spatial representations. To be more explicit, let $v_g \in T_g G$ and consider

$$\xi^b = T_g L_{g^{-1}} v_g \quad \text{and} \quad \xi^s = T_g R_{g^{-1}} v_g.$$

The relationship between spatial and body velocities can be written in terms of the infinitesimal version of the adjoint action of G on itself, which is called the adjoint action of the Lie group on its Lie algebra.

Definition 2.3.5. *The adjoint action of G on $\mathfrak{g}$ is defined as the map $Ad : G \times \mathfrak{g} \longrightarrow \mathfrak{g}$ given by $Ad(g, \xi) = Ad_g \xi = T_{g^{-1}} L_g (T_e R_{g^{-1}} \xi).$*

A simple computation shows that $\xi^b = Ad_{g^{-1}}\xi^s$. Similarly, given $\alpha_g \in T^*G$, we may define

$$p_b = T_eL_g^*\alpha_g \quad \text{and} \quad p_s = T_eR_g^*\alpha_g \, .$$

The relation between the spatial and body momenta is given by means of the coadjoint action.

Definition 2.3.6. *The coadjoint action of G on $\mathfrak{g}^*$ is defined as the map* $CoAd : G \times \mathfrak{g}^* \longrightarrow \mathfrak{g}^*$ *given by* $CoAd(g,p) = (Ad_{g^{-1}})^*p = T_e^*L_{g^{-1}}(T_{g^{-1}}^*R_gp)$.

The body momentum is related to the spatial momentum via $p_s = CoAd_gp_b$.

Lie groups are mathematical objects that have been, and still are, intensively studied in their own right. For us, they will also be interesting because they are the natural geometrical setting for describing the symmetries (translational, rotational,...) that many mechanical systems exhibit. Their presence will generally allow us to develop reduction methods to simplify the description of (and, in some cases, help integrate) the dynamics of the given mechanical system.

This notion of symmetry or invariance of the system is formally expressed through the concept of action.

Definition 2.3.7. *A (left) action of a Lie group G on a manifold Q is a smooth mapping* $\Phi : G \times Q \longrightarrow Q$ *such that,*

1. $\Phi(e, q) = q$, *for all* $q \in Q$.

2. $\Phi(g, \Phi(h, q)) = \Phi(gh, q)$ *for all* $g, h \in G$, $q \in Q$.

The same definition can be stated for right actions, but we consider here left actions, which is the usual convention in Mechanics.

We will normally only be interested in the action as a mapping from Q to Q, and so will write the action as $\Phi_g : Q \longrightarrow Q$, where $\Phi_g(q) = \Phi(g, q)$, for $g \in G$. In some cases, we shall make a slight abuse of notation and write gq instead of $\Phi_g(q)$. The orbit of the G-action through a point q is $\mathrm{Orb}_G(q) = \{gq \mid g \in G\}$.

An action is said to be *free* if all its isotropy groups are trivial, that is, the relation $\Phi_g(q) = q$ implies $g = e$, for any $q \in Q$ (note that, in particular, this implies that there are no fixed points). An action is said to be *proper* if $\tilde{\Phi} : G \times Q \longrightarrow Q \times Q$ defined by $\tilde{\Phi}(g,q) = (q, \Phi(g,q))$ is a proper mapping, i.e., if $K \subset Q \times Q$ is compact, then $\tilde{\Phi}^{-1}(K)$ is compact. Finally, an action is said to be *simple* or *regular* if the set Q/G of orbits has a differentiable manifold structure such that the canonical projection of Q onto Q/G is a submersion.

If Φ is a free and proper action, then Φ is simple, and therefore Q/G is a smooth manifold and $\pi : Q \longrightarrow Q/G$ is a submersion [1, 253]. We will deal with simple Lie group actions.

Let ξ be an element of the Lie algebra $\mathfrak{g}$. Consider the $\mathbb{R}$-action on Q defined by

$$\Phi^\xi : \mathbb{R} \times Q \longrightarrow Q$$
$$(t, q) \longmapsto \Phi(\exp(t\xi), q) .$$

It is easy to verify that this indeed satisfies the defining properties of an action. Alternatively, we can interpret Φ^ξ as a flow on the manifold Q. Consequently, it determines a vector field on Q, given by

$$\xi_Q(q) = \left.\frac{d}{dt}\right|_{t=0} \left(\Phi(\exp(t\xi), q)\right) ,$$

which is called the *fundamental vector field* or *infinitesimal generator* of the action corresponding to ξ. These vector fields generate the tangent space of the orbits of the G-action, that is

$$T_q(\mathrm{Orb}_G(q)) = \{\xi_Q(q) \mid \xi \in \mathfrak{g}\} .$$

The basic properties of infinitesimal generators are,

$-$ $(Ad_g\xi)_Q(q) = T\Phi_g\xi_Q(\Phi_{g^{-1}}(q))$, for any $g \in G$, $q \in Q$ and $\xi \in \mathfrak{g}$,

$-$ $[\xi_Q, \eta_Q] = -[\xi, \eta]_Q$, for $\xi, \eta \in \mathfrak{g}$.

Given a Lie group G, we can consider the natural action of G on itself by left multiplication

$$\Phi : G \times G \longrightarrow G$$
$$(g, h) \longmapsto gh .$$

For any $\xi \in \mathfrak{g}$, the corresponding fundamental vector field of the action is given by

$$\xi_G(h) = \left.\frac{d}{dt}\right|_{t=0} \left(\exp(t\xi) \cdot h\right) = T_e R_h \xi ,$$

that is, the right-invariant vector field defined by ξ.

An action Φ of G on a manifold Q induces an action of the Lie group on the tangent bundle of Q, $\hat{\Phi} : G \times TQ \longrightarrow TQ$ defined by $\hat{\Phi}(g, v_q) = T\Phi_g(v_q)(= \Phi_{g_*}(v_q))$ for any $g \in G$ and $v_q \in T_qQ$. $\hat{\Phi}$ is called the *lifted action* of Φ.

2.4 Principal connections

IN this section, we briefly review the notion of a principal connection on a principal fiber bundle. For details we refer to [119] (note that the actions considered there are right actions).

Let Ψ be a Lie group action on the configuration manifold Q. Assuming that Ψ is free and proper, we can endow the quotient space $Q/G = N$ with a manifold structure such that the canonical projection $\pi : Q \longrightarrow N$ is a surjective submersion. Note that the kernel of $\pi_*(= T\pi)$ consists of the vertical tangent vectors, i.e. the vectors tangent to the orbits of G in Q. We shall denote the bundle of vertical vectors by $\mathcal{V}_\pi$, with $(\mathcal{V}_\pi)_q = T_q(\mathrm{Orb}_G(q))$, $q \in Q$.

In the framework of the mechanics of (coupled) rigid bodies, robotic locomotion, etc., the quotient manifold N is commonly called the *shape space* of the system under consideration and the Lie group G is called the *pose* or *fiber space*. We then have that $Q(N, G, \pi)$ is a *principal fiber bundle* with bundle space Q, base space N, structure group G and projection π.

Note that the bundle space Q is locally trivial, that is, for every point $q \in Q$ there is a neighborhood U of $\pi(q)$ in N such that there exists a diffeomorphism $\psi : \pi^{-1}(U) \longrightarrow G \times U$, $\psi(q) = (\varphi(q), \pi(q))$, for which $\varphi : \pi^{-1}(U) \longrightarrow G$ satisfies $\varphi(\Psi_g q) = L_g \varphi(q)$, for all $g \in G$ and $q \in \pi^{-1}(U)$. Under the identification provided by this diffeomorphism, the action of the Lie group on Q can be simply read as left multiplication in the fiber, that is, $\Psi_g(h, n) = (gh, n) \in G \times N$.

In problems of locomotion it is most often the case that the splitting of the bundle space can be written globally, $Q = G \times N$. This corresponds to the notion of trivial principal fiber bundle. The pose coordinates $g \in G$ describe the position and orientation of the system, whereas the shape coordinates $n \in N$ describe the internal shape.

A *principal connection* on $Q(N, G, \pi)$ can be defined as a distribution $\mathcal{H}$ on Q satisfying the following properties

1. $T_q Q = \mathcal{H}_q \oplus (\mathcal{V}_\pi)_q, \forall q \in Q$,

2. $\mathcal{H}_{gq} = T_q \Psi_g(\mathcal{H}_q)$, i.e. the distribution $\mathcal{H}$ is G-invariant,

3. $\mathcal{H}_q$ depends smoothly on q.

The subspace $\mathcal{H}_q$ of $T_q Q$ is called the *horizontal subspace* at q determined by the connection. Alternatively, a principal connection can be characterized by a $\mathfrak{g}$-valued 1-form γ on Q satisfying the following conditions

1. $\gamma(\xi_Q(q)) = \xi$, for all $\xi \in \mathfrak{g}$,

2. $\gamma(T\Psi_g X) = \mathrm{Ad}_g(\gamma(X))$, for all $X \in TQ$.

The horizontal subspace at q is then given by $\mathcal{H}_q = \{v_q \in T_qQ \mid \gamma(v_q) = 0\}$. A vector field X on Q is called horizontal if $X(q) \in \mathcal{H}_q$ at each point q.

Given a principal connection, property (i) above implies that every vector $v \in T_qQ$ can be uniquely written as

$$v = v_1 + v_2,$$

with $v_1 \in \mathcal{H}_q$ and $v_2 \in (V_\pi)_q$. We denote by $\mathbf{h} : TQ \longrightarrow \mathcal{H}$ and $\mathbf{v} : TQ \longrightarrow V_\pi$ the corresponding horizontal and vertical projectors, respectively. The *horizontal lift* of a vector field Y on N is the unique vector field Y^h on Q which is horizontal and projects onto Y, $\pi_*(Y^h) = Y \circ \pi$.

The *curvature* Ω of the principal connection is the g-valued 2-form on Q defined as follows: for each $q \in Q$ and $u, v \in T_qQ$

$$\Omega(u, v) = d\gamma(\mathbf{h}u, \mathbf{h}v) = -\gamma([U^h, V^h]_q),$$

where U^h and V^h are the horizontal lifts of any two (local) vector fields U and V on N for which $U^h(q) = \mathbf{h}u$ and $V^h(q) = \mathbf{h}v$, respectively. The curvature measures the lack of integrability of the horizontal distribution and plays a fundamental role in the theory of holonomy (see [119] for a comprehensive treatment).

2.5 Riemannian geometry

THE subject of Riemannian geometry is a very vast one and here we shall present only that part of it that will be used later on. A detailed discussion of Riemannian geometry can be found in [58, 119].

A Riemannian metric $\mathcal{G}$ is a $(0, 2)$-tensor on a manifold Q which is symmetric and positive-definite. This means that

1. $\mathcal{G}(v_q, w_q) = \mathcal{G}(w_q, v_q)$, for all $v_q, w_q \in TQ$,

2. $\mathcal{G}(v_q, v_q) \geq 0$, and $\mathcal{G}(v_q, v_q) = 0$ if and only if $v_q = 0$.

A Riemannian manifold is a pair $(Q, \mathcal{G})$, where Q is a differentiable manifold and $\mathcal{G}$ is a Riemannian metric.

Given a Riemannian manifold, we may consider the "musical" isomorphisms

$$\flat_\mathcal{G} : TQ \longrightarrow T^*Q, \quad \sharp_\mathcal{G} : T^*Q \longrightarrow TQ,$$

defined as $\flat_{\mathcal{G}}(v) = \mathcal{G}(v, \cdot)$ and $\sharp_{\mathcal{G}} = \flat_{\mathcal{G}}^{-1}$. If $f \in C^{\infty}(Q)$, we define its *gradient* as the vector field given by $\operatorname{grad} f = \sharp_{\mathcal{G}}(df)$.

A vector field $X \in \mathfrak{X}(Q)$ is said to be *Killing* if its flow leaves invariant the metric, that is, $\mathcal{L}_X \mathcal{G} = 0$.

Every Riemannian manifold is endowed with a canonical affine connection, called the Levi-Civita connection. In general, an *affine connection* [1, 119] is defined as an assignment

$$\nabla : \mathfrak{X}(Q) \times \mathfrak{X}(Q) \longrightarrow \mathfrak{X}(Q)$$
$$(X, Y) \longmapsto \nabla_X Y$$

which satisfies the following properties for any $X, Y, Z \in \mathfrak{X}(Q)$, $f \in C^{\infty}(Q)$,

1. it is $\mathbb{R}$-bilinear,

2. $\nabla_{fX + gY} Z = f \nabla_X Z + g \nabla_Y Z$,

3. $\nabla_X(fY) = f \nabla_X Y + \mathcal{L}_X(f) Y$.

We shall call $\nabla_X Y$ the *covariant derivative of Y with respect to X*. In local coordinates (q^A) on Q, we have that

$$\nabla_X Y = \left(\frac{\partial Y^A}{\partial q^B} X^B + \Gamma^A_{BC} X^B Y^C \right) \frac{\partial}{\partial q^A},$$

where $\Gamma^A_{BC}(q)$ are the *Christoffel symbols* of the affine connection defined by

$$\nabla_{\frac{\partial}{\partial q^B}} \frac{\partial}{\partial q^C} = \Gamma^A_{BC} \frac{\partial}{\partial q^A}. \tag{2.1}$$

A curve $c : [a, b] \longrightarrow Q$ is a *geodesic* for ∇ if $\nabla_{\dot{c}(t)} \dot{c}(t) = 0$. Locally, the condition for a curve $t \mapsto (q^1(t), \ldots, q^n(t))$ to be a geodesic can be expressed as

$$\ddot{q}^A + \Gamma^A_{BC} \dot{q}^B \dot{q}^C = 0, \quad 1 \leq A \leq n. \tag{2.2}$$

The geodesic equation (2.2) is a second-order differential equation on Q, which can obviously be written as a first-order differential equation on TQ. The vector field corresponding to this first-order equation is given in coordinates by

$$Z_{\nabla} = v^A \frac{\partial}{\partial q^A} - \Gamma^A_{BC} v^B v^C \frac{\partial}{\partial v^A},$$

and is called the *geodesic spray* associated with the affine connection ∇. Hence, the integral curves of the geodesic spray Z_{∇}, $(q^A(t), \dot{q}^A(t))$ are the solutions of the geodesic equation. Other important objects related to an affine connection are the *torsion* tensor, which is defined by

$$T : \mathfrak{X}(Q) \times \mathfrak{X}(Q) \longrightarrow \mathfrak{X}(Q)$$
$$(X, Y) \longmapsto \nabla_X Y - \nabla_Y X - [X, Y],$$

and the *curvature* tensor , given by

$$R : \mathfrak{X}(Q) \times \mathfrak{X}(Q) \times \mathfrak{X}(Q) \longrightarrow \mathfrak{X}(Q)$$
$$(X, Y, Z) \longmapsto \nabla_X \nabla_Y Z - \nabla_Y \nabla_X Z - \nabla_{[X,Y]} Z .$$

Locally, if we write

$$T(\frac{\partial}{\partial q^A}, \frac{\partial}{\partial q^B}) = T^C_{AB} \frac{\partial}{\partial q^C} , \quad R(\frac{\partial}{\partial q^A}, \frac{\partial}{\partial q^B}, \frac{\partial}{\partial q^C}) = R^D_{ABC} \frac{\partial}{\partial q^D} ,$$

we obtain

$$T^C_{AB} = \Gamma^C_{AB} - \Gamma^C_{BA} , \quad R^D_{ABC} = \frac{\partial \Gamma^D_{BC}}{\partial q^A} - \frac{\partial \Gamma^D_{AC}}{\partial q^B} + \Gamma^E_{BC} \Gamma^D_{AE} - \Gamma^E_{AC} \Gamma^D_{BE} .$$

The *Levi-Civita connection* $\nabla^{\mathcal{G}}$ associated with the metric $\mathcal{G}$ is determined by the formula

$$2\mathcal{G}(Z, \nabla_X Y) = X(\mathcal{G}(Z, Y)) + Y(\mathcal{G}(Z, X)) - Z(\mathcal{G}(Y, X))$$
$$+ \mathcal{G}(X, [Z, Y]) + \mathcal{G}(Y, [Z, X]) - \mathcal{G}(Z, [Y, X]), \quad (2.3)$$

where $X, Y, Z \in \mathfrak{X}(Q)$. One can compute the Christoffel symbols of $\nabla^{\mathcal{G}}$ to be

$$\Gamma^A_{BC} = \frac{1}{2} \mathcal{G}^{AD} \left(\frac{\partial \mathcal{G}_{DB}}{\partial q^C} + \frac{\partial \mathcal{G}_{DC}}{\partial q^B} - \frac{\partial \mathcal{G}_{BC}}{\partial q^D} \right) ,$$

where $(\mathcal{G}^{AD})$ denotes the inverse matrix of $(\mathcal{G}_{DA} = \mathcal{G}(\frac{\partial}{\partial q^D}, \frac{\partial}{\partial q^A}))$.

The Levi-Civita connection is torsion-free, i.e. $T(X, Y) = 0$ for all X, $Y \in \mathfrak{X}(Q)$.

2.5.1 Metric connections

In this section, we want to collect some simple facts about metric connections that will be useful for the study of generalized Chaplygin systems in Chapter 5.

Definition 2.5.1. *An affine connection ∇ is called metric with respect to $\mathcal{G}$ if $\nabla \mathcal{G} = 0$, that is,*

$$Z(\mathcal{G}(X, Y)) = \mathcal{G}(\nabla_Z X, Y) + \mathcal{G}(X, \nabla_Z Y),$$

for all $X, Y, Z \in \mathfrak{X}(Q)$.

The Levi-Civita connection $\nabla^{\mathcal{G}}$ can alternatively be defined as the unique torsion-free affine connection which is metric with respect to $\mathcal{G}$.

Let ∇ be a metric connection with respect to $\mathcal{G}$. The following proposition asserts that ∇ is fully determined by its torsion T.

Proposition 2.5.2. *Let T be a skew-symmetric (1,2)-tensor on Q. Then there exists a unique metric connection ∇ whose torsion is precisely T.*

Proof. Let us suppose that there exists such metric connection ∇. Then we have that

$$Z(\mathcal{G}(X,Y)) = \mathcal{G}(\nabla_Z X, Y) + \mathcal{G}(X, \nabla_Z Y),$$
$$X(\mathcal{G}(Z,Y)) = \mathcal{G}(\nabla_X Z, Y) + \mathcal{G}(Z, \nabla_X Y),$$
$$Y(\mathcal{G}(X,Z)) = \mathcal{G}(\nabla_Y X, Z) + \mathcal{G}(X, \nabla_Y Z),$$

for all $X, Y, Z \in \mathfrak{X}(Q)$. Now

$$
\begin{aligned}
Z&(\mathcal{G}(X,Y)) + X(\mathcal{G}(Z,Y)) - Y(\mathcal{G}(X,Z)) \\
&= \mathcal{G}(\nabla_X Z + \nabla_Z X, Y) + \mathcal{G}(\nabla_Z Y - \nabla_Y Z, X) + \mathcal{G}(\nabla_X Y - \nabla_Y X, Z) \\
&= \mathcal{G}(2\nabla_X Z + T(Z,X) + [Z,X], Y) \\
&\qquad + \mathcal{G}(T(Z,Y) + [Z,Y], X) + \mathcal{G}(T(X,Y) + [X,Y], Z) \\
&= 2\mathcal{G}(\nabla_X Z, Y) + \mathcal{G}(T(Z,X) + [Z,X], Y) \\
&\qquad + \mathcal{G}(T(Z,Y) + [Z,Y], X) + \mathcal{G}(T(X,Y) + [X,Y], Z).
\end{aligned}
$$

Consequently, the connection ∇ is uniquely determined by the formula

$$
\mathcal{G}(\nabla_X Z, Y) = \mathcal{G}(\nabla_X^{\mathcal{G}} Z, Y) - \frac{1}{2}(\mathcal{G}(Y, T(X,Z)) + \mathcal{G}(X, T(Z,Y)) + \mathcal{G}(Z, T(X,Y))).
$$

$\square$

This proposition implies that the Christoffel symbols $\bar{\Gamma}^A_{BC}$ of the metric connection ∇ in a local chart (q^A) are given by

$$
\bar{\Gamma}^A_{BC} = \Gamma^A_{BC} - \frac{1}{2}\mathcal{G}^{AK}\left(\mathcal{G}_{KM}T^M_{BC} + \mathcal{G}_{BM}T^M_{CK} + \mathcal{G}_{CM}T^M_{BK}\right),
$$

where Γ^A_{BC} are the Christoffel symbols of the connection $\nabla^{\mathcal{G}}$ and $T = T^C_{AB}dq^A \otimes dq^B \otimes \frac{\partial}{\partial q^C}$.

Another way to characterize metric connections is the following. In general, the (1,2)-tensor field S which encodes the difference between an affine connection ∇ on a Riemannian manifold and the Levi-Civita connection corresponding to the Riemannian metric, is called the *contorsion* of ∇ (cf. [215]), that is,

$$\nabla_X Y = \nabla_X^{\mathcal{G}} Y + S(X, Y).$$

If ∇ is a metric connection, then

$$\begin{aligned}
Z(\mathcal{G}(X, Y)) &= \mathcal{G}(\nabla_Z X, Y) + \mathcal{G}(X, \nabla_Z Y) \\
&= \mathcal{G}(\nabla_Z^{\mathcal{G}} X + S(Z, X), Y) + \mathcal{G}(X, \nabla_Z^{\mathcal{G}} Y + S(Z, Y)) \\
&= Z(\mathcal{G}(X, Y)) + \mathcal{G}(S(Z, X), Y) + \mathcal{G}(X, S(Z, Y)),
\end{aligned}$$

which implies that $\mathcal{G}(S(Z, X), Y) + \mathcal{G}(X, S(Z, Y)) = 0$. Herewith we have proved the following

Proposition 2.5.3. ∇ *is a metric connection if and only if*

$$\mathcal{G}(S(Z, X), X) = 0, \quad \forall X, Z \in \mathfrak{X}(Q). \tag{2.4}$$

As a consequence of the two characterizations we have obtained for metric connections, we can establish the next result.

Corollary 2.5.4. *There is a one-to-one correspondence between (1,2)-tensors S verifying (2.4) and skew-symmetric (1,2)-tensors T. This correspondence is given by*

$$S \longrightarrow T,$$

where $T(X, Y) = S(X, Y) - S(Y, X)$ and

$$T \longrightarrow S,$$

where $\mathcal{G}(S(X, Z), Y) = -\dfrac{1}{2}\left((\mathcal{G}(Y, T(X, Z)) + \mathcal{G}(X, T(Y, Z)) + \mathcal{G}(Z, T(X, Y)))\right).$

The equations for the geodesics of a metric connection can be written

$$\nabla_{\dot{c}(t)} \dot{c}(t) = 0 \iff \nabla_{\dot{c}(t)}^{\mathcal{G}} \dot{c}(t) = -S(\dot{c}(t), \dot{c}(t)),$$

or, in local coordinates,

$$\ddot{q}^A + \Gamma_{BC}^A \dot{q}^B \dot{q}^C = \sum_{B < C} \mathcal{G}^{AK}\left(\mathcal{G}_{BM} T_{CK}^M + \mathcal{G}_{CM} T_{BK}^M\right) \dot{q}^B \dot{q}^C,$$

for each $A = 1, \ldots, n$.

Finally, it is important to note that metric connections obviously preserve the kinetic energy associated with the metric $\mathcal{G}$, that is, if $c(t)$ is a geodesic of ∇, we have that

$$\frac{d}{dt}\left(\frac{1}{2}\mathcal{G}(\dot{c}(t), \dot{c}(t))\right) = \mathcal{G}(\nabla_{\dot{c}(t)} \dot{c}(t), \dot{c}(t)) = 0.$$

2.6 Symplectic manifolds

$\mathbf{W}$HEN studying mechanics, a basic mathematical tool is the notion of symplectic manifold. The exposition here follows [149].

Definition 2.6.1. *An almost-symplectic manifold is a pair, (P, ω), where P is a differentiable manifold and ω is a nondegenerate 2-form on P. An almost-symplectic manifold is symplectic if the ω is closed, $d\omega = 0$.*

Associated with a symplectic manifold, there are two canonical "musical" isomorphisms

$$\flat_\omega : \mathfrak{X}(P) \longrightarrow \Omega^1(P), \quad \sharp_\omega : \Omega^1(P) \longrightarrow \mathfrak{X}(P),$$

defined as $\flat_\omega(X) = i_X \omega$ and $\sharp_\omega = \flat_\omega^{-1}$. Given a function $f \in C^\infty(P)$, we define the corresponding *Hamiltonian vector field* by

$$X_f = \sharp_\omega(df).$$

The flow of a Hamiltonian vector field leaves the symplectic form invariant, that is $\mathcal{L}_{X_f} \omega = 0$. Any vector field with this property is called a *locally Hamiltonian vector field*. This terminology has the following explanation: if X is a locally Hamiltonian vector field, then the 1-form $i_X \omega$ is closed and hence, by Poincaré's Lemma, it is locally exact, i.e. there locally exists a function f_X such that $i_X \omega = df_X$.

Every symplectic manifold is naturally equipped with a bracket of functions defined by

$$\begin{aligned} C^\infty(P) \times C^\infty(P) &\longrightarrow C^\infty(P) \\ (f, g) &\longmapsto \{f, g\} = \omega(X_f, X_g). \end{aligned}$$

This bracket is a Poisson bracket (see Section 2.8) and, hence, every symplectic manifold is a Poisson manifold.

Now we turn to identifying some important distributions on symplectic manifolds. Given a distribution $\mathcal{D}$ on P, we define its orthogonal complement with respect to ω by

$$\mathcal{D}_p^\perp = \{v \in T_p P \mid \omega(p)(v, u) = 0, \, \forall u \in \mathcal{D}_p\}, \quad p \in P.$$

We say that $\mathcal{D}$ is isotropic if $\mathcal{D}^\perp \subset \mathcal{D}$, coisotropic if $\mathcal{D} \subset \mathcal{D}^\perp$, Lagrangian if $\mathcal{D} = \mathcal{D}^\perp$ and symplectic if $\mathcal{D} \cap \mathcal{D}^\perp = \{0\}$. The same definitions are valid for submanifolds of P, imposing the corresponding requirements on the tangent spaces of the submanifold. For instance, a symplectic submanifold W will be a submanifold of P such that the pullback $i^*\omega$ of the 2-form ω by the inclusion map $i : W \hookrightarrow P$ is a symplectic form on W. Hence, $(W, i^*\omega)$ is a symplectic manifold.

The cotangent bundle of any manifold is equipped with a canonical symplectic form, as we will now describe. Let Q be a manifold. Consider the 1-form on T^*Q defined by

$$
\begin{aligned}
\Theta_Q : T^*Q &\longrightarrow T^*(T^*Q) \\
\alpha_q &\longmapsto \Theta_Q(\alpha_q) : T(T^*Q) \longrightarrow \mathbb{R} \\
& \qquad\qquad\qquad\quad v_{\alpha_q} \longmapsto \langle \alpha_q, T\pi_Q(v_{\alpha_q}) \rangle.
\end{aligned}
$$

Θ_Q is called the *Liouville 1-form*. In local coordinates (q^A, p_A) on T^*Q, it reads $\Theta_Q = p_A dq^A$. It has some nice properties, such as, for instance, $\beta^* \Theta_Q = \beta$, for any 1-form β on Q. It also allows us to define the 2-form

$$
\omega_Q = -d\Theta_Q .
$$

Obviously, ω_Q is closed. In addition, it is also nondegenerate, and therefore (T^*Q, ω_Q) is a symplectic manifold. It is called the *canonical symplectic form on T^*Q*. In local coordinates, one can see that

$$
\omega_Q = dq^A \wedge dp_A .
$$

In fact, every symplectic manifold is locally isomorphic to a cotangent bundle. This is a consequence of the following result.

Theorem 2.6.2 (Darboux's theorem). *Let (P, ω) be a symplectic manifold of dimension $2r$. Every point $p \in P$ has an open neighborhood U, which is the domain of a chart (U, φ) with local coordinates $x^1, \ldots, x^{2r}$ such that the 2-form ω has the local expression*

$$
\omega = \sum_{i=1}^{r} dx^i \wedge dx^{r+i} .
$$

2.7 Symplectic and Hamiltonian actions

LET $\Phi : G \times P \longrightarrow P$ be an action of a Lie group G on a symplectic manifold (P, ω). The action Φ is called *symplectic* if $\Phi_g^* \omega = \omega$ for all $g \in G$, that is, for every $X, Y \in T_p P$, we have that

$$
\omega(p)(X, Y) = \omega(\Phi(g, p))(\Phi_{g_*}X, \Phi_{g_*}Y) ,
$$

for all $g \in G$ and $p \in P$. The invariance of the symplectic form under the action readily implies that all the infinitesimal generators are locally Hamiltonian vector fields, namely,

$$\mathcal{L}_{\xi_P}\omega = 0 , \quad \text{for each } \xi \in \mathfrak{g} .$$

Assume that the infinitesimal generators are indeed globally Hamiltonian. This means that for each $\xi \in \mathfrak{g}$, there exists a function $J_\xi \in C^\infty(P)$ such that

$$i_{\xi_P}\omega = dJ_\xi .$$

Note that J_ξ is determined up to a constant on each connected component of the manifold P. The set $\{J_\xi\}_{\xi \in \mathfrak{g}}$ allows us to construct the *momentum mapping*

$$
\begin{aligned}
J : P &\longrightarrow \mathfrak{g}^* \\
x &\longmapsto J(x) : \mathfrak{g} \longrightarrow \mathbb{R} \\
&\qquad\qquad \xi \longmapsto J_\xi(x)
\end{aligned}
$$

The actions Φ which admit a momentum mapping J are called *Hamiltonian actions*. A momentum mapping J is called *CoAd*-equivariant provided

$$J(\Phi_g(p)) = CoAd_g(J(p)), \quad \forall g \in G, \ p \in P .$$

Assume P is connected. Given a momentum mapping, fix $g \in G$ and $\xi \in \mathfrak{g}$. Then we can verify that the function $\Psi_{g,\xi} : P \longrightarrow \mathbb{R}$, $p \longmapsto \langle J(\Phi_g(x)), \xi \rangle - \langle J(x), Ad_g(\xi) \rangle$ is constant on P. This allows us to define the map $\sigma : G \longrightarrow \mathfrak{g}^*$ as $\langle \sigma(g), \xi \rangle = \Psi_{g,\xi}(p)$. This map σ is called the *coadjoint cocycle* associated with J. This cocycle defines a cohomology class $[\sigma]$ which can be proved to be uniquely determined by the action Φ admitting the momentum mapping [1, 149]. Note that *CoAd*-equivariant momentum mappings have $\sigma \equiv 0$. In general, for a given momentum mapping J, we can define the action on $\mathfrak{g}^*$,

$$
\begin{aligned}
\Phi : G \times \mathfrak{g}^* &\longrightarrow \mathfrak{g}^* \\
(g, \mu) &\longmapsto CoAd_g\mu + \sigma(g) ,
\end{aligned}
$$

such that the following diagram is commutative

$$
\begin{array}{ccc}
P & \xrightarrow{\ \Phi_g\ } & P \\
{\scriptstyle J}\big\downarrow & & \big\downarrow{\scriptstyle J} \\
\mathfrak{g}^* & \xrightarrow[\ \Phi_g\ ]{} & \mathfrak{g}^*
\end{array}
$$

for all $g \in G$, i.e., J is Φ-equivariant.

2.8 Almost-Poisson manifolds

THE concept of a Poisson manifold is a generalization of the concept of symplectic manifold, as we will see below. We refer the reader to [149, 237] for a thorough discussion of this topic.

Definition 2.8.1. *An almost-Poisson bracket is a mapping* $\{\cdot,\cdot\} : C^\infty(P) \times C^\infty(P) \longrightarrow C^\infty(P)$ *satisfying the following properties,*

1. *bilinearity over* $\mathbb{R}$*:* $\{\alpha^i f_i, \beta^j g_j\} = \alpha^i \beta^j \{f_i, g_j\}$*, for all* α^i*,* $\beta^j \in \mathbb{R}$*,* f_i*,* $g_j \in C^\infty(P)$*,*

2. *skew-symmetry:* $\{f, g\} = -\{g, f\}$*, for all* f*,* $g \in C^\infty(P)$*,*

3. *the Leibniz rule:* $\{h, fg\} = \{h, f\}g + f\{h, g\}$*, for all* f*,* g *and* $h \in C^\infty(P)$*.*

A Poisson bracket is an almost-Poisson bracket that additionally verifies

(iv) the Jacobi identity: $\{f, \{g, h\}\} + \{g, \{h, f\}\} + \{h, \{f, g\}\} = 0$*, for all* f*,* g *and* $h \in C^\infty(P)$*.*

An (almost-)Poisson manifold is a pair $(P, \{\cdot, \cdot\})$, where P is a differentiable manifold and $\{\cdot, \cdot\}$ is a (almost-)Poisson bracket.

As a consequence of the Leibniz rule, for any function f on P we have that $\{\cdot, f\}$ is a derivation of $C^\infty(P)$. Hence, there exists a well defined vector field X_f given by

$$X_f(g) = \{g, f\},$$

which will be called the *Hamiltonian vector field* of f.

On a (almost-)Poisson manifold we have a unique $(2, 0)$-tensor or bivector field Λ such that

$$\{f, g\} = \Lambda(df, dg).$$

We call Λ the (almost-)Poisson bivector of $(P, \{\cdot, \cdot\})$.

Almost-Poisson manifolds can also be defined as those manifolds admitting a bivector field Λ. It can be seen [237] that the Jacobi identity is equivalent to $[\Lambda, \Lambda] = 0$, where $[\cdot, \cdot]$ denotes the Schouten-Nijenhuis bracket . Therefore, Poisson manifolds can alternatively be defined as almost-Poisson manifolds whose bivector field Λ satisfies $[\Lambda, \Lambda] = 0$.

In the usual way, Λ determines a bundle homomorphism over the identity,

$$\sharp_\Lambda : T^*P \longrightarrow TP,$$

defined by $\beta(\sharp_\Lambda(\alpha)) = \Lambda(\alpha, \beta)$, α, $\beta \in T^*P$. Notice, in particular, that one has $\sharp_\Lambda(df) = X_f$. If the manifold is Poisson and $\sharp_\Lambda$ is bijective, then we can define a symplectic 2-form whose associated Poisson bracket is $\{\cdot, \cdot\}$.

Let $(P, \{\cdot, \cdot\})$ be a Poisson manifold. The image of T^*P under the homomorphism $\sharp_\Lambda$ defines a generalized distribution $\mathcal{D}^\Lambda$ by

$$\mathcal{D}_p^\Lambda = \sharp_\Lambda(T_p^*P), \quad p \in P.$$

The rank of the Poisson structure at p is the dimension of the space $\mathcal{D}_p^\Lambda$. $\mathcal{D}^\Lambda$ is called the *characteristic distribution* of the Poisson structure. Note that $\mathcal{D}^\Lambda$ is generated by the Hamiltonian vector fields and consequently is a differentiable distribution as defined in Section 2.2.

The following result can be deduced from the theory of integrability of generalized distributions [237].

Theorem 2.8.2. *The characteristic distribution $\mathcal{D}^\Lambda$ is completely integrable, and the Poisson structure induces symplectic structures on the leaves of $\mathcal{D}^\Lambda$.*

The leaves of $\mathcal{D}^\Lambda$ are called the *symplectic leaves* of the Poisson manifold and $\mathcal{D}^\Lambda$ is also said to be the *symplectic foliation* of P. Note that, since $\mathcal{D}^\Lambda$ is a generalized distribution, the symplectic leaves can have different dimensions.

The local structure of Poisson manifolds is given by the following result.

Theorem 2.8.3 (Weinstein [255]). *Let (P, Λ) be a n-dimensional Poisson manifold, and $p \in P$ a point where the rank of the Poisson structure is $2r$, $0 \leq 2r \leq n$. Then there exists a chart (U, φ) of P, whose domain contains p, with local coordinates $(x^1, \ldots, x^r, y^1, \ldots, y^r, z^1, \ldots, z^{n-2r})$ such that, on U,*

$$\Lambda = \sum_{i=1}^r \frac{\partial}{\partial y^i} \wedge \frac{\partial}{\partial x^i} + \sum_{1 \leq k < l \leq n-2r} b^{kl}(z^1, \ldots, z^{n-2r}) \frac{\partial}{\partial z^k} \wedge \frac{\partial}{\partial z^l},$$

and $b^{kl}(p) = 0$.

2.8.1 Almost-Poisson reduction

Almost-Poisson brackets will be very useful in Chapter 4, when we discuss the reduction and reconstruction of the dynamics of nonholonomic systems with symmetry. Here, we briefly recall the main results of Poisson reduction as developed in [162], but rephrased for almost-Poisson manifolds.

Definition 2.8.4. *Let (P, Λ_P) be an almost-Poisson manifold. Then a pair (N, E) consisting of a submanifold $j : N \hookrightarrow P$, and a vector subbundle E of $TP_{|N}$ will be called a reductive structure of (P, Λ_P) if the following conditions are satisfied,*

1. $E \cap TN$ *is tangent to a foliation* $\mathcal{F}$ *whose leaves are the fibers of a submersion* $\pi : N \longrightarrow S$;

2. *for all* φ, $\psi \in C^\infty(P)$ *such that* $d\varphi$ *and* $d\psi$ *vanish on* E, $d\{\varphi, \psi\}_P$ *also vanishes on* E.

Furthermore, if S, *defined by* (i), *has an almost-Poisson structure* Λ_S *such that for any local* C^∞ *functions* f, g *on* S, *and any local extensions* φ, ψ *of* $\pi^* f$, $\pi^* g$, *with* $d\varphi_{|E} = d\psi_{|E} = 0$, *the relation*

$$\{\varphi, \psi\}_P \circ j = \{f, g\}_S \circ \pi$$

holds, we say that (P, N, E) *is a reducible triple, and* (S, Λ_S) *is the reduced almost-Poisson manifold of* (P, Λ_P) *via* (N, E).

The bundle E is sometimes called the *control bundle*. The following theorem characterizes reducible triples.

Theorem 2.8.5. *Let* (N, E) *be a reductive structure of the almost-Poisson manifold* (P, Λ_P). *Then* (P, N, E) *is a reducible triple iff*

$$\sharp_P(E^o) \subseteq TN + E.$$

2.9 The geometry of the tangent bundle

IN this section we want to collect some facts about the geometry of the tangent bundle that will be useful in the symplectic formulation of Lagrangian systems. The interested reader may consult [78, 138].

First of all, we define what we shall understand by a Lagrangian system.

Definition 2.9.1. *A Lagrangian system consists of a* n-*dimensional manifold* Q, *representing the space of all possible configurations of the system, and a function on the tangent bundle of* Q, $L : TQ \longrightarrow \mathbb{R}$, *called the Lagrangian of the system.*

This can be taken as a purely mathematical definition, although its motivation is clearly physical: it is generally assumed that the dynamical model associated with the Lagrangian function describes the real behavior of the problem under consideration. For a mechanical system, the Lagrangian is given by

$$L(q, \dot{q}) = T(q, \dot{q}) - V(q),$$

where $T : TQ \longrightarrow \mathbb{R}$ denotes the kinetic energy of the system and $V :$ $Q \longrightarrow \mathbb{R}$ the potential energy. This type of Lagrangian is called *natural* or *mechanical*. In any case, the geometric treatment of Lagrangian systems exposed here is valid for a wide class of Lagrangians.

In the previous definition of Lagrangian system we are implicitly assuming that there are no external forces acting on it. Mathematically, an external force field F is usually modeled by a bundle map $F : TQ \longrightarrow T^*Q$ over the identity (although there may be more general types of forces which depend also on accelerations: see [141] for a complete exposition). Using the dual mapping of $T\tau_q : TTQ \longrightarrow TQ$, we will often identify F with the one-form on TQ given by $T^*\tau_Q \circ F$. In Chapter 8, we shall also consider systems with forces that depend only on configurations, rather than on configurations and velocities. In that case, the force field is just given by a one-form on Q, $F' : Q \longrightarrow T^*Q$, so that $F = F' \circ \tau_Q$.

There are two canonical geometric objects associated with the tangent bundle of a manifold, which we describe next. On the one hand, one has the *dilation* or *Liouville vector field*[1]. Consider the 1-parameter group of dilations

$$\phi_t : TQ \longrightarrow TQ$$
$$v_q \longmapsto e^t v_q .$$

The Liouville vector field is the infinitesimal generator of this 1-parameter group, i.e.

$$\Delta : TQ \longrightarrow T(TQ)$$
$$v_q \longmapsto \frac{d}{dt}\big|_{t=0} (\phi_t(v_q)) .$$

In local coordinates, one can check that $\Delta = \dot{q}^A \frac{\partial}{\partial \dot{q}^A}$. On the other hand, there exists a $(1,1)$-tensor field on TQ

$$S : T(TQ) \longrightarrow T(TQ)$$
$$w_{v_q} \longmapsto \frac{d}{dt}\big|_{t=0} \left(v_q + t T\tau_Q(w_{v_q})\right) ,$$

which is known as the *vertical endomorphism* or the *almost tangent structure* of TQ. The local expression of S is given by $S = \frac{\partial}{\partial \dot{q}^A} \otimes dq^A$.

Both objects play a fundamental role in the geometrical description of the dynamics of Lagrangian systems. The role of the Liouville vector field will be shown in Chapter 3 when defining the energy function, and in Chapter 8 when dealing with dissipative mechanical control systems. The vertical endomorphism will enable us to construct the bundle of reaction forces associated with a constrained system in Chapter 3.

By means of the vertical endomorphism we can also define the following relevant additional structures on the tangent bundle on which a Lagrangian

---

[1] In fact, the Liouville vector field can be defined on any vector bundle: see [149].

function is given. The *Poincaré-Cartan 1-form* is the pull-back by S of the differential of the Lagrangian function, i.e.

$$\Theta_L = S^*(dL).$$

In local coordinates, one can see that $\Theta_L = \frac{\partial L}{\partial \dot{q}^A} dq^A$. From Θ_L, one can construct the *Poincaré-Cartan 2-form*

$$\omega_L = -d\Theta_L.$$

Obviously ω_L is closed, but in general need not be nondegenerate. If the Lagrangian is regular, i.e. if for any coordinates $(q^A, \dot{q}^A)$ on TQ we have that the Hessian matrix

$$\left(\frac{\partial^2 L}{\partial \dot{q}^A \partial \dot{q}^B} \right)$$

is invertible, then ω_L is nondegenerate, and hence symplectic. For a mechanical system, L is always regular (indeed, the Hessian matrix is positive-definite). If L is non-regular (or singular), then ω_L is just presymplectic, meaning that the musical mapping $\flat_{\omega_L} : TTQ \longrightarrow T^*TQ$ is not an isomorphism. In local coordinates, we have

$$\omega_L = dq^A \wedge d\left(\frac{\partial L}{\partial \dot{q}^A} \right) = \frac{\partial^2 L}{\partial \dot{q}^A \partial q^B} dq^A \wedge dq^B + \frac{\partial^2 L}{\partial \dot{q}^A \partial \dot{q}^B} dq^A \wedge d\dot{q}^B.$$

Another way of constructing the Poincaré-Cartan forms is through the use of the Legendre transformation. Let us define the latter. Let $v_q \in T_q Q$. Since $T_q Q$ is a vector space, we can identify $T_q Q \equiv T_{v_q}(T_q Q)$ and regard $T_q Q$ as contained in $T_{v_q}(TQ)$. Then, for each v_q, we can consider the linear mapping $(dL)_{v_q} \circ i : T_q Q \hookrightarrow T_{v_q}(TQ) \longrightarrow \mathbb{R}$, or, stated otherwise, $(dL)_{v_q} \circ i \in T_q^* Q$. Thus we get the Legendre transformation

$$\mathcal{F}L : TQ \longrightarrow T^*Q$$
$$v_q \longmapsto (dL)_{v_q} \circ i.$$

Now, it is easy to verify that in local coordinates this map reads

$$\mathcal{F}L(q^A, \dot{q}^A) = \left(q^A, \frac{\partial L}{\partial \dot{q}^A}(q, \dot{q}) \right).$$

Some simple additional computations then yield that

$$\Theta_L = \mathcal{F}L^*\Theta_Q \quad \text{and} \quad \omega_L = \mathcal{F}L^*\omega_Q.$$

If L is regular, the Legendre transformation is a local diffeomorphism. The Lagrangian is called *hyperregular* if $\mathcal{F}L$ is a global diffeomorphism. In that case, $\mathcal{F}L$ indeed defines a symplectomorphism between the symplectic manifolds (TQ, ω_L) and (T^*Q, ω_Q).

If the Lagrangian is singular, then the Legendre transformation is not a local diffeomorphism. Often, it is assumed that L is almost regular, which means that ω_L has constant rank $2r$ (that is, $\flat_{\omega_L}$ has constant rank $2r$), the image $\mathcal{F}L(TQ)$ is a submanifold of T^*Q and the fibers of $\mathcal{F}L$ are connected. Under these assumptions, one can develop a Hamiltonian description of the dynamics of the system, making use of Dirac's theory of constraints [82] and its geometrization due to Gotay and Nester [93, 94, 95] (see [56] for a comprehensive overview).

Let $\Phi : G \times Q \longrightarrow Q$ be an action of a Lie group G on Q. Consider the lifted action $\hat{\Phi} : G \times TQ \longrightarrow TQ$. We say that the Lagrangian L is G-invariant if $L \circ \hat{\Phi}_g = L$, for all $g \in G$. The invariance of the Lagrangian implies that the lifted action is symplectic, namely

$$\hat{\Phi}_g^* \omega_L = \omega_L, \quad \forall g \in G.$$

Moreover, we can define the map,

$$
\begin{aligned}
J^L : TQ &\longrightarrow \mathfrak{g}^* \\
v_q &\longmapsto J^L(v_q) : \mathfrak{g} \longrightarrow \mathbb{R} \\
&\qquad\qquad\quad \xi \longmapsto \langle \mathcal{F}L(v_q), \xi_Q \rangle = \langle \Theta_L(v_q), \xi_{TQ} \rangle.
\end{aligned}
$$

A straightforward computation yields

$$i_{\xi_{TQ}} \omega_L = dJ_\xi^L,$$

where $J_\xi^L : TQ \longrightarrow \mathbb{R}$ is defined by $J_\xi^L(v) = \langle J^L(v), \xi \rangle$, and hence we have that the lifted action $\hat{\Phi}$ is indeed Hamiltonian.

Finally, other notions that will be used later are that of *vertical lift* and *complete lift* of a vector field [138]. Let $X \in \mathfrak{X}(Q)$, the vertical lift of X is the vector field X^{lift} on TQ defined by

$$X^{\text{lift}}(v_q) = \left.\frac{d}{dt}\right|_{t=0} (v_q + tX(q)).$$

To define the complete lift of a vector field, we first have to define the complete lift of a function. The *complete lift* to TQ of a function $f \in C^\infty(Q)$ is the function $f^c \in C^\infty(TQ)$ given by $f^c = df$. The complete lift to TQ of $X \in \mathfrak{X}(Q)$ is the unique vector field $X^c \in \mathfrak{X}(TQ)$ such that $X^c(f^c) = (Xf)^c$, $\forall f \in C^\infty(Q)$.

3 Nonholonomic systems

IN this chapter, we present the class of systems which are the subject of study of the monograph: nonholonomic systems. Classical books such as Appell [6], Painlevé [198], Pars [201], Rosenberg [207] and Whittaker [257] account for much of the developments of the analytical mechanics of nonholonomic problems. The book by Neimark and Fufaev [188] remains as a basic reference in the area. The beginning of the study of nonholonomic mechanics from a differential geometric perspective is marked by the work by Vershik and Faddeev [243, 244]. Since then, many authors have contributed to this growing body of research, which has experimented a huge thrust in the 90's (see [25, 59] and references therein).

The chapter is organized as follows. In Section 3.1 we present Hamilton's principle for unconstrained systems and a geometric formalization using symplectic geometry. Section 3.2 contains several classical and modern examples of nonholonomic systems. In Section 3.3 we state the Lagrange-d'Alembert principle and derive the equations of motion. Finally in Section 3.4 we introduce two alternative intrinsic formulations of the dynamics of nonholonomic systems which will be key in our exposition along the rest of the book.

3.1 Variational principles in Mechanics

SOME authors prefer to first derive the Lagrange or Hamilton equations starting from a Newtonian formulation and then obtain variational principles as theorems. Others assume variational principles and derive the Hamiltonian and Lagrange equations as theorems. We prefer the second approach, since this seems to be in better accordance with the fundamental role that variational principles have played in the evolution of mathematical models in Mechanics.

3.1.1 Hamilton's principle

In this section we give a brief account of the variational principles involved in the derivation of the equations of motion in Classical Mechanics.

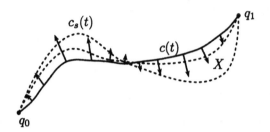

Fig. 3.1. Illustration of a variation c_s and an infinitesimal variation X of a curve c with endpoints q_0 and q_1.

We will typically consider curves $c : [a, b] \longrightarrow Q$, which connect two given points q_0, q_1 in the configuration manifold Q. These curves may be subject to some constraints, but in this section we shall focus our attention on the unconstrained case. The set of all such curves which are twice differentiable will be denoted by

$$C^2(q_0, q_1, [a, b]) = \{c : [a, b] \longrightarrow Q \mid c \text{ is } C^2, \ c(a) = q_0 \text{ and } c(b) = q_1\}.$$

It can be proved that this set is a smooth infinite-dimensional manifold [21]. Let c be a curve in $C^2(q_0, q_1, [a, b])$. As is well known, the tangent space of $C^2(q_0, q_1, [a, b])$ at c is given by

$$T_c C^2(q_0, q_1, [a, b]) = \{X : [a, b] \longrightarrow TQ \mid X \text{ is } C^1, \ X(t) \in T_{c(t)}Q, \\ X(a) = 0 \text{ and } X(b) = 0\}.$$

A tangent vector X at c is then a vector field along the curve c which vanishes at the end points, $c(a)$ and $c(b)$.

Since X is a tangent vector to the manifold $C^2(q_0, q_1, [a, b])$, we may write it as the tangent vector at $s = 0$ of a curve in $C^2(q_0, q_1, [a, b])$, $s \in (-\epsilon, \epsilon) \subset \mathbb{R} \longmapsto c_s$ which passes through c at $s = 0$, $c_0 = c$, that is,

$$X = \frac{dc_s}{ds}\bigg|_{s=0}.$$

The curve c_s is called a *variation* of c and the vector X is called an *infinitesimal variation* of c.

Next, we consider the classical action functional [1, 10] associated with a Lagrangian $L : TQ \longrightarrow \mathbb{R}$, defined by

$$\mathcal{J} : C^2(q_0, q_1, [a, b]) \longrightarrow \mathbb{R}$$
$$c \longmapsto \int_a^b L(\dot{c}(t))\, dt.$$

We then have

$$dJ(c)(X) = \frac{d}{ds}\Big|_{s=0} (J(c_s)) = \int_a^b \frac{d}{ds}\Big|_{s=0} L(c_s, \dot{c}_s) dt .$$

A direct computation using integration by parts [1] shows that in local coordinates we can write

$$dJ(c)(X) = \int_a^b \left(\frac{\partial L}{\partial q^A} - \frac{d}{dt}\left(\frac{\partial L}{\partial \dot{q}^A} \right) \right) X^A \, dt , \qquad (3.1)$$

for $c \in C^2(q_0, q_1, [a, b])$ and $X \in T_c C^2(q_0, q_1, [a, b])$.

Definition 3.1.1 (Hamilton's principle). *A curve $c \in C^2(q_0, q_1, [a, b])$ is a motion of the Lagrangian system defined by L if and only if c is a critical point of J, that is, $dJ(c) = 0$.*

Therefore, a motion of the Lagrangian system extremizes the functional J among all its possible variations.

Being a critical point of the functional J means that $dJ(c)(X) = 0$ for all $X \in T_c C^2(q_0, q_1, [a, b])$. In view of (3.1) this is the same as

$$\int_a^b \left(\frac{\partial L}{\partial q^A} - \frac{d}{dt}\left(\frac{\partial L}{\partial \dot{q}^A} \right) \right) X^A \, dt = 0, \quad \forall X^A .$$

Then, using the fundamental lemma of the Calculus of Variations, it is easy to prove that for a curve c the condition of being critical is equivalent to being a solution of the Euler-Lagrange equations for the Lagrangian L,

$$\frac{d}{dt}\left(\frac{\partial L}{\partial \dot{q}^A} \right) - \frac{\partial L}{\partial q^A} = 0, \quad 1 \leq A \leq n . \qquad (3.2)$$

In the presence of external forces, say $F^1, \ldots, F^m$, we must consider the total work done by these forces along the motion, which is given by

$$W(c) = \sum_{j=1}^m \int_a^b F^j(\dot{c}(t)) dt .$$

Definition 3.1.2 (Integral Lagrange-d'Alembert principle). *Let c be a curve in $C^2(q_0, q_1, [a, b])$. Then c is a motion of the Lagrangian system defined by L, and with external forces $F^1, \ldots, F^m$ if and only if c is such that $dJ(c) = W(c)$.*

In coordinates, the integral Lagrange-d'Alembert principle can be expressed in the following way: we have that $dJ(c) = W(c)$ if and only if $dJ(c)(X) = W(c)(X)$ for all $X \in T_c C^2(q_0, q_1)$, that is,

$$\int_a^b \left(\frac{\partial L}{\partial q^A} - \frac{d}{dt} \left(\frac{\partial L}{\partial \dot{q}^A} \right) \right) X^A \, dt = \sum_{j=1}^m \int_a^b F_A^j(\dot{c}(t)) X^A dt \, , \quad \forall X^A \, .$$

Using again the fundamental lemma of the Calculus of Variations, we obtain the classical forced Euler-Lagrange equations,

$$\frac{d}{dt} \left(\frac{\partial L}{\partial \dot{q}^A} \right) - \frac{\partial L}{\partial q^A} = \sum_{j=1}^m F_A^j \, , \quad 1 \leq A \leq n \, . \tag{3.3}$$

3.1.2 Symplectic formulation

In this section we present an intrinsic formulation of the equations of motion (3.2) for a Lagrangian system that may subjected to external forces. For more details see [1, 138].

First of all, we must define the energy of a system with Lagrangian function L. If we interpret the Liouville vector field Δ on TQ as a derivation on functions, the energy of the system is given by

$$E_L = \Delta(L) - L \, .$$

In the mechanical case, this exactly corresponds to $E_L = T + V$, that is, kinetic plus potential energy.

Now, consider the equation

$$i_X \omega_L = dE_L \, , \tag{3.4}$$

in which X is interpreted as the unknown. If the Lagrangian L is regular, then ω_L is symplectic and $X = X_{E_L} = \Gamma_L$, the Hamiltonian vector field associated with the energy function, is the unique solution of (3.4). In such a case, Γ_L is a second order differential equation (SODE for short), meaning that its integral curves $c(t) \in TQ$ are of the form $c(t) = \frac{d}{dt}\tau_Q(c(t))$. Intrinsically, this is encoded in the equality

$$S(\Gamma_L) = \Delta \, ,$$

where S is the almost-tangent structure on TQ (see Section 2.9). In local coordinates, this reads as $c(t) = (q^A(t), \dot{q}^A(t))$. The curves $q(t) = \tau_Q(c(t))$ are called the solution curves of Γ_L and are just the solutions of the Euler-Lagrange equations (3.2).

If L is not regular, then it does not exist in general a vector field X verifying (3.4). Even if it does, it will not be unique nor a SODE. The Gotay-Nester's algorithm [93, 94, 95], which is a geometrization of Dirac's algorithm of constraints [82], treats this interesting situation. In any case, assuming that we restrict our attention to SODE vector fields, what one can easily compute is that the solution curves of an hypothetical X, if it exists, are precisely those solutions of the Euler-Lagrange equations. So, equation (3.4) is the intrinsic geometric writing of the equations (3.2).

The presence of external forces is easily considered in this formalism. Define the sum of all external forces as the one-form on TQ, $F = \sum_{j=1}^{m} F^j$. Then, equations (3.3) are written as

$$i_X \omega_L = dE_L + F. \tag{3.5}$$

3.2 Introducing constraints

SO far, we have been dealing with mechanical systems for which any of their parts were allowed to move in any direction. However, in many cases, the very construction of the physical system under consideration precludes an arbitrary motion of its various parts; their motions and positions are somehow interrelated and must satisfy a number of conditions. In mechanics, this exactly corresponds to the presence of *constraints* imposed on the system.

The actual form of these constraints may be varied. If they impose restrictions on the possible configurations of the individual parts of the system, we shall call them geometric constraints. If they restrict the kinematically possible motions of the system, i.e. the possible values of the velocities of the individual parts, we call them kinematic. Every geometric constraint gives rise to a certain kinematic constraint by differentiation. But the converse does not hold in general.

The existence of kinematic constraints that impose no restrictions on the possible configurations of a mechanical system was recognized in comparatively recent times. Lagrange himself overlooked them in his celebrated *Mécanique Analytique*. It was only in 1894 that Hertz introduce the distinction between holonomic and nonholonomic constraints, that we describe in the following.

Let (q^A), $A = 1, \ldots, n$, be coordinates on the configuration manifold Q. The conditions on the motions and positions of the system may be expressed by a system of inequalities of the form

$$f_i(q^A, \dot{q}^A, t) \geq 0, \quad 1 \leq i \leq m.$$

These constraints are termed one-sided and nonlimiting. Alternatively, we may have equations of the form

$$f_i(q^A, \dot{q}^A, t) = 0, \quad 1 \leq i \leq m. \tag{3.6}$$

These are called two-sided and limiting. In addition, limiting constraints are classified as time-dependent (classically, *rheonomic*) or time-independent (*scleronomic*), depending on whether or not they contain time explicitly, and geometric or kinematic. The constraints are geometric if they are expressed by equations of the form

$$f_i(q^A, t) = 0, \quad 1 \leq i \leq m.$$

Otherwise, they are kinematic. The kinematic constraints are integrable if the corresponding system of differential equations (3.6) is integrable. Integrable or *holonomic* kinematic constraints are essentially geometric constraints, and in this sense impose restrictions on the possible configurations. In contrast, nonintegrable or *nonholonomic* constraints cannot be reduced to geometric constraints. We give below a precise mathematical formulation of these concepts, since the classification of Hertz in necessarily local in nature due to the choice of coordinates.

The nonholonomic constraints usually encountered in Mechanics are of rolling or non-sliding type, and are linear or affine in the velocities. Examples can be found in wheeled robots, locomotion devices, etc. There are also situations in which the constraints are nonlinear, although in this case there does not exist an unanimous consensus about the physical principle which gives the correct equations of motion. We shall return to this point in the following section.

We remark that one can find in the literature another kind of nonholonomic constraints which are called *dynamic* nonholonomic constraints, that is, constraints preserved by the basic Euler-Lagrange or Hamiltonian equations, such as angular or linear momentum. These "constraints" are not externally imposed on the system, but are obtained as consequences of the equations of motion, i.e. *a posteriori*. Hence, they must be treated as conservation laws, rather than constraints.

Throughout the book, we will assume that the constraints imposed on the mechanical system can be globally described by a submanifold M of the phase space TQ.

Definition 3.2.1. *A nonholonomic Lagrangian system on a manifold Q consists of a pair (L, M), where $L : TQ \longrightarrow \mathbb{R}$ is the Lagrangian of the system and M is a submanifold of TQ.*

We will tacitly assume that $\tau_Q(M) = Q$.

The motions of the system are forced to take place on M, that is, the allowed velocities for the nonholonomic Lagrangian system are those belonging to M. This requires the introduction of some, in principle unknown, "reaction forces" (the constraint forces). This kind of problems include systems in robotics, wheeled vehicular dynamics and motion generation, and are the subject of research of what is commonly known as *Nonholonomic Mechanics*.

In case M is a vector subbundle of TQ, we are dealing with the case of linear constraints. We shall refer then to M as $\mathcal{D}$. If, in addition, this subbundle corresponds to an integrable distribution, we are precisely reduced to the case of holonomic constraints. If $\mathcal{F}$ denotes the foliation on Q defined by $\mathcal{D}$, for any initial condition $v_q \in \mathcal{D}_q$, the problem is then reduced to an unconstrained system living in the leaf of $\mathcal{F}$ containing q (and thus, with less degrees of freedom than the original problem). In case M is an affine subbundle modeled on a vector bundle $\mathcal{D}$, we are treating affine constraints. We will assume then that there exists a globally defined vector field $\gamma \in \mathfrak{X}(Q)$ such that $v_q \in M_q$ if and only if $v_q - \gamma(q) \in \mathcal{D}_q$.

In Chapter 6 we shall deal with a more general situation in which M is contained in TQ, but it is not equipped with a manifold structure.

An important class of nonholonomic systems are those for which the dilation vector field Δ is tangent to the constraint submanifold M. In local coordinates, this condition implies that the constraint functions $\phi_i = 0$ describing M should satisfy

$$\left(\dot{q}^A \frac{\partial \phi_i}{\partial \dot{q}^A} \right)_{|M} = 0 \,.$$

This type of nonholonomic constraints are called *homogeneous*. This condition will be verified in particular if the constraints are homogeneous in the velocities, as is the case of linear constraints.

Classically, there have been many examples of nonholonomic systems studied. The book of Neimark and Fufaev [188] accounts for quite a number of them. Some are the example of a uniform sphere rolling on a surface of revolution, the rolling disk (inclined, vertical), the bicycle, the plate with a knife edge on an inclined plane, the rolling of a solid of revolution on an horizontal plane, the wobblestone or rattleback, etc. In the following, we present some of these examples together with related modern ones which have been in the development of the geometric study of nonholonomic mechanics.

3.2.1 The rolling disk

Consider a disk that rolls without sliding on an horizontal plane. Fix a coordinate system Oxy on the plane. The position of the disk is given by the coordinates (x, y) of the point of contact P with the floor, the angle ψ measured from a chosen point of the rim S to the point of contact P or rotation angle, the angle φ between the tangent to the disk at the point P and the Ox axis or heading angle, and the angle of inclination ϑ between the plane of disk and the floor (see Figure 3.2). The configuration space is hence $Q = \mathbb{R}^2 \times \mathbb{S}^1 \times \mathbb{S}^1 \times \mathbb{S}^1$.

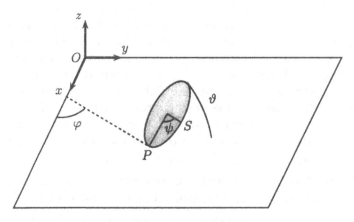

Fig. 3.2. The rolling disk

The Lagrangian of this mechanical system is of natural type, $L = T - V$. The kinetic energy is given by

$$
\begin{aligned}
T = {}& \frac{1}{2}m \left(\dot{x}^2 + \dot{y}^2 + R^2\dot{\vartheta}^2 + R^2\dot{\varphi}^2 \sin^2 \vartheta \right) \\
& - mR \left(\dot{\vartheta} \cos \vartheta (\dot{x} \sin \varphi - \dot{y} \cos \varphi) + \dot{\varphi} \sin \vartheta (\dot{x} \cos \varphi + \dot{y} \sin \varphi) \right) \\
& + \frac{1}{2}I_1 \left(\dot{\vartheta}^2 + \dot{\varphi}^2 \cos^2 \vartheta \right) + \frac{1}{2}I_2 \left(\dot{\psi} + \dot{\varphi} \sin \vartheta \right)^2 \, ,
\end{aligned}
$$

where m is the mass, and I_1, I_2 are the principal moments of inertia of the disk. The potential energy of the disk is

$$
V = mgR \cos \vartheta \, .
$$

The condition that the disk rolls without sliding on the horizontal plane means that the instantaneous velocity of the point of contact of the disk is equal to zero at all times, otherwise the disk would necessarily slip. This gives rise to the following constraints

$$
\phi_1 = \dot{x} - (R \cos \varphi)\dot{\psi} = 0, \quad \phi_2 = \dot{y} - (R \sin \varphi)\dot{\psi} = 0, \tag{3.7}
$$

where R is the radius of the disk. Note that both ϕ_1 and ϕ_2 are nonholonomic in the sense that they cannot be "integrated" and expressed in terms of

$(x, y, \psi, \varphi, \vartheta)$ only. Both constraints are also linear in the velocities. They determine the following distribution of allowed velocities,

$$D = \mathrm{span}\left\{R\cos\psi\frac{\partial}{\partial x} + R\sin\psi\frac{\partial}{\partial y} + \frac{\partial}{\partial\psi}, \frac{\partial}{\partial\varphi}, \frac{\partial}{\partial\vartheta}\right\}.$$

3.2.2 A homogeneous ball on a rotating table

A homogeneous sphere of radius r and unit mass $(m = 1)$ rolls without sliding on a horizontal table which rotates with non constant angular velocity $\Omega(t)$ about a vertical axis through one of its points. Apart from the constant gravitational force, no other external forces are assumed to act on the sphere.

Choose a Cartesian reference frame with origin at the center of rotation of the table and z-axis along the rotation axis. Let (x, y) denote the position of the point of contact of the sphere with the table. The configuration space is $Q = \mathbb{R}^2 \times SO(3)$, where $SO(3)$ may be parameterized by the Eulerian angles (φ, θ, ψ) (see Figure 3.3).

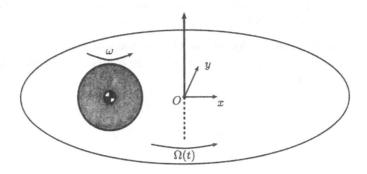

Fig. 3.3. A ball on a rotating table

The potential energy is constant, so we may put $V = 0$. In addition, since we do not consider external forces, the Lagrangian is given by the kinetic energy of the sphere, i.e.

$$L(= T) = \frac{1}{2}\left(\dot{x}^2 + \dot{y}^2 + k^2(\dot{\theta}^2 + \dot{\varphi}^2 + \dot{\psi}^2 + 2\dot{\varphi}\dot{\psi}\cos\theta)\right).$$

The constraint equations, expressing the condition of rolling without slipping, are

$$\dot{x} - r\dot{\theta}\sin\psi + r\dot{\varphi}\sin\theta\cos\psi = -\Omega(t)y$$
$$\dot{y} + r\dot{\theta}\cos\psi + r\dot{\varphi}\sin\theta\sin\psi = \Omega(t)x.$$

Note that in this case the constraints are nonholonomic, time-dependent and affine in the velocities. They determine the following space of allowed velocities,

$$M = \text{span}\left\{ r\sin\psi\frac{\partial}{\partial x} - r\cos\psi\frac{\partial}{\partial y} + \frac{\partial}{\partial\theta}, \right.$$
$$\left. r\sin\theta\cos\psi\frac{\partial}{\partial x} + r\sin\theta\sin\psi\frac{\partial}{\partial y} + \frac{\partial}{\partial\varphi}, \frac{\partial}{\partial\psi} \right\} + \gamma,$$

where γ is the time-dependent vector field defined by

$$\gamma = -\Omega(t)y\frac{\partial}{\partial x} + \Omega(t)x\frac{\partial}{\partial y}.$$

It is well known that the problem of the rolling ball on a rotating table can be most elegantly treated exploiting the symmetry of the problem by means of the formalism of "quasi-coordinates", with the angular velocity playing the role of "quasi-velocity". For a classical treatment of the theory of quasi-coordinates, we refer to [188, 201]. A more recent discussion of the use of quasi-coordinates in the study of nonholonomic systems with symmetry, within a differential geometrical setting, can be found in [120].

In the following we briefly discuss the approach of "quasi-coordinates". In terms of the Euler's angles we have that the components of the angular velocity of the sphere read

$$\omega_x = \dot\theta\cos\psi + \dot\varphi\sin\theta\sin\psi,$$
$$\omega_y = \dot\theta\sin\psi - \dot\varphi\sin\theta\cos\psi,$$
$$\omega_z = \dot\psi + \dot\varphi\cos\theta.$$

Let us take $(x, y, \theta, \varphi, \psi, \dot x, \dot y, \omega_x, \omega_y, \omega_z)$ as coordinates on the tangent bundle, with the components of the angular velocity now being regarded as the "velocities" associated with some quasi-coordinates. Following the classical treatments, such as [188, 201], we put $\omega_x = \dot q^1, \omega_y = \dot q^2, \omega_z = \dot q^3$, with q^1, q^2, q^3 denoting quasi-coordinates. The latter merely have a symbolic meaning in the sense that in the present example, for instance, the partial derivative operators $\partial/\partial q^i$ should be interpreted as linear combinations of the partial derivatives with respect to Euler's angles. Also to the differential forms dq^i one should attach the appropriate meaning, i.e. they do not represent exact differentials but, instead, we should read them as $dq^1 = \cos\psi\, d\theta + \sin\theta\sin\psi\, d\varphi$, etc. The Lagrangian is now given by

$$L = \frac{1}{2}\left(\dot x^2 + \dot y^2 + k^2(\omega_x^2 + \omega_y^2 + \omega_z^2)\right)$$

and the nonholonomic constraints read simply

$$\dot{x} - r\omega_y = -\Omega(t)y, \qquad \dot{y} + r\omega_x = \Omega(t)x \ .$$

Finally, the space M is described as

$$M = \text{span}\left\{ r\frac{\partial}{\partial x} + \frac{\partial}{\partial q^2}, -r\frac{\partial}{\partial y} + \frac{\partial}{\partial q^1}, \frac{\partial}{\partial q^3} \right\} + \gamma \ .$$

3.2.3 The Snakeboard

The Snakeboard [29, 148, 195] is a variant of the skateboard in which the passive wheel assemblies can pivot freely about a vertical axis. By coupling the twisting of the human torso with the appropriate turning of the wheels (where the turning is controlled by the rider's foot movement), the rider can generate a snake-like locomotion pattern without having to kick off the ground.

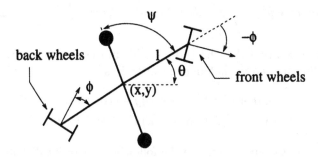

Fig. 3.4. The Snakeboard model. Figure courtesy of Jim Ostrowski.

A simplified model is shown in Figure 3.4. We assume that the front and rear wheel axles move through equal and opposite rotations. This is based on the observations of human Snakeboard riders who use roughly the same phase relationship. A momentum wheel rotates about a vertical axis through the center of mass, simulating the motion of a human torso. Figure 3.5 shows a robotic prototype of the Snakeboard built in Caltech [195].

The position and orientation of the Snakeboard is determined by the coordinates of the center of mass (x, y) and its orientation θ. The shape variables are (ψ, ϕ), so the configuration space is $Q = SE(2) \times \mathbb{S}^1 \times \mathbb{S}^1$. The physical parameters for the system are the mass of the board, m; the inertia of the rotor, J_r; the inertia of the wheels about the vertical axes, J_w; and the half-length of the board, l. A key component of the Snakeboard is the use of the rotor inertia to drive the body. To keep the rotor and body inertias on similar scales, we make the additional simplifying assumption [29, 196] that the inertias of the system satisfy $J + J_r + 2J_w = ml^2$.

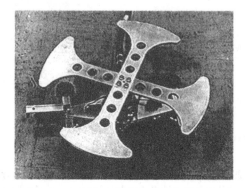

Fig. 3.5. A prototype robotic Snakeboard. Figure courtesy of Jim Ostrowski.

The Lagrangian of the system is given by its kinetic energy

$$L = \frac{1}{2}m(\dot{x}^2 + \dot{y}^2) + \frac{1}{2}(J + J_r + 2J_w)\dot{\theta}^2 + J_r\dot{\theta}\dot{\psi} + \frac{1}{2}J_r\dot{\psi}^2 + J_w\dot{\phi}^2 \,.$$

The assumption that the wheels do not slip in the direction of their axles yields the following two nonholonomic constraints

$$-\sin(\theta + \phi)\dot{x} + \cos(\theta + \phi)\dot{y} - l\cos\phi\,\dot{\theta} = 0\,,$$
$$-\sin(\theta - \phi)\dot{x} + \cos(\theta - \phi)\dot{y} + l\cos\phi\,\dot{\theta} = 0\,.$$

They both are linear and determine the following constraint distribution

$$D = \mathrm{span}\left\{-r\cos\phi\cos(\theta - \phi)\frac{\partial}{\partial x} - r\cos\phi\sin(\theta - \phi)\frac{\partial}{\partial y} + \sin\phi\frac{\partial}{\partial \theta}, \frac{\partial}{\partial \psi}, \frac{\partial}{\partial \phi}\right\}.$$

3.2.4 A variation of Benenti's example

To end with this series of simple examples, we will treat some variants considered in [103] of an example proposed by Benenti [19].

Consider the problem of two point masses forced to move on a plane with parallel velocities. The configuration space is $Q = \mathbb{R}^2 \times \mathbb{R}^2$. We denote by (x_1, y_1) the position of the particle of mass m_1 and by (x_2, y_2) the position of the particle of mass m_2. The Lagrangian of the system is the kinetic energy,

$$L = \frac{1}{2}m_1(\dot{x}_1^2 + \dot{y}_1^2) + \frac{1}{2}m_2(\dot{x}_2^2 + \dot{y}_2^2)\,.$$

The constraint on the velocities is given by the function on the tangent bundle TQ,

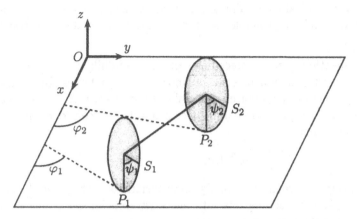

Fig. 3.6. A variation of Benenti's system.

$$\Psi = \dot{x}_1 \dot{y}_2 - \dot{x}_2 \dot{y}_1 = 0 \,.$$

The constraint Ψ is a genuine nonlinear nonholonomic constraint.

A variation of this problem is the following: let us substitute the point masses by disks rolling without sliding on the plane (see Figure 3.6) . Then the system consists of two copies of the system discussed in Section 3.2.1, now rolling *vertically* and with the additional nonlinear constraint on the velocities,

$$\Psi = \mathbf{v}_1 \wedge \mathbf{v}_2 = 0 \,, \qquad (3.8)$$

where $\mathbf{v}_i$ denotes the velocity of the center of mass of the ith disk. Here, we are identifying the vector $\mathbf{v}_1 \wedge \mathbf{v}_2$ orthogonal to the (x, y)-plane with its projection onto the z-axes, so that the constraint Ψ takes real values.

Replacing in (3.8) the constraints given by the rolling conditions, eq. (3.7), this can be rewritten as,

$$\Psi = R_1 R_2 \sin(\varphi_2 - \varphi_1) \dot{\psi}_1 \dot{\psi}_2 = 0 \,.$$

The constraint means that either one of the disks is not rolling ($\dot{\psi}_i = 0$) and the angles φ_i are arbitrary or, if the disks are rolling ($\dot{\psi}_i \neq 0$), then $\varphi_1 = \varphi_2 = \varphi$ and the nonlinear constraint Ψ is redundant: the two disks roll independently and freely keeping parallel directions.

3.3 The Lagrange-d'Alembert principle

IN this section, we derive the equations of motion for nonholonomic systems subject to affine constraints. We start by explaining important physical notions such as ideal constraint and virtual displacement. After stating the Lagrange-d'Alembert principle, we turn to the mathematical derivation of the equations.

As we mentioned in the preceding section, the fulfilling of the constraints requires the introduction of some unknown reaction forces. In connection with the problem of eliminating this unknown character, it is customary to introduce the concept of virtual displacements. Let us consider a number of nonholonomic constraints

$$\sum a_{iA}\dot{q}^A + a_{i0} = 0, \quad 1 \leq i \leq k.$$

Virtual variations are infinitesimally small variations of the generalized coordinates (q^A) that are compatible with the constraints imposed on the system, that is, satisfy the equations

$$\sum a_{iA}\delta q^A = 0.$$

Displacements of a system corresponding to virtual variations of its generalized coordinates are called *virtual displacements*. We must emphasize that we are speaking of virtual displacements of the system at a given instant of time and for a given configuration of the system.

Definition 3.3.1 (Principle of Virtual Work). *Nonholonomic constraints on a nonholonomic Lagrangian system are said to be ideal if the reaction forces associated with them perform no work in any virtual displacement of the system.*

Throughout the book, we always assume that we are dealing with ideal constraints. This is the case, for example, for constraints of non-sliding type.

Classically, an *admissible path* is a trajectory of the system consistent with the constraints. This corresponds to curves $c : [a, b] \longrightarrow Q$ connecting two points in Q which satisfy the constraints determined by M, meaning that the tangent vectors of the curve belong to M, $\dot{c}(t) \in M_{c(t)}$. The set of all such curves which are twice differentiable will be denoted by

$$\tilde{C}^2(q_0, q_1, [a, b]) = \{c : [a, b] \longrightarrow Q \mid c \in C^2(q_0, q_1, [a, b]) \text{ and } \dot{c}(t) \in M_{c(t)}\}.$$

Note that $\tilde{C}^2(q_0, q_1, [a, b])$ is a subset of $C^2(q_0, q_1, [a, b])$ introduced in Section 3.1.1. Given a curve $c \in \tilde{C}^2(q_0, q_1, [a, b])$, the set of all possible virtual variations along c is the following subset of $T_c C^2(q_0, q_1, [a, b])$,

$$\mathcal{V}_c = \{X \in T_c C^2(q_0, q_1, [a, b]) \mid X(t) \in \mathcal{D}_{c(t)}, \forall t \in [a, b]\},$$

where $M = (\mathcal{D}, \gamma)$. The relevant equations describing the dynamic behavior of systems subject to general ideal constraints are obtained through the Lagrange-d'Alembert principle, which we state next.

Definition 3.3.2 (Lagrange-d'Alembert principle). *Let c be an admissible path in $\tilde{C}^2(q_0, q_1, [a, b])$. Then c is a motion of the nonholonomic Lagrangian system (L, M) if*

$$d\mathcal{J}(c)(X) = 0 \, , \quad \text{for all } X \in \mathcal{V}_c \, .$$

Remark 3.3.3. This formulation of the Lagrange-d'Alembert principle is known as Hölder principle. There is another equivalent formulation due to Gauss, the Gauss' principle of least constraint (see [10]).

Remark 3.3.4. Note that the Lagrange-d'Alembert principle is not *truly* variational, since the motion of the nonholonomic system is not a critical point of any functional in the sense of the Calculus of Variations. If we followed a variational approach to the dynamics of the constrained Lagrangian system, we could consider the restriction of the action functional $\mathcal{J}$ to the submanifold $\tilde{C}^2(q_0, q_1, [a, b])$ of $C^2(q_0, q_1, [a, b])$, $\tilde{\mathcal{J}} : \tilde{C}^2(q_0, q_1, [a, b]) \longrightarrow \mathbb{R}$ and try to find extremals of this restricted functional.

This point of view, that is, extremizing the action functional among all the curves satisfying the constraints, gives rise to the so-called *Vakonomic Mechanics*, which will not be treated here (see [10]). This mechanics is thus obtained through a purely variational principle by imposing the fulfilling of the constraints on the variations themselves, not on the infinitesimal variations, as it does the Lagrange-d'Alembert principle. The resulting equations are not equivalent in general to the nonholonomic equations of motion, although there exist cases in which the nonholonomic solutions can be regarded as vakonomic solutions: for instance, the holonomic case or the rolling disk [26, 70, 145]. It is generally assumed that the relevant equations describing the dynamical behavior of systems subject to constraints of non-sliding type are obtained through the Lagrange-d'Alembert principle [145, 260], though vakonomic dynamics or the constrained variational problem is the natural setting of many optimization problems encountered in Economic Growth theories, Engineering, Control Theory, motion of micro-organisms, etc.

The derivation of the constrained variational equations can present some anomalies: in some cases, it could happen that a given admissible path does not admit enough variations in $\tilde{C}^2(q_0, q_1, [a, b])$ and at the same time extremizes the restricted action functional $\tilde{\mathcal{J}}$. Such solutions are called singular or abnormal, in contrast with the regular or normal case. A classical example of the abnormal situation is due to Carathéodory (see [10]). In recent years, there has been several works devoted to the existence of such type of solutions, in connection with sub-Riemannian geometry (see [132, 153, 179]).

Now, we derive the equations of motion for a nonholonomic system. In a local description, a constraint submanifold M of codimension k can be defined by the vanishing of k independent functions ϕ_i (the constraint functions). If the constraints are affine, the functions ϕ_i can be taken to be of the form $\phi_i(q, \dot{q}) = \mu_{iA}(q)\dot{q}^A + \mu_{i0}(q)$, $i = 1, \ldots, k$. Therefore, for the virtual displacements, we have by definition that

$$X \in \mathcal{V}_c \iff \mu_{iA}X^A = 0,\ 1 \le i \le k.$$

The Lagrange-d'Alembert principle asserts that $c \in \tilde{C}^2(q_0, q_1, [a, b])$ is a motion of the nonholonomic system if $d\mathcal{J}(c)(X) = 0$, for all $X \in \mathcal{V}_c$. From (3.1) and using the Lagrange multipliers technique, this is equivalent to

$$\frac{d}{dt}\left(\frac{\partial L}{\partial \dot{q}^A}\right) - \frac{\partial L}{\partial q^A} = \lambda^i \mu_{iA}, \qquad (3.9)$$

which, together with the constraint equations $\mu_{iA}\dot{q}^A + \mu_{i0}(q) = 0$ ($i = 1, \ldots, k$), determine the dynamics of the nonholonomic system. Here, the λ^i are the Lagrange multipliers to be determined. The right-hand side of eq. (3.9) precisely represents the "reaction force" induced by the constraints.

External forces can be incorporated into the discussion in the same way we exposed for the unconstrained case. If $F^1, \ldots, F^m$ are external forces acting on the nonholonomic system, define $F = \sum_{j=1}^m F^j = F_A dq^A$. Then the equations of motion are given by

$$\begin{cases} \dfrac{d}{dt}\left(\dfrac{\partial L}{\partial \dot{q}^A}\right) - \dfrac{\partial L}{\partial q^A} = \lambda^i \mu_{iA} + F_A \\ \mu_{iA}\dot{q}^A + \mu_{i0}(q) = 0. \end{cases} \qquad (3.10)$$

Before ending this section, we make a remark concerning nonlinear nonholonomic constraints. They do not occur frequently in real physical problems and there is little agreement in the literature about the right mathematical model to incorporate them. The most widely employed model makes use of the so-called "Chetaev's rule". In classical terms, this rule can be interpreted as extending the definition of the concept of virtual displacements to the case where nonlinear constraints are present. The assumption then is that the analogous version of the Lagrange-d'Alembert principle remains valid. Most authors accept this model, but criticism about its physical correctness has been formulated for instance in [157, 202]. Its geometrical implementation is practically the same as in the linear case, so this can be seen as a good reason for examining the Chetaev model from a purely mathematical perspective.

If $\phi_i(q^A, \dot{q}^A) = 0$ determine locally the nonlinear submanifold M of TQ, then the Chetaev's rule implies that the equations of motion for the constrained Lagrangian system are

$$\frac{d}{dt}\left(\frac{\partial L}{\partial \dot{q}^A}\right) - \frac{\partial L}{\partial q^A} = \lambda^i \frac{\partial \phi_i}{\partial \dot{q}^A}. \qquad (3.11)$$

together with the constraint equations.

3.4 Geometric formalizations

A S mentioned in the Introduction, considerable efforts have been made to adapt and extend several ideas and techniques from the geometric treatment of unconstrained problems to the study of systems with nonholonomic constraints. The list of contributions is extensive and it would be nearly impossible to make a complete account of them all here. Roughly speaking, we can say that the subject has been approached from quite some different points of view, including a Lagrangian approach (using reduction theory [29, 195], tangent bundle geometry [57, 137] and jet theory [126, 213]), a Hamiltonian approach [18, 35, 80, 122, 156] and a formulation in terms of (almost-)Poisson structures [53, 123, 241]. In this section, we review two interesting approaches to nonholonomic dynamics: the symplectic approach, which in particular is well suited for the treatment of nonholonomic systems with symmetry, and the affine connection approach.

3.4.1 Symplectic approach

In the presence of nonholonomic constraints, the formalization of the equations of motion must be modified in order to incorporate the constraints into the picture. We shall always assume that the constraints verify the so-called *admissibility condition* (see e.g. [137]), i.e. for all $x \in M$

$$\dim (T_x M)^o = \dim S^* ((T_x M)^o) \, ,$$

where the annihilator of $T_x M$ is taken in $T_x^* TQ$. In coordinates, if M is locally defined by the annihilation of $\phi_1, \ldots, \phi_k$, this condition has the interpretation that the rank of the matrix

$$\frac{\partial (\phi_1, \ldots, \phi_k)}{\partial (\dot{q}^1, \ldots, \dot{q}^n)}$$

is k for any choice of coordinates $(q^A, \dot{q}^A)$ in TQ.

Next, we define the bundle of reaction forces or *Chetaev bundle*, the geometric representation for the constraint forces. This is given by $S^*(TM^o) \subset T^* TQ_{|M}$. Consider then the distribution F on TQ along M whose annihilator is precisely this bundle, $F^o = S^*((TM)^o)$.

The equations of motion for the nonholonomic system are then given by

$$\begin{cases} (i_X \omega_L - dE_L)_{|M} \in F^o \, , \\ X_{|M} \in TM \, . \end{cases} \tag{3.12}$$

The nonholonomic system will have a unique solution X if it satisfies the *compatibility condition* $F^\perp \cap TM = 0$. If the Hessian of L with respect to the velocities is definite, then this condition is automatically satisfied, and this is the usual case in Mechanics since $L = T - V$, where T is the kinetic energy of a Riemannian metric on Q, and V is the potential energy. If the compatibility condition is not satisfied, then the Gotay-Nester algorithm [93, 94] can be adapted to deal with the nonholonomic case and obtain a submanifold where the dynamics is consistent (see [137]).

If $M = (\mathcal{D}, \gamma)$ is locally described by equations of the form

$$\phi_i = \mu_{iA}(q)\dot{q}^A + \mu_{i0}(q) = 0, \qquad i = 1, \ldots k,$$

with $k = \operatorname{codim}(M)$, putting

$$C_{ij} = -\mu_{iA} W^{AB} \mu_{jB}, \quad i, j = 1, \ldots k,$$

where (W^{AB}) is the inverse of the Hessian matrix

$$\left(\frac{\partial^2 L}{\partial \dot{q}^A \partial \dot{q}^B} \right)$$

compatibility locally translates into regularity of the matrix (C_{ij}) (cf. [137], where the compatibility condition for a nonholonomic system was called the regularity condition).

If we define $Z_i = \sharp_L(S^*(d\phi_i))$, $1 \le i \le k$, it is easy to see that $F^\perp$ is locally generated by $Z_1, \ldots, Z_k$. We have the expression in local coordinates

$$Z_i = \mu_{iA} W^{AB} \frac{\partial}{\partial \dot{q}^B}.$$

Under the compatibility condition, a simple counting of dimensions yields that $T_x TQ = F_x^\perp \oplus T_x M$, $\forall x \in M$, which gives rise to two complementary projectors

$$\mathcal{P}_x : T_x TQ \longrightarrow T_x M, \qquad \mathcal{Q}_x : T_x TQ \longrightarrow F_x^\perp.$$

A direct calculation shows that the constrained dynamics X is obtained by projecting the unconstrained Euler-Lagrange vector field Γ_L (restricted to M) to TM with respect to this decomposition, i.e. $X = \mathcal{P}(\Gamma_{L|M})$. Therefore, we have that $X = \Gamma_{L|M} - \mathcal{Q}(\Gamma_{L|M}) = \Gamma_{L|M} - D^j Z_j$ (the summation in j is understood). Since X is tangent to M, then

$$X(\phi_i) = \Gamma_{L|M}(\phi_i) - D^j Z_j(\phi_i) = 0, \quad 1 \le i \le k.$$

As $Z_j(\phi_i) = -C_{ij}$, we conclude that $D^j = -C^{ji} \Gamma_{L|M}(\phi_i)$, where (C^{ij}) is the inverse of the matrix (C_{ij}).

It should be pointed out that the solution X of (3.12) satisfies automatically the SODE condition along M, i.e. $S(X)_{|M} = \Delta_{|M}$. This implies that, in local coordinates, the integral curves of X on M are of the form $(q^A(t), \dot{q}^A(t) \equiv \frac{dq^A}{dt}(t))$, whereby the $q^A(t)$ are solutions of the system of differential equations (3.9), together with the constraint equations $\mu_{iA}(q)\dot{q}^A + \mu_{i0}(q) = 0$, $i = 1, \ldots, k$. The local coordinate expression for X reads

$$X = \dot{q}^A \frac{\partial}{\partial q^A} + W^{AB} \left(\frac{\partial L}{\partial q^B} - \frac{\partial \hat{p}_B}{\partial q^C} \dot{q}^C \right.$$

$$+ C^{ij} \frac{\partial \mu_{iD}}{\partial q^C} \mu_{jB} \dot{q}^C \dot{q}^D + W^{CD} \left(\frac{\partial L}{\partial q^D} - \dot{q}^E \frac{\partial \hat{p}_D}{\partial q^E} \right) C^{ij} \mu_{jB} \mu_{iC} \right) \frac{\partial}{\partial \dot{q}^A} \,,$$

where, for ease of writing, we have put $\hat{p}_A = \dfrac{\partial L}{\partial \dot{q}^A}$, $A = 1, \ldots, n$.

In [18, 80, 136], the following alternative approach has been proposed. The compatibility condition is equivalent to the condition that the distribution $F \cap TM$ on M determines a symplectic vector bundle on M. Then, $T_x TQ = (F_x \cap T_x M) \oplus (F_x \cap T_x M)^\perp$, $\forall x \in M$, with induced projectors

$$\bar{\mathcal{P}}_x : T_x TQ \longrightarrow (F_x \cap T_x M), \quad \bar{\mathcal{Q}}_x : T_x TQ \longrightarrow (F_x \cap T_x M)^\perp \,.$$

It should be noted that, in general, the projection of the unconstrained dynamics Γ_L by $\bar{\mathcal{P}}$ will not produce the constrained dynamics X. However, in the case of homogeneous constraints, we have

$$\bar{\mathcal{P}}(\Gamma_L) = \mathcal{P}(\Gamma_L) = X$$

along M. The dynamics X is called the *distributional* Hamiltonian vector field of E_L with respect to $(\omega_L, M, F \cap TM)$.

By means of the projector $\bar{\mathcal{P}}$, we can define the so-called *nonholonomic bracket on* M, $\{\cdot, \cdot\}_M$, in the following manner (see [53, 50, 123, 157, 241]). Consider $\lambda, \sigma : M \longrightarrow \mathbb{R}$ and take $\tilde{\lambda}, \tilde{\sigma}$ arbitrary extensions to TQ, $\tilde{\lambda} \circ j_M = \lambda$, $\tilde{\sigma} \circ j_M = \sigma$, with $j_M : M \hookrightarrow TQ$. Then

$$\{\lambda, \sigma\}_M = \omega_L(\bar{\mathcal{P}}(X_{\tilde{\lambda}}), \bar{\mathcal{P}}(X_{\tilde{\sigma}})) \circ j_M \,.$$

It is a routine to verify that this bracket is well-defined and is indeed an almost-Poisson bracket. In general, $\{\cdot, \cdot\}_M$ does not verify the Jacobi identity, except if the constraints are holonomic [123, 241]. This almost-Poisson bracket is very important because, in the case of homogeneous constraints, it gives the evolution of the constrained dynamics in the following sense: for any function $f \in C^\infty(M)$, its evolution along integral curves of X on M is given by

$$\dot{f} = X(f) = \{f, E_L\}_M \,.$$

Note that the homogeneity of the constraints and the SODE character of X, implies automatically that $X \in F$, since for $\alpha \in F^o$, we have that

$\alpha = S^*\tilde{\alpha}$, with $\tilde{\alpha} \in TM^o$, and therefore $\alpha(X) = \tilde{\alpha}(SX) = \tilde{\alpha}(\Delta) = 0$. Making use of this fact, we can prove the *nonholonomic Noether theorem* [80, 136, 220], which ensures us when a function φ is a constant of the motion of the dynamics.

Theorem 3.4.1. *Consider a nonholonomic Lagrangian system with homogeneous constraints. A function $\varphi : TQ \longrightarrow \mathbb{R}$ is a constant of the motion of the dynamics X if and only if the energy is constant along the integral curves of the vector field $\bar{\mathcal{P}}(X_\varphi)$, that is, $\bar{\mathcal{P}}(X_\varphi)(E_L) = 0$.*

Proof. Note that along M

$$X(\varphi) = d\varphi(X) = \omega_L(X_\varphi, X) = \omega_L(\bar{\mathcal{P}}(X_\varphi), X),$$

where in the last equality we have used the fact that $X \in F \cap TM$. By skew-symmetry, $\omega_L(\bar{\mathcal{P}}(X_\varphi), X) = -(i_X\omega_L)\bar{\mathcal{P}}(X_\varphi)$. Now, from equation (3.12), we have that $i_X\omega_L = dE_L + \beta$, with $\beta \in F^o$. Since $\bar{\mathcal{P}}(X_\varphi) \in F$, we conclude

$$X(\varphi) = -(dE_L)\bar{\mathcal{P}}(X_\varphi) = -\bar{\mathcal{P}}(X_\varphi)(E_L).$$

$\square$

Before ending this section, we recall that for nonholonomic Lagrangian systems with constraints which are linear (or, more general, homogeneous) in the velocities, the energy E_L is a conserved quantity. This can be deduced as follows. From eq. (3.12), we can write

$$i_X\omega_L = dE_L + \beta, \tag{3.13}$$

with $\beta \in F^o$. Contracting the latter equation with the dynamics vector field and using the fact that $X \in F$ due to the homogeneity of the constraints, we get $X(E_L) = 0$.

Another interesting property to which we will refer in Chapter 7 is the following: from (3.13), the evolution of the symplectic form along the trajectories of the system is given by

$$\mathcal{L}_X\omega_L = i_X d\omega_L + d i_X\omega_L = d\beta. \tag{3.14}$$

In the absence of constraints, the symplectic form is preserved by the flow of the equations, a fact already discovered by Lagrange himself.

3.4.2 Affine connection approach

In this section we will describe from a different point of view the dynamics of nonholonomic systems of mechanical type, namely, in terms of the theory of affine connections. This formalism will be useful later in the study of Chaplygin systems in Chapter 5 and the control aspects of simple mechanical systems in Chapter 8. Here we restrict our attention to the case of linear constraints.

Let Q be the configuration space of a mechanical system with Lagrangian

$$L(v) = \frac{1}{2}g(v,v) - V \circ \tau_Q(v), \ v \in T_qQ,$$

where g is a Riemannian metric on Q and $V : Q \to \mathbb{R}$ is the potential energy function. We denote by ∇^g the (covariant derivative operator of the) Levi-Civita connection associated with the metric g. Let $\mathcal{F} = \{F^1, \ldots, F^m\}$ be a set of m linearly independent 1-forms on Q, which physically correspond to forces or torques depending on configurations only (recall the convention adopted in Section 2.9). We shall denote by $Y_1, \ldots, Y_m$ the vector fields defined as $Y_i = b_g^{-1}(F^i)$.

It is well known that a curve $c : I \to Q$ is a solution of the forced Euler-Lagrange equations (3.3) for the Lagrangian L and the forces $\mathcal{F}$ iff

$$\nabla^g_{\dot{c}(t)}\dot{c}(t) = -\operatorname{grad} V(c(t)) + \sum_{i=1}^m Y_i(c(t)), \tag{3.15}$$

where the gradient is also considered with respect to the metric g. For non-holonomic systems there is a similar description. The second-order differential equations (3.10) for the mechanical system with Lagrangian L, forces $\mathcal{F}$ and constraints $\mathcal{D}$, can be written intrinsically as

$$\nabla^g_{\dot{c}(t)}\dot{c}(t) + \operatorname{grad} V(c(t)) - \sum_{i=1}^m Y_i(c(t)) \in \mathcal{D}^\perp_{\dot{c}(t)}, \ \dot{c}(t) \in \mathcal{D}_{c(t)}, \tag{3.16}$$

where $\mathcal{D}^\perp$ here denotes the g-orthogonal complement to $\mathcal{D}$ (see e.g. [27, 143, 144, 235, 244]). Alternatively, if we denote by $\mathcal{P} : TQ \longrightarrow \mathcal{D}$, $\mathcal{Q} : TQ \longrightarrow \mathcal{D}^\perp$ the complementary g-orthogonal projectors, we can define

$$\overline{\nabla}_X Y = \nabla^g_X Y + (\nabla^g_X \mathcal{Q})(Y),$$

and verify that it is indeed an affine connection [143]. Now, writing (3.16) as

$$\nabla^g_{\dot{c}(t)}\dot{c}(t) + \operatorname{grad} V(c(t)) - \sum_{i=1}^m Y_i(c(t)) = \lambda(t) \in \mathcal{D}^\perp_{\dot{c}(t)},$$

and applying $\mathcal{Q}$, we get

$$\lambda(t) = \mathcal{Q}(\nabla^g_{\dot{c}(t)}\dot{c}(t)) + \mathcal{Q}(\operatorname{grad} V(c(t))) - \mathcal{Q}(\sum_{i=1}^m Y_i(c(t))).$$

In addition, we have that

$$(\nabla^g_{\dot{c}(t)}\mathcal{Q})(\dot{c}(t)) = \nabla^g_{\dot{c}(t)}\mathcal{Q}(\dot{c}(t)) - \mathcal{Q}(\nabla^g_{\dot{c}(t)}\dot{c}(t)) = -\mathcal{Q}(\nabla^g_{\dot{c}(t)}\dot{c}(t)),$$

since $\dot{c}(t) \in \mathcal{D}_{c(t)}$. Therefore, the nonholonomic equations of motion (3.16) can be rewritten as

$$\overline{\nabla}_{\dot{c}(t)}\dot{c}(t) = -\mathcal{P}(\operatorname{grad} V(c(t))) + \sum_{i=1}^{m} \mathcal{P}(Y_i(c(t))), \qquad (3.17)$$

and where we select the initial velocity in $\mathcal{D}$ (cf. [144]).

It can be easily deduced from its definition that the connection $\overline{\nabla}$ restricts to $\mathcal{D}$, that is,

$$\overline{\nabla}_X Y = \mathcal{P}(\nabla^g_X Y) \in \mathcal{D},$$

for all $Y \in \mathcal{D}$ and $X \in \mathfrak{X}(Q)$. The class of affine connections that restrict to a given distribution has been studied in [143]. In particular, such a behavior implies that the distribution $\mathcal{D}$ is *geodesically invariant*, that is, for every geodesic $c(t)$ of $\overline{\nabla}$ starting from a point in $\mathcal{D}$, $\dot{c}(0) \in \mathcal{D}_{c(0)}$, we have that $\dot{c}(t) \in \mathcal{D}_{c(t)}$. In [143], a nice property is derived which characterizes geodesically invariant distributions in terms of the so-called symmetric product of vector fields. The *symmetric product* of two vector fields $X, Y \in \mathfrak{X}(Q)$ is defined by

$$\langle X : Y \rangle = \nabla_X Y + \nabla_Y X.$$

This property asserts that $\mathcal{D}$ is geodesically invariant if and only if we have that $\langle X : Y \rangle \in \mathcal{D}$, for all $X, Y \in \mathcal{D}$. Note in passing that the symmetric product of vector fields is a differential geometric concept, first appeared in the study of gradient dynamical systems [79, 238], with important applications to control theory. For instance, as we shall see in Chapter 8, it plays a fundamental role in the controllability analysis and series expansions results of mechanical control systems.

We conclude this section by presenting some properties of the nonholonomic affine connection $\overline{\nabla}$.

Proposition 3.4.2. *For all $Z \in \mathfrak{X}(Q)$ and $X, Y \in \mathcal{D}$, we have that*

$$Z(g(X,Y)) = g(\overline{\nabla}_Z X, Y) + g(X, \overline{\nabla}_Z Y).$$

That is, $\overline{\nabla}$ is a metric connection with respect to g.

Proof. In view of the definition of $\overline{\nabla}$, we have that

$$\begin{aligned}
g(\overline{\nabla}_Z X, Y) + g(X, \overline{\nabla}_Z Y) &= g(\nabla^g_Z X, Y) + g(X, \nabla^g_Z Y) \\
&\quad + g((\nabla^g_Z \mathcal{Q})(X), Y) + g(X, (\nabla^g_Z \mathcal{Q})(Y)) = Z(g(X,Y)),
\end{aligned}$$

since $(\nabla^g_Z \mathcal{Q})(X), (\nabla^g_Z \mathcal{Q})(Y) \in \mathcal{D}^\perp$ (see Proposition 6.1 in [143]). $\qquad \square$

We derive from this proposition that the connection $\overline{\nabla}$ also has the following property: parallel transport is an isometry along the distribution $\mathcal{D}$.

A direct computation shows that the torsion of $\overline{\nabla}$ is the skew-symmetric $(1, 2)$-tensor field

$$\overline{T}(X, Y) = (\nabla^g_X \mathcal{Q})(Y) - (\nabla^g_Y \mathcal{Q})(X).$$

Observe that if $X, Y \in \mathcal{D}$, we have that

$$\overline{T}(X, Y) = \nabla_X(\mathcal{Q}(Y)) - \mathcal{Q}(\nabla_X Y) - \nabla_Y(\mathcal{Q}(X)) + \mathcal{Q}(\nabla_Y X)$$
$$= -\mathcal{Q}(\nabla_X Y - \nabla_Y X) = -\mathcal{Q}([X, Y])$$

and $\overline{T}(X, Y) \in \mathcal{D}^\perp$. It is easy now to conclude that if $\overline{\nabla}$ is a torsion-free connection, then $\mathcal{D}$ is an integrable distribution.

4 Symmetries of nonholonomic systems

IN this chapter, we develop a theory of reduction and reconstruction of the dynamics for nonholonomic systems with symmetry, making use of the symplectic formalism described in Chapter 3.

Nonholonomic systems with symmetry have been studied in the classical books of Mechanics such as [6, 198, 201, 257]. The first modern treatment is due to Koiller [120] (see also [244]), who considered the case of Chaplygin systems. His work produced a renewed interest in the subject, with contributions that are based on different viewpoints, such as the Hamiltonian formalism [18], Lagrangian reduction [29], the geometry of the tangent bundle [50, 68] or Poisson methods [156, 241], among others. Several relevant results contained in these works will be reviewed along the chapter.

The exposition is organized as follows. In Section 4.1 we introduce the basic notation used along the chapter. A classification of systems with symmetry is presented, depending on the relative position of the constraints and the symmetry directions. In Section 4.2 we study the purely kinematic or principal case, in which none of the symmetries is compatible with the constraints. In Section 4.3 we treat the other extreme situation: the horizontal case. Here, all the symmetries fulfill the constraints. Section 4.4 deals with the general case. We present an almost-Poisson formulation of the reduction in this situation. Finally, Section 4.5 presents a special subcase of the general one, in which two subsequent reductions can be performed, first a horizontal one, and then a kinematic one. Examples are included throughout the exposition to illustrate the various reduction schemes.

4.1 Nonholonomic systems with symmetry

LET us consider a nonholonomic Lagrangian system with symmetry. More precisely, the given data are a regular Lagrangian function $L : TQ \longrightarrow \mathbb{R}$, a homogeneous constraint submanifold $M \subset TQ$ and a Lie group action Ψ of G on Q, such that both L and M are G-invariant with respect to the lifted action on TQ, $\hat{\Psi} : G \times TQ \longrightarrow TQ$. In addition, we will assume that $\hat{\Psi}$ is free and proper and that the system verifies the compatibility condition (cf. Section 3.4.1). As a consequence, the solution X of (3.12) is unique. Notice also that this implies that X is G-invariant, since it is easy to verify that $T\hat{\Psi}_g(X)$ is again a solution of (3.12) for all $g \in G$.

Taking into account the available symmetries, we can reduce the number of degrees of freedom and isolate important geometric objects driving the dynamics of the system.

Note that from the assumptions on the lifted action $\hat{\Psi}$, we have that Ψ is also free and proper and therefore $\pi : Q \longrightarrow Q/G$ has a principal G-bundle structure. Moreover, $\hat{\Psi}$ is symplectic with respect to ω_L, due to the G-invariance of the Lagrangian (see Section 2.9). For any $\xi \in \mathfrak{g}$, the infinitesimal generators ξ_{TQ} and ξ_Q of $\hat{\Psi}$ and Ψ, respectively, are τ_Q-related,

$$\tau_{Q_*} \circ \xi_{TQ} = \xi_Q \circ \tau_Q . \tag{4.1}$$

Let us denote by $\rho : TQ \longrightarrow \overline{TQ} = TQ/G$ the natural projection. From the given assumptions it follows that the energy E_L and the vector subbundle F defined in Section 3.4.1 are G-invariant. The induced action of G on M, i.e. the restriction of $\hat{\Psi}$ to $G \times M$, is still free and proper and we can regard the orbit space $\overline{M} = M/G$ as a submanifold of $\overline{TQ}$. We shall denote the fundamental vector fields associated with this action by a subscript M, so that $\xi_M = \xi_{TQ|M}$ for $\xi \in \mathfrak{g}$. Moreover, the energy E_L will induce a function $\overline{E}_L$ on $\overline{TQ}$.

In the sequel, we will denote by $\mathcal{V}$ the subbundle of TTQ whose fibers are the tangent spaces to the G-orbits, i.e. $\mathcal{V}_x = T_x(Gx)$, $\forall x \in TQ$ or, equivalently, $\mathcal{V} = \ker T\rho$. Note that $\mathcal{V}_x \subset T_x M$ for all $x \in M$, i.e. $\mathcal{V}_{|M} \subset TM$. For simplicity, we will also usually write $\mathcal{V}$, instead of $\mathcal{V}_{|M}$, when referring to its restriction to M (the precise meaning should be clear from the context).

For all unconstrained systems that admit group symmetries, Noether's theorem [1, 138] states that the invariance of the Lagrangian implies a momentum conservation law. In other words, the system has a first integral, for example, conservation of linear and angular momentum. Many systems in Mechanics (the falling cat, the satellite with rotors,...) obey these conservation laws.

This can be easily seen in the symplectic formalism explained in the preceeding chapter. Let $J : TQ \longrightarrow \mathfrak{g}^*$ be the canonical momentum mapping associated with the G-action, i.e. $J = J^L$ (cf. Section 2.9). For each $\xi \in \mathfrak{g}$, denote by J_ξ the function on TQ defined by means of the pairing $\langle J(\cdot), \xi \rangle$. Then we have that

$$\Gamma_L(J_\xi) = dJ_\xi(\Gamma_L) = (i_{\xi_{TQ}}\omega_L)(\Gamma_L) = -(i_{\Gamma_L}\omega_L)(\xi_{TQ}) = -\xi_{TQ}(E_L) .$$

But the invariance of the Lagrangian function precisely implies that $\xi_{TQ}(E_L) = 0$ and hence the momentum gives us conserved quantities.

In the presence of nonholonomic constraints, however, these conservation laws must be modified to account for the effect of constraint forces. These effects are exactly the reason why the Snakeboard, for instance, can build up momentum, even though the external forces of constraint do no work on the system. Indeed, taking the solution X of equation (3.12), we have

$$X(J_\xi) = -i_X\omega_L(\xi_M) = -\xi_M(E_L) + \beta(\xi_M),$$

with $\beta \in F^o$. Therefore, the invariance of the system implies only that

$$X(J_\xi) = \beta(\xi_M). \qquad (4.2)$$

This motivates the introduction of the notion of *horizontal symmetry* of the nonholonomic system [18, 29], which is just an element of the Lie algebra, $\xi \in \mathfrak{g}$, such that ξ_M is a section of $\mathcal{V} \cap F$. This kind of symmetries provides us with conserved quantities, since they automatically verify $\beta(\xi_M) = 0$, for $\beta \in F^o$ and hence $X(J_\xi) = 0$. Some authors refer to this result as a version of Noether's theorem for nonholonomic systems [29, 80, 220].

In general, the situation can be quite involved. It is possible that some Lie algebra elements are horizontal symmetries (but not the whole Lie algebra) or even none. In order to deal with general nonholonomic systems with symmetry, we will identify three types of situations following [29, 51]. This classification arises from considering carefully the intersection $\mathcal{V} \cap F$, which points out how well the symmetries fit in the constrained system.

1. *The purely kinematic case:* $\mathcal{V}_x \cap F_x = \{0\}$ and $T_x M = \mathcal{V}_x + (F_x \cap T_x M)$, for all $x \in M$. That is, there are no horizontal symmetries.

2. *The case of horizontal symmetries:* $\mathcal{V}_x \cap F_x = \mathcal{V}_x$, for all $x \in M$, which is equivalent to $\mathcal{V}_x \subset F_x$, for all $x \in M$.

3. *The general case:* $\{0\} \subsetneqq \mathcal{V}_x \cap F_x \subsetneqq \mathcal{V}_x$, for all $x \in M$.

Remark 4.1.1. In the above classification, we are assuming a regularity condition, namely, that the rank of the intersection $\mathcal{V}_x \cap F_x$ remains constant for all $x \in M$. The study of the singular case, that is, when this does not hold true, is still an open issue. A different singular situation, which is receiving increasing attention [15, 221], is when the action $\hat{\Psi}$ is not free and hence we do not have a manifold structure on the quotient space.

Remark 4.1.2. In the third case of the classification, although the intersection $\mathcal{V} \cap F$ is non-trivial, it is possible to have no horizontal symmetry. For instance, the fact that an element $\xi \in \mathfrak{g}$ verifies $\xi_M(x) \in F_x$ for a given $x \in M$ does not imply in general that $\xi_M(y) \in F_y$ for all $y \in M$. This observation will be more clear when we reach Section 4.4, where we shall treat it carefully.

Remark 4.1.3. This classification can be extended to a more general class of systems, namely, constrained systems with symmetry [51, 68], which may include, among others, singular Lagrangian systems. The subsequent discussion is also valid for this type of systems, but here we will restrict our attention to nonholonomic Lagrangian systems.

In the following sections, we are going to treat specifically each of these cases. Before proceeding, we would like to point out that, unlike in the unconstrained case where, within the Hamiltonian [166] and Poisson formulations [162], one can prove that the system obtained after reduction is of the same type as the original one, here the situation changes: the reduction process leads us in general to a different category of systems. For instance, in some cases of purely kinematic systems, we shall see that one ends up with an unconstrained reduced system.

An alternative viewpoint is the one proposed by Bates and Śniatycki [18]. Their treatment has the advantage of fully respecting the category of systems under consideration. For the sake of completeness, we shall recall here the symplectic reduction established by them for nonholonomic systems (see also [122]). We point out that this reduction has been generalized to general constrained systems with symmetry by Cantrijn *et al.* in [51] (see also [48, 68]).

Recall from Section 3.4.1 that the homogeneity of the constraints and the SODE character of X implies automatically that $X \in F$. We define a (generalized) vector subbundle $\mathcal{U}$ of $T_M TQ$ by

$$\mathcal{U} = (F \cap TM) \cap (\mathcal{V} \cap F)^{\perp}, \tag{4.3}$$

where $(\mathcal{V} \cap F)^{\perp}$ is the ω_L-complement of $\mathcal{V} \cap F$ in $T_M TQ$.

There are two main reasons to consider this, in principle rather "strange", bundle. On the one hand, from equation (3.12) it is straightforward to verify that indeed the solution of the dynamics belongs to $\mathcal{U}$, namely $X \in \mathcal{U}$. Denoting by $\omega_{\mathcal{U}}$ the restriction of ω_L to $\mathcal{U}$, the nonholonomic equations of motion (3.12) can by rewritten as

$$i_X \omega_{\mathcal{U}} = d_{\mathcal{U}} E_L, \tag{4.4}$$

where $d_{\mathcal{U}} E_L$ denotes the restriction of dE_L to $\mathcal{U}$.

To explain clearly the other reason, we need some more derivations. Indeed, it is not hard to see that $\mathcal{U}$ is G-invariant and, hence, projects under $T\rho$ onto a subbundle $\bar{\mathcal{U}}$ of $T_{\overline{M}} \overline{TQ}$. In general, this bundle need not be of constant rank, i.e. it determines a generalized distribution on $\overline{TQ}$ along $\overline{M}$. In the sequel, however, we will always tacitly assume that $\mathcal{U}$ is a genuine vector bundle over $\overline{M}$.

We can see that $\omega_{\mathcal{U}}$ is also G-invariant and since, moreover, $i_{\tilde{\xi}}\omega_{\mathcal{U}} = 0$ for all $\tilde{\xi} \in \mathcal{V} \cap \mathcal{U}$, it pushes down to a 2-form $\omega_{\overline{\mathcal{U}}}$ on $\overline{\mathcal{U}}$ (i.e. $\omega_{\overline{\mathcal{U}}}$ only acts on vectors belonging to $\overline{\mathcal{U}}$). This is precisely the other reason to consider the bundle $\mathcal{U}$: it allows us to project the symplectic form to the reduced bundle $\overline{\mathcal{U}}$.

Similarly, $d_{\mathcal{U}}E_L$ pushes down to a 1-form $d_{\overline{\mathcal{U}}}\overline{E}_L$ on $\overline{\mathcal{U}}$, which is simply the restriction of $d\overline{E}_L$ to $\overline{\mathcal{U}}$. Note that neither $\omega_{\overline{\mathcal{U}}}$ nor $d_{\overline{\mathcal{U}}}\overline{E}_L$ are differential forms on $\overline{M}$; they are exterior forms on a vector bundle over $\overline{M}$, with smooth dependence on the base point.

Proposition 4.1.4 ([18, 51]). *Let X be the G-invariant solution of (4.4). Then, the projection $\overline{X}$ of X onto $\overline{M}$ is a section of $\overline{\mathcal{U}}$ satisfying the equation*

$$i_{\overline{X}}\omega_{\overline{\mathcal{U}}} = d_{\overline{\mathcal{U}}}\overline{E}_L .$$

Proof. It readily follows from the symmetry assumptions and the previous considerations. □

It is important to observe that, in general, the 2-form $\omega_{\overline{\mathcal{U}}}$ may be degenerate. However, under the compatibility condition, $F^{\perp} \cap TM = 0$, one can prove that $\omega_{\overline{\mathcal{U}}}$ is nondegenerate, such that $(\overline{\mathcal{U}}, \omega_{\overline{\mathcal{U}}})$ becomes a symplectic vector bundle over $\overline{M}$ (see [18]). The reduced dynamics is then *uniquely determined* by the equation mentioned in Proposition 4.1.4.

Indeed, let us suppose that $Y \in \overline{\mathcal{U}}$ is such that $\omega_{\overline{\mathcal{U}}}(Y, Z) = 0$, for all $Z \in \overline{\mathcal{U}}$. Then, there exists a $Y' \in \mathcal{U}$ such that $T\rho(Y') = Y$ and $\omega_L(Y', Z') = 0$, for all $Z' \in \mathcal{U}$. Otherwise said, $Y' \in \mathcal{U} \cap \mathcal{U}^{\perp}$. Then Y' can be written as $Y' = Y_1 + Y_2$, where $Y_1 \in (F \cap TM)^{\perp}$ and $Y_2 \in \mathcal{V} \cap F$. Since $\mathcal{V} \cap F \subset F \cap TM$ and $Y' \in \mathcal{U}$, we have that $Y_1 \in F \cap TM \cap (F \cap TM)^{\perp}$. Now, the compatibility condition implies that this latter intersection yields the zero section, so $Y_1 = 0$. Finally, we have that $Y = T\rho(Y') = T\rho(Y_2) = 0$, since $Y_2 \in \mathcal{V}$. Consequently, $\omega_{\overline{\mathcal{U}}}$ is nondegenerate.

Now, we turn to the discussion of each of the cases established above. We start by the purely kinematic or vertical case.

4.2 The purely kinematic case

SUPPOSE that $\mathcal{V}_x \cap F_x = \{0\}$ and $T_x M = \mathcal{V}_x + (F_x \cap T_x M)$, for all $x \in M$. In principle, this leads us to think that the symmetries do not play an important role in the reduction, because none of them is compatible with the bundle of reaction forces. However, we will show in the following that the symplectic scheme explained above takes a nice form here due to the particular geometry involved in the system.

4.2.1 Reduction

In the vertical case, we have that $T_x M = V_x \oplus (F_x \cap T_x M)$, for all $x \in M$. Moreover, $\mathcal{U} = F \cap TM$, so $TM = V_{|M} \oplus \mathcal{U}$. Since $\mathcal{U}$ is G-invariant, this decomposition defines a principal connection Υ on the principal G-bundle $\rho_{|M} : M \to \overline{M}$, with horizontal subspace $\mathcal{U}_x$ at $x \in M$. Note, in passing, that $\mathcal{U}$ here represents a vector bundle of constant rank. In addition, X is horizontal, i.e. $X \in \mathcal{U}$.

Denote by $\mathbf{h}_\Upsilon : TM \longrightarrow \mathcal{U}$ and $\mathbf{v}_\Upsilon : TM \longrightarrow V$ the horizontal and vertical projectors associated with the decomposition $TM = V_{|M} \oplus \mathcal{U}$, respectively. The curvature of Υ is the tensor field of type (1,2) on M given by

$$R = \frac{1}{2}[\mathbf{h}_\Upsilon, \mathbf{h}_\Upsilon] ,$$

where $[\cdot, \cdot]$ denotes the Nijenhuis bracket of (1,1)-tensor fields [20]. Taking into account that $\overline{\mathcal{U}} = T\overline{M}$, we obtain on $\overline{M}$ a 2-form $\overline{\omega}_L$ (which is now a genuine differential form on $\overline{M}$) and a function $\overline{E}_L$ such that, following Proposition 4.1.4, the projection $\overline{X}$ of X verifies

$$i_{\overline{X}}\overline{\omega}_L = d\overline{E}_L . \tag{4.5}$$

It should be pointed out that the reduced 2-form $\overline{\omega}_L$ in general need not be closed, i.e. it is an almost symplectic form. We will show, however, that one can construct a reduced equation, equivalent to (4.5), but now in terms of a closed 2-form on $\overline{M}$.

Denote by θ' the 1-form on M defined by $\theta' = j_M^* \theta_L$, where $j_M : M \hookrightarrow TQ$ is the canonical inclusion. By means of the given solution X of (3.12) we can construct a 1-form α_X on M as follows:

$$\alpha_X = i_X(\mathbf{h}_\Upsilon^* d\theta' - d\mathbf{h}_\Upsilon^* \theta') , \tag{4.6}$$

with the usual convention that, for an arbitrary p-form β, $\mathbf{h}_\Upsilon^* \beta$ is the p-form defined by the prescription $\mathbf{h}_\Upsilon^* \beta(X_1, \ldots, X_p) = \beta(\mathbf{h}_\Upsilon(X_1), \ldots, \mathbf{h}_\Upsilon(X_p))$.

Proposition 4.2.1. *[51] The 1-forms $\mathbf{h}_\Upsilon^* \theta'$ and α_X are projectable. Moreover, the projection $\overline{X}$ of the dynamics X, which is a solution of (4.5), also satisfies*

$$i_{\overline{X}} d\overline{\theta'}_h = d\overline{E}_L + \overline{\alpha_X} , \tag{4.7}$$

where $\overline{\theta'}_h$ and $\overline{\alpha_X}$ are the projections of the 1-forms $\mathbf{h}_\Upsilon^* \theta'$ and α_X, resp.

Remark 4.2.2. The sign of $\overline{\alpha_X}$ in (4.7) differs from the one in the corresponding expression stated in [51, 68], where the discussion took place in a more general symplectic framework with an exact symplectic structure $\omega = d\theta$ (whereas here we have $\omega_L = -d\theta_L$). The signs would agree if we would have defined θ' as $-j_M^* \theta_L$. The formulation of Proposition 4.2.1 for general constrained systems is straightforward (see [51, 68]) and we shall make use of it in Section 4.5.

Remark 4.2.3. Proposition 4.2.1 describes a situation where a nonholonomic system (3.12) with symmetry, admits a reduction to an unconstrained system (4.7), but with an additional nonconservative force represented by $\overline{\alpha_X}$. Indeed, by construction, the 1-form α_X satisfies $i_X \alpha_X = 0$, which implies

$$i_{\overline{X}} \overline{\alpha_X} = 0 \,.$$

As a consequence, we have that the unconstrained system (4.7) has the energy $\overline{E}_L$ conserved.

Chaplygin systems In the following, we specialize our discussion to a class of nonholonomic systems with symmetry that fall into the purely kinematic case, the so-called *Chaplygin systems*.

The given data are a principal G-bundle $\pi : Q \longrightarrow Q/G$, associated with a free and proper action Ψ of G on Q, a Lagrangian $L : TQ \longrightarrow \mathbb{R}$ which is G-invariant with respect to the lifted action on TQ, and linear nonholonomic constraints determined by the horizontal distribution (here denoted as $\mathcal{D}$) of a principal connection γ on π.

Remark 4.2.4. Classically, a mechanical system with Lagrangian $L(q^A, \dot{q}^A)$, $A = 1, \ldots, n$, subject to k linear nonholonomic constraints, is said to be of Chaplygin type if coordinates (q^a, q^α) can be found, with $a = 1, \ldots, k$ and $\alpha = k+1, \ldots, n$, such that the constraints can be written in the form $\dot{q}^a = B_\alpha^a(q^{k+1}, \ldots, q^n)\dot{q}^\alpha$ and such that L does not depend on the coordinates q^a (see e.g. [188]). Such a system can be (locally) interpreted as a special case of the generalized Chaplygin systems introduced above, with $Q = \mathbb{R}^n$ and with an action defined by the Abelian group $G = \mathbb{R}^k$. Koiller [120] refers to the more general case, considered here, as "non-Abelian Chaplygin systems".

Note that there exists a natural identification $\mathcal{D} \cong Q \times_{Q/G} T(Q/G)$ as principal G-bundles over $T(Q/G)$. The isomorphism is obtained by mapping $v_q \in \mathcal{D}$ onto $(q, \pi_*(v_q))$. It then follows that $\mathcal{D}/G$ can be naturally identified with $T(Q/G)$ and we have, in particular,

$$\rho_{|\mathcal{D}} = \pi_{*|\mathcal{D}} \,. \tag{4.8}$$

Henceforth, the restriction of ρ to $\mathcal{D}$ will also be simply denoted by ρ. We can summarize the situation for Chaplygin systems in the following diagram,

$$TD = \mathcal{U} \oplus \mathcal{V}_\rho \xrightarrow{\ \rho_*\ } T\overline{D} \cong \overline{\mathcal{U}}$$

$$\mathcal{D} \xrightarrow{\ \rho\ } T(Q/G) \cong \overline{\mathcal{D}}$$

Apart from the symplectic action $\hat{\Psi}$ and the 1-form α_X, we have some additional geometric objects involved in the symplectic approach to the reduction process for this type of systems.

The connections. The principal connection Υ that we obtained above is obviously related to the original connection γ of the Chaplygin system. Indeed, take $w \in T_{v_q}\mathcal{D}$ and consider $\tau_{Q_*}w \in T_qQ$. Then, we can write

$$\tau_{Q_*}w = \big(\tau_{Q_*}w - (\gamma(\tau_{Q_*}w))_Q(q)\big) + (\gamma(\tau_{Q_*}w))_Q(q)\,,$$

where $\tau_{Q_*}w - (\gamma(\tau_{Q_*}w))_Q(q) \in \mathcal{D}_q$ and $(\gamma(\tau_{Q_*}w))_Q(q) \in (\mathcal{V}_\pi)_q$. Putting $\gamma(\tau_{Q_*}w) = \xi \in \mathfrak{g}$, a direct computation shows that $w - \xi_{TQ}(v_q) \in \mathcal{U}$ and, consequently, $w = (w - \xi_{TQ}(v_q)) + \xi_{TQ}(v_q)$ is the $(\mathcal{U}, \mathcal{V}_\rho)$ decomposition of w. Herewith we have proved the following property.

Proposition 4.2.5. *The connection 1-forms Υ and γ are related by $\Upsilon = \tau_Q^*\gamma$, i.e. $\Upsilon_{v_q}(w) = \gamma_q(\tau_{Q_*}w)$ for any $v_q \in \mathcal{D}$ and $w \in T_{v_q}\mathcal{D}$.*

Let us denote the horizontal projectors, associated with γ, resp. Υ, by $\mathbf{h}_\gamma : TQ \longrightarrow \mathcal{D}(\subset TQ)$, resp. $\mathbf{h}_\Upsilon : T\mathcal{D} \longrightarrow \mathcal{U}(\subset T\mathcal{D})$. Likewise, the vertical projectors onto $\mathcal{V}_\pi$, resp. $\mathcal{V}_\rho$, will be denoted by $\mathbf{v}_\gamma$, resp. $\mathbf{v}_\Upsilon$. In order not to further overload the notations, we will use the same superscript ^{h} for the horizontal lifts of vectors (vector fields) with respect to either γ or Υ; in principle it should always be clear from the context which horizontal lift operation is being used.

We now have that

$$\tau_{Q_*|T\mathcal{D}} \circ \mathbf{h}_\Upsilon = \mathbf{h}_\gamma \circ \tau_{Q_*|T\mathcal{D}}\,, \tag{4.9}$$

i.e. the following diagram is commutative,

$$(TTQ \supset)\ T\mathcal{D} \xrightarrow{\ \mathbf{h}_\Upsilon\ } T\mathcal{D}\ (\subset TTQ)$$

$$\tau_{Q_*} \downarrow \qquad\qquad \downarrow \tau_{Q_*}$$

$$TQ \xrightarrow{\ \mathbf{h}_\gamma\ } TQ$$

Indeed, taking into account Proposition 4.2.5, we see that for any $w \in T_{v_q}\mathcal{D}$, $\gamma_q(\tau_{Q_*}\mathbf{h}_\Upsilon(w)) = \Upsilon_{v_q}(\mathbf{h}_\Upsilon(w)) = 0$ and, hence, $\tau_{Q_*}\mathbf{h}_\Upsilon(w)$ is horizontal with respect to γ, i.e. $\tau_{Q_*}\mathbf{h}_\Upsilon(w) \in \mathcal{D}_q$. By definition we also have $\mathbf{h}_\gamma(\tau_{Q_*}w) \in \mathcal{D}_q$. Using the fact that $\pi \circ \tau_Q = \tau_{Q/G} \circ \pi_*$ we obtain

$$\pi_*(\mathbf{h}_\gamma(\tau_{Q_*}w)) = \pi_*\tau_{Q_*}w = \tau_{Q/G_*}(\pi_*)_*w = \tau_{Q/G_*}\rho_*w,$$

where the last equation follows from (4.8). Similarly, we have:

$$\pi_*(\tau_{Q_*}(\mathbf{h}_\Upsilon w)) = \tau_{Q/G_*}(\pi_*)_*(\mathbf{h}_\Upsilon w) = \tau_{Q/G_*}\rho_*(\mathbf{h}_\Upsilon w) = \tau_{Q/G_*}\rho_*w.$$

We thus see that the γ-horizontal tangent vectors at q, $\tau_{Q_*}(\mathbf{h}_\Upsilon w)$ and $\mathbf{h}_\gamma(\tau_{Q_*}w)$, have the same projection under π_* and, therefore, they are equal. This completes the proof of (4.9). Denoting the curvature tensors of the principal connections γ and Υ by Ω^γ and Ω^Υ, respectively, one can easily deduce from Proposition 4.2.5 and (4.9) the relation

$$\Omega^\Upsilon = \tau_Q^*\Omega^\gamma. \tag{4.10}$$

The Lagrangians. The Lagrangian L of the given mechanical system induces a Lagrangian $L^* : T(Q/G) \longrightarrow \mathbb{R}$ on the quotient space $\overline{\mathcal{D}} \cong T(Q/G)$, given by $L^*(\overline{q}, v_{\overline{q}}) = L(q, v_q^h)$ for any $q \in \pi^{-1}(\overline{q})$, and where v_q^h denotes the γ-horizontal lift of $v_{\overline{q}}$ at q. This is well-defined because of the G-invariance of L. Moreover, under the compatibility condition, one can show that L^* is a regular Lagrangian on $T(Q/G)$ (cf. [137]). A quick set of calculations shows that

$$\overline{E}_L = E_{L^*}$$

under the identification $\overline{\mathcal{D}} \equiv T(Q/G)$.

Now, we are in a position to state that Proposition 4.2.1 above takes the following form for nonholonomic Chaplygin systems (see [51, 120]).

Proposition 4.2.6. *The dynamics X of the generalized Chaplygin system projects onto $\overline{\mathcal{D}}$, and its projection $\overline{X}$ is determined by the equation*

$$i_{\overline{X}}\omega_{L^*} = dE_{L^*} + \overline{\alpha_X}, \tag{4.11}$$

where $\overline{\alpha_X}$ is the projection of the 1-form α_X, defined by (4.6).

It can be easily verified that the form $\overline{\alpha_X}$ is a semi-basic 1-form on $\overline{\mathcal{D}}$ (see also Section 5.3), from which it then follows that the vector field $\overline{X}$, defined by (4.11), is a SODE vector field. Moreover, one can show that not only the contraction of $\overline{\alpha_X}$ with $\overline{X}$ vanishes, but that, more generally, $i_Y\overline{\alpha_X} = 0$ for any SODE Y on $\overline{\mathcal{D}}$. Thus, a generalized Chaplygin system reduces to an unconstrained mechanical system, with an external nonconservative force of "gyroscopic" type, which is geometrically represented by the 1-form $\overline{\alpha_X}$ (see also [120, 188]). The "gyroscopic" character of this force is also in agreement with the fact that the projected energy function E_{L^*} is a conserved quantity of the reduced dynamics. But there is more to be said about it.

Proposition 4.2.7. *The 2-form* $\Sigma = \mathbf{h}_{\Upsilon}^* d\theta' - d\mathbf{h}_{\Upsilon}^* \theta'$ *on* $\mathcal{D}$ *projects onto a 2-form* $\overline{\Sigma}$ *on* $\overline{\mathcal{D}}$ *and the 1-form* $\overline{\alpha_X}$ *satisfies*

$$\overline{\alpha_X} = i_{\overline{X}}\overline{\Sigma}.$$

Proof. Let $\xi_{\mathcal{D}}$ be the fundamental vector field of the G-action on $\mathcal{D}$, induced by an arbitrary element $\xi \in \mathfrak{g}$. We must prove that $\xi_{\mathcal{D}}$ belongs to the characteristic distribution of the 2-form Σ. First, we have that

$$i_{\xi_{\mathcal{D}}}\Sigma = -i_{\xi_{\mathcal{D}}}d\mathbf{h}_{\Upsilon}^*\theta'.$$

For any vector field Y on $\mathcal{D}$, $i_{\xi_{\mathcal{D}}}d\mathbf{h}_{\Upsilon}^*\theta'(Y) = \xi_{\mathcal{D}}(\theta'(\mathbf{h}_{\Upsilon}Y)) - \theta'(\mathbf{h}_{\Upsilon}[\xi_{\mathcal{D}}, Y])$. Now, if Y is vertical, we readily see that $i_{\xi_{\mathcal{D}}}d\mathbf{h}_{\Upsilon}^*\theta'(Y) = 0$. If Y is horizontal, we have

$$i_{\xi_{\mathcal{D}}}d\mathbf{h}_{\Upsilon}^*\theta'(Y) = \mathcal{L}_{\xi_{\mathcal{D}}}\theta'(Y) = 0,$$

because of the G-invariance of θ'. It therefore remains to prove that $i_{\xi_{\mathcal{D}}}d\Sigma = 0$. For any two vector fields Y, Z we have that

$$\begin{aligned}
i_{\xi_{\mathcal{D}}}d\Sigma(Y, Z) &= \left(i_{\xi_{\mathcal{D}}}d\mathbf{h}_{\Upsilon}^* d\theta'\right)(Y, Z) \\
&= \xi_{\mathcal{D}}\left(\mathbf{h}_{\Upsilon}^* d\theta'\right)(Y, Z) - \mathbf{h}_{\Upsilon}^* d\theta'([\xi_{\mathcal{D}}, Y], Z) + \mathbf{h}_{\Upsilon}^* d\theta'([\xi_{\mathcal{D}}, Z], Y).
\end{aligned}$$

If at least one of the vector fields Y and Z is vertical, then $i_{\xi_{\mathcal{D}}}d\Sigma(Y, Z) = 0$. Taking Y and Z both horizontal, we find, taking into account the G-invariance of $d\theta'$ and $\xi_{\mathcal{D}} \in \mathfrak{X}(\mathcal{D})$,

$$\begin{aligned}
i_{\xi_{\mathcal{D}}}d\Sigma(Y, Z) &= \xi_{\mathcal{D}}\left(d\theta'(Y, Z)\right) - d\theta'([\xi_{\mathcal{D}}, Y], Z) - d\theta'(Y, [\xi_{\mathcal{D}}, Z]) \\
&= (\mathcal{L}_{\xi_{\mathcal{D}}}d\theta')(Y, Z) = 0.
\end{aligned}$$

The last part of the proposition now immediately follows from (4.6) and the projectability of X. $\square$

We will make use of this result in the following chapter. To end this section, we write some local expressions for some of the geometrical objects just introduced which will be useful later.

Consider a local trivialization $U \times G$ of π, with *adapted bundle coordinates* (r^α, g^a), where $a = 1, \ldots, k = \dim G$ and $\alpha = 1, \ldots, n - k$. Choosing a basis e_a $(a = 1, \ldots, k)$ of the Lie algebra $\mathfrak{g}$, and using the left trivialization $TG \cong G \times \mathfrak{g}$, a tangent vector $v \in T_{(x,g)}(U \times G) \cong T_x U \times \mathfrak{g}$ can be represented by a pair (w, ξ), whereby $w \in T_x U$ and $\xi = \xi^a e_a \in \mathfrak{g}$. In terms of the coordinates $(r^\alpha, g^a, \dot{r}^\alpha, \xi^a)$ on $T(U \times G)$ the G-invariant Lagrangian can then be written as

$$L = \ell(r^\alpha, \dot{r}^\alpha, \xi^a).$$

Strictly speaking, ℓ represents the reduction of L to TQ/G. With respect to the given local trivialization, we further denote the connection coefficients of the given principal connection γ by $\mathcal{A}^a_\alpha = \mathcal{A}^a_\alpha(r^1, \ldots, r^{n-k})$, and then the constraints take the form $\xi^a = -\mathcal{A}^a_\alpha \dot{r}^\alpha$ or, equivalently,

$$w = g^{-1} dg + \mathcal{A}^b_\beta(r) dr^\beta e_b. \tag{4.12}$$

In particular, it follows that the reduced Lagrangian L^* is given by $L^*(r^\alpha, \dot{r}^\alpha) = \ell(r^\alpha, \dot{r}^\alpha, -\mathcal{A}^a_\beta \dot{r}^\beta e_a)$.

With all the above, one can now derive the following coordinate expression for the reduced dynamics (see also [29, 120]),

$$X = \dot{r}^\alpha \frac{\partial}{\partial r^\alpha} + \hat{W}^{\alpha\beta} \left(\frac{\partial L^*}{\partial r^\beta} - \dot{r}^\varsigma \frac{\partial \hat{p}_\beta}{\partial r^\varsigma} - \alpha_\beta \right) \frac{\partial}{\partial \dot{r}^\alpha}, \tag{4.13}$$

where $(\hat{W}^{\alpha\beta})$ is the inverse of the Hessian matrix

$$\left(\frac{\partial^2 L^*}{\partial \dot{r}^\alpha \partial \dot{r}^\beta} \right),$$

$\hat{p}_\alpha = \dfrac{\partial L^*}{\partial \dot{r}^\alpha}$, and $\alpha_\beta dr^\beta$ is the local expression for the gyroscopic 1-form $\overline{\alpha_X}$. The α_β are explicitly given by

$$\alpha_\beta = -\left(\frac{\partial \ell}{\partial \xi^a} \right)^* \left(\frac{\partial \mathcal{A}^a_\varsigma}{\partial r^\beta} - \frac{\partial \mathcal{A}^a_\beta}{\partial r^\varsigma} - c^a_{bc} \mathcal{A}^b_\beta \mathcal{A}^c_\varsigma \right) \dot{r}^\varsigma, \tag{4.14}$$

where the $*$ on the right-hand side indicates that, after computing the derivative of ℓ with respect to ξ^a, one replaces the ξ^b everywhere by $-\mathcal{A}^b_\alpha \dot{r}^\alpha$. The constants c^a_{bc} appearing in the last term on the right-hand side are the structure constants of $\mathfrak{g}$ with respect to the chosen basis, i.e. $[e_b, e_c] = c^a_{bc} e_a$. Note in passing that the expressions

$$\frac{\partial \mathcal{A}^a_\varsigma}{\partial r^\beta} - \frac{\partial \mathcal{A}^a_\beta}{\partial r^\varsigma} - c^a_{bc} \mathcal{A}^b_\beta \mathcal{A}^c_\varsigma$$

are the coefficients of the curvature of γ in local form.

The interested reader may now consult the original coordinate treatment of Chaplygin in Chapter III, Section 3 of [188], which is also reviewed in [120].

Vertical rolling disk Consider the example of the disk introduced in Section 3.2.1, but now rolling vertically on a horizontal plane. This system is called the *vertical rolling disk* [29, 51].

The configuration space is $Q = \mathbb{R} \times \mathbb{S}^1 \times \mathbb{S}^1$. The dynamics of this mechanical system is described by the regular Lagrangian,

$$L = \frac{1}{2}\left(m\dot{x}^2 + m\dot{y}^2 + I_1\dot{\theta}_1^2 + I_2\dot{\theta}_2^2\right),$$

where m is the mass, and I_1, I_2 are moments of inertia; and the nonholonomic constraints,

$$\phi_1 = \dot{x} - (R\cos\theta_1)\dot{\theta}_2 = 0, \quad \phi_2 = \dot{y} - (R\sin\theta_1)\dot{\theta}_2 = 0,$$

where R is the radius of the disk.

Consider the group $G = \mathbb{R}^2$ and its trivial action by translations on Q,

$$\Phi: \quad \begin{aligned} G \times Q &\longrightarrow Q \\ (r,s) \times (x,y,\theta_1,\theta_2) &\longmapsto (x+r, y+s, \theta_1, \theta_2). \end{aligned}$$

Note that $\rho : Q \to \mathbb{S}^1 \times \mathbb{S}^1$ is a principal G-bundle and M, the constraint manifold, is the horizontal subbundle of a principal connection, so that the given system is a Chaplygin system. Following the above analysis we then obtain,

$$L^* = \frac{1}{2}\left(I_1\dot{\theta}_1^2 + (mR^2 + I_2)\dot{\theta}_2^2\right),$$
$$\omega_{L^*} = I_1\, d\theta_1 \wedge d\dot{\theta}_1 + (mR^2 + I_2)\, d\theta_2 \wedge d\dot{\theta}_2.$$

In this particular case the gyroscopic 1-form $\overline{\alpha}_X = 0$. So the reduced equation (4.11) becomes

$$i_X\omega_{L^*} = dE_{L^*}.$$

That is, an unconstrained Lagrangian system!

After the above reduction procedure, the system can still have some symmetries we have not taken into account. This is precisely the case here, where there is still some symmetry of the system to be considered. Denote the Lie group $\mathbb{S}^1 \times \mathbb{S}^1$ by K and let us define:

$$\Phi: \quad \begin{aligned} K \times Q/G &\longrightarrow Q/G \\ ((\lambda_1, \lambda_2), (\theta_1, \theta_2)) &\longmapsto (\theta_1 + \lambda_1, \theta_2 + \lambda_2). \end{aligned}$$

If we consider the lifted action $\hat{\Phi}$ of Φ to $T(Q/G)$, it is clear that the Lagrangian L^* is K-invariant. Then, we can make a further reduction.

Thus, in general, the reduced system (4.7) can still possess more symmetries which have to be considered. Let $\Psi : K \times \overline{M} \longrightarrow \overline{M}$ be an action on $\overline{M}$ that leaves invariant the reduced energy function $\overline{E}_L$ and the 1-form $\overline{\theta'}_h$. Denote by $\mathfrak{k}$ the Lie algebra of K and define a momentum mapping, $J : \overline{M} \longrightarrow \mathfrak{k}^*$, in the usual manner: $\langle J(\overline{m}), \eta \rangle = -\langle \overline{\theta'}_h(\overline{m}), \eta_{\overline{M}}(\overline{m}) \rangle$ for $\overline{m} \in \overline{M}$ and $\eta \in \mathfrak{k}$. It is easy to see that $i_{\eta_{\overline{M}}} d\overline{\theta'}_h = dJ_\eta$ for all $\eta \in \mathfrak{k}$. Using equation (4.7) and the K-invariance of $\overline{E}_L$, we obtain a momentum equation,

$$\overline{X}(J_\eta) = \overline{\alpha_X}(\eta_{\overline{M}}) . \tag{4.15}$$

In Section 4.5 we will develop this idea for a special subcase of the general case mentioned above: for a number of problems, we will be able to "split up" the reduction process into two known steps. First, a horizontal reduction and secondly a kinematic reduction.

Chapter 5 will be devoted to study the integrability aspects of mechanical Chaplygin systems.

The nonholonomic free particle We will discuss here an instructive example due to Rosenberg [207] which has been extensively treated also in [17, 18, 29]. Consider a particle moving in space, so the configuration space is given by $Q = \mathbb{R}^3$, subject to the nonholonomic constraint

$$\phi = \dot{z} - y\dot{x} = 0 .$$

The Lagrangian function is the kinetic energy of the particle

$$L = \frac{1}{2} \left(\dot{x}^2 + \dot{y}^2 + \dot{z}^2 \right) ,$$

and the Poincaré-Cartan two-form is

$$\omega_L = dx \wedge d\dot{x} + dy \wedge d\dot{y} + dz \wedge d\dot{z} .$$

The constraint manifold is the distribution determined by the annihilation of the linear constraint ϕ

$$M = \text{span} \left\{ \frac{\partial}{\partial x} + y\frac{\partial}{\partial z}, \frac{\partial}{\partial y} \right\} .$$

Choose local coordinates $(x, y, z, \dot{x}, \dot{y})$ on M. We find that the distribution $F_{|M}$ is generated by the vectors fields,

$$F_{|M} = \text{span} \left\{ \frac{\partial}{\partial x} + y\frac{\partial}{\partial z}, \frac{\partial}{\partial y}, \frac{\partial}{\partial \dot{x}}, \frac{\partial}{\partial \dot{y}}, \frac{\partial}{\partial \dot{z}} \right\} .$$

The symplectic vector bundle $F \cap TM$ is given by

$$F \cap TM = \text{span} \left\{ \frac{\partial}{\partial x} + y \frac{\partial}{\partial z}, \frac{\partial}{\partial y} + \dot{x} \frac{\partial}{\partial \dot{z}}, \frac{\partial}{\partial \dot{x}} + y \frac{\partial}{\partial \dot{z}}, \frac{\partial}{\partial \dot{y}} \right\} ,$$

with symplectic orthogonal complement

$$(F \cap TM)^{\perp} = \text{span} \left\{ \frac{\partial}{\partial \dot{z}} - y \frac{\partial}{\partial \dot{x}}, \frac{\partial}{\partial z} + \dot{x} \frac{\partial}{\partial \dot{y}} - y \frac{\partial}{\partial x} \right\} .$$

Consider the Lie group $G = \mathbb{R}^2$ and its action on Q,

$$\Psi : \quad \begin{array}{ccc} G \times Q & \longrightarrow & Q \\ ((r, s), (x, y, z)) & \longmapsto & (x + r, y, z + s) . \end{array}$$

It is a simple verification to see that L and M are G-invariant.

If we consider the lifted action $\hat{\Psi}$ of Ψ to TQ, the infinitesimal generators of this action are $\left\{ \frac{\partial}{\partial x}, \frac{\partial}{\partial z} \right\}$. For each $m = (x, y, z, \dot{x}, \dot{y}) \in M$, we have

$$\mathcal{V}_m \cap F_m = \text{span} \left\{ \left(\frac{\partial}{\partial x} + y \frac{\partial}{\partial z} \right) \Big|_m \right\} .$$

Therefore the nonholonomic free particle falls into the general case. We shall come again to this example in Section 4.4, but here we will show now how it can also be seen as a Chaplygin system.

Consider the Lie group $G = \mathbb{R}$ and its trivial action by translation on Q,

$$\Phi : \quad \begin{array}{ccc} G \times Q & \longrightarrow & Q \\ (s, (x, y, z)) & \longmapsto & (x, y, z + s) . \end{array}$$

Note that M is the horizontal subspace of a connection γ on the principal fiber bundle $Q \longrightarrow Q/G$, where $\gamma = (dz - y dx)e$, with $\{e\}$ the infinitesimal generator of the translation. Therefore this is a Chaplygin system.

Following the above analysis, we obtain that

$$L^* = \frac{1}{2} \left((1 + y^2) \dot{x}^2 + \dot{y}^2 \right) ,$$

and the reduced system

$$i_X \omega_{L^*} = dE_{L^*} + \overline{\alpha_X} ,$$

where $\overline{\alpha_X} = -\dot{x} \dot{y} y \, dx + y \dot{x}^2 \, dy$.

4.2.2 Reconstruction

A natural problem related to the reduction of mechanical systems with symmetry concerns the reverse procedure: once the solutions of the reduced dynamics have been obtained, how can one recover from it the solutions of the original system. This is called the "reconstruction problem" of the dynamics. This problem is intimately related to the concepts of geometric and dynamic phases, which play an important role in various aspects of mechanics [159] and in the study of locomotion systems (for example, in the generation of net motion by cyclic changes in shape space [117, 216]). Alternatively, some authors have pursued the following idea: given a system whose equations of motion are complex, they try to find another one with equations of motion that are easier to integrate, even tough the dimension may be greater, and such that its reduction by a certain group of symmetries yields the original system. This method has been used, for instance, in [115] to prove the complete integrability of certain Hamiltonian systems.

We now discuss the reconstruction of the dynamics on the constraint submanifold M from the reduced dynamics on $\overline{M}$. Suppose that the flow of the reduced system $\overline{X}$ is known. Take $\overline{c}(t)$ an integral curve of $\overline{X}$ starting from a point $\overline{x} \in \overline{M}$, and fix $x \in \rho^{-1}(\overline{x})$. We want to find the corresponding integral curve $c(t)$ of X starting from x which projects on $\overline{c}(t)$, i.e. $\rho(c(t)) = \overline{c}(t)$. But we must realize that the curve $c(t)$ is just the horizontal lift of $\overline{c}(t)$ starting from x with respect to the principal connection Υ. We prove this simple fact in the following

Proposition 4.2.8. *The integral curve $c(t)$ of X, starting at $x \in M$, is the horizontal lift with respect to the principal connection Υ of the integral curve $\overline{c}(t)$ of $\overline{X}$ starting at $\overline{x} = \rho(x)$.*

Proof. Let $d(t)$ denote the horizontal lift of $\overline{c}(t)$ starting from x. Then $\rho(d(t)) = \overline{c}(t)$ and $d(0) = x$. Since X and $\overline{X}$ are ρ-related, we have that $T\rho(X(d(t))) = \overline{X}\rho(d(t)) = \overline{X}(\overline{c}(t)) = \dot{\overline{c}}(t) = T\rho(\dot{d}(t))$. Therefore $\dot{d}(t) - X(d(t))$ is vertical. But it is also horizontal, because $X \in \mathcal{U}$. Then we deduce that $\dot{d}(t) = X(d(t))$ and therefore $d(t) = c(t)$.   □

Thus, in the vertical case, the reconstruction problem is just a horizontal lift operation with respect to the induced connection Υ living on M.

We recall now briefly the concepts of geometric, dynamic and total phases for the reconstruction process [159]. The *geometric phase* is just the holonomy of a closed path $\overline{c}(t)$ with respect to the connection Υ, that is, the Lie group element g so that $d(1) = g \cdot d(0)$. In general, we will have that $c(t)$, the integral curve projecting on $\overline{c}(t)$, is not exactly $d(t)$, the horizontal lift of $\overline{c}(t)$, but a shift of this curve, $c(t) = g(t) \cdot d(t)$. We call the Lie group element $g(1)$ the *dynamic phase*. And the *total phase* will stand for $h = g(1) \cdot g$.

Corollary 4.2.9. *In the purely kinematic case, the geometric phase coincides with the total phase.*

Concerning Chaplygin systems, the above description remains valid, of course, but we can say a little more about the holonomy of the two connections, γ and Υ.

Let $\bar{c}(t)$ be the integral curve of $\overline{X}$ starting from $\bar{x}$. Fix $x \in \rho^{-1}(\bar{x})$ and consider its horizontal lift, $c(t)$, with respect to Υ starting from x. We have proved that $c(t)$ is precisely the integral curve of X starting from x which projects on $\bar{c}(t)$. Let $\bar{q}(t)$ be the projection of $\bar{c}(t)$ to Q/G, $\bar{q}(t) = \pi_{Q/G}(\bar{c}(t))$. We will denote by $q^M(t)$ its horizontal lift with respect to γ. Finally, we write $q(t) = \pi_Q(c(t))$. Then we have $\pi(q(t)) = \pi \circ \pi_Q(c(t)) = \pi_{Q/G} \circ \rho(c(t)) = \pi_{Q/G}(\bar{c}(t)) = \bar{q}(t)$. Since $c(t)$ is an integral curve of a SODE, we have $c(t) = \dot{q}(t) \in M$. So we have proved that $q(t)$ is just the horizontal lift of $\bar{q}(t)$, i.e. $q(t) = q^M(t)$.

Now, we study the holonomy of $\bar{c}(t)$. Let us suppose that $\bar{c}(t)$ is a closed loop. We have $\bar{c}(0) = \bar{c}(1) = \bar{x}$ and $c(0) = x$. Consequently, $c(1) = gx$ and g is the geometric phase, which is, in the vertical case, the total phase. As $c(t) = \dot{q}^M(t)$, we have that $\dot{q}^M(1) = g\dot{q}^M(0)$ which in particular implies that $q^M(1) = gq^M(0)$. Then we have proved the following result.

Proposition 4.2.10. *The geometric phase (with respect to Υ) of a closed integral curve of $\overline{X}$ is the same as the geometric phase (with respect to γ) of its projection to Q/G.*

Plate with a knife edge on an inclined plane The configuration space of the plate with a knife edge on an inclined plane is $Q = \mathbb{R}^2 \times \mathbb{S}^1$ with coordinates (x, y, θ) (see Figure 4.1). The center of mass of the plate is assumed to coincide with the point (x, y) of contact of the knife edge and the plane [188].

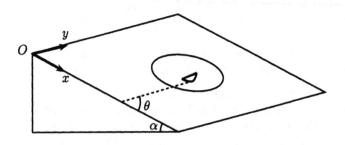

Fig. 4.1. Plate with a knife edge on an inclined plane

This system is determined by the following data: the regular Lagrangian function,

$$L: \quad TQ \quad \longrightarrow \mathbb{R}$$
$$(x, y, \theta, \dot{x}, \dot{y}, \dot{\theta}) \longmapsto \frac{1}{2}(\dot{x}^2 + \dot{y}^2) + \frac{1}{2}k^2\dot{\theta}^2 + gx\sin\alpha \,,$$

where the mass of the plate is assumed equal to unity; and the nonholonomic constraint function

$$\phi = \dot{y} - \dot{x}\tan\theta = 0 \,.$$

Consider the Lie group $G = \mathbb{R}$ and its trivial action by translation on Q,

$$\Phi: \quad \mathbb{R} \times Q \quad \longrightarrow \quad Q$$
$$(r, (x, y, \theta)) \longmapsto (x, y + r, \theta) \,,$$

with associated fibration

$$\rho: \quad Q \quad \longrightarrow \quad \mathbb{R} \times \mathbb{S}^1$$
$$(x, y, \theta) \longmapsto (x, \theta) \,.$$

Note that $\rho: Q \longrightarrow \mathbb{R} \times \mathbb{S}^1$ is a principal bundle, with structure group G, and M, the constraint submanifold, is the horizontal distribution of a principal connection, γ. The connection 1-form is $\gamma = dy - \tan\theta \, dx$. Therefore, this is a Chaplygin system.

The corresponding reduced system (4.11) is described by the reduced Lagrangian,

$$L^*: \quad T(\mathbb{R} \times \mathbb{S}^1) \quad \longrightarrow \quad \mathbb{R}$$
$$(x, \theta, \dot{x}, \dot{\theta}) \longmapsto \frac{1}{2}\left(\sec^2\theta \dot{x}^2 + k^2\dot{\theta}^2\right) + gx\sin\alpha \,,$$

and the gyroscopic 1-form

$$\overline{\alpha_X} = \tan\theta \sec^2\theta \left[(\dot{x})^2 \, d\theta - \dot{x}\dot{\theta} \, dx\right] \,.$$

After some calculations, one finds the following equations of motion,

$$\ddot{x} = -\dot{x}\dot{\theta}\tan\theta + g\sin\alpha\cos^2\theta \,, \quad \ddot{\theta} = 0 \,.$$

We obtain that $\theta = \omega t + \theta_0$, where ω and θ_0 are constants. Consequently, a solution for the initial conditions $\theta_0 = x_0 = \dot{x}_0 = 0$ and $\dot{\theta}_0 = \omega$ is

$$x = \frac{g}{2\omega^2}\sin\alpha\sin^2\omega t \,, \quad \theta = \omega t \,.$$

This curve $\overline{q}(t) = (x(t), \theta(t))$ is closed since

$$\bar{q}(0) = \bar{q}(2\pi/\omega) \, .$$

The horizontal lift $q(t) = q^M(t)$ of the curve $\bar{q}(t)$ with initial conditions $\theta_0 = x_0 = \dot{x}_0 = y_0 = \dot{y}_0 = 0$ and $\dot{\theta}_0 = \omega$ is

$$x = \frac{g}{2\omega^2} \sin\alpha \sin^2\omega t \, , \quad y = \frac{g}{2\omega^2} \sin\alpha \left[\omega t - \frac{1}{2}\sin\omega t\right] \, , \quad \theta = \omega t \, .$$

Observe that $q(0) = (0,0,0)$ and $q(2\pi/\omega) = (0, (g\pi/\omega^2)\sin\alpha, 0)$. Therefore, the geometric phase of the curve $\bar{q}(t)$ is $(g\pi/\omega^2)\sin\alpha$.

4.3 The case of horizontal symmetries

IN this section we focus our attention on the extreme case complementary to the purely kinematic one. The assumption now is that $V_x \cap F_x = V_x$, for all $x \in M$ or, equivalently, $V_{|M} \subset F$. In particular, every infinitesimal generator of the given group action then yields a horizontal symmetry. Thus, in this case, all the symmetries are compatible with the bundle F. This leads us to suspect that we can perform a reduction procedure similar to the unconstrained case. Note also that an unconstrained system with symmetry can be regarded as a special subcase of this situation, since we then have $M = TQ$, $F = T(TQ)$ and, obviously, $V \subset T(TQ)$.

4.3.1 Reduction

Taking into account that, by assumption, $V_{|M} \subset F$, we find that for the solution X of (3.12), we have along the constraint submanifold M

$$X(J_\xi) = 0 \, ,$$

i.e. the components of the momentum mapping are conserved quantities for the constrained dynamics.

Let $\mu \in \mathfrak{g}^*$ be a weakly regular value of $J = J^L$ (cf. Section 2.9). Since the action, $\hat{\Psi}$, of G on TQ is free and proper, we have that the isotropy group G_μ acts freely and properly on the level set $J^{-1}(\mu)$. It is known (see [1, 149, 166, 178]) that under these conditions $((TQ)_\mu = J^{-1}(\mu)/G_\mu, \omega_\mu)$ is a symplectic manifold, where ω_μ is the 2-form defined by

$$\pi_\mu^* \omega_\mu = j_\mu^* \omega_L \, ,$$

with $\pi_\mu : J^{-1}(\mu) \longrightarrow (TQ)_\mu$ the canonical projection and $j_\mu : J^{-1}(\mu) \hookrightarrow TQ$ the natural inclusion.

Imposing a condition of clean intersection of M and $J^{-1}(\mu)$, we have that $M' = M \cap J^{-1}(\mu)$ is a submanifold of $J^{-1}(\mu)$ which is G_μ-invariant. Passing to the quotient, we then obtain a submanifold $M_\mu = M'/G_\mu$ of $(TQ)_\mu$. This submanifold can be identified, via the adequate embedding, with $\overline{M} \cap (TQ)_\mu$.

Next, we can define a distribution F' on TQ along M' by putting

$$F'_{x'} = T_{x'}(J^{-1}(\mu)) \cap F_{x'} , \ \forall x' \in M' ,$$

and we make the further simplifying assumption that F' has constant rank. It is obvious that F' is a G_μ-invariant subbundle of $T_{M'}TQ$ and, hence, it projects onto a subbundle F_μ of $T(TQ)_\mu$ along M_μ. Finally, since the restriction of the energy E_L to $J^{-1}(\mu)$ is also G_μ-invariant, it induces a function E_{L_μ} on $(TQ)_\mu$.

Theorem 4.3.1. *[51] Suppose that X is the (G-invariant) solution of (3.12). Then, X induces a vector field X_μ on M_μ, such that*

$$\begin{cases} (i_{X_\mu}\omega_\mu - dE_{L_\mu})_{|M_\mu} \in F_\mu^o , \\ X_\mu \in TM_\mu . \end{cases} \tag{4.16}$$

In the case of horizontal symmetries we have therefore that, under the appropriate assumptions, the given nonholonomic constrained problem on (TQ, ω_L) reduces to a constrained problem on $((TQ)_\mu, \omega_\mu)$.

4.3.2 Reconstruction

The parallelism of the horizontal case with unconstrained systems can also be tracked in the process of reconstruction of the dynamics, as we see next.

We start with $c_\mu(t)$ an integral curve of X_μ with initial condition $c_\mu(0) = m_\mu \in M_\mu$. Choose $m \in (\pi_\mu)^{-1}(m_\mu)$. We would like to find the unique integral curve $c(t)$ of X which satisfies $c(0) = m$. As X and X_μ are π_μ-related, $c(t)$ projects on $c_\mu(t)$.

Let $d(t)$ be a curve in TQ such that $\pi_\mu(d(t)) = c_\mu(t)$ (later, we will discuss how to obtain such curves). Put $c(t) = g(t)d(t)$, for some curve $g(t)$ in G_μ, with $g(0) = e$. As $c(t)$ is an integral curve of X, we have that $X(c(t)) = \dot{c}(t)$, i.e.

$$X(g(t)d(t)) = \frac{d}{dt}(g(t)d(t)) = g(t)\dot{d}(t) + g(t)((g^{-1}(t)\dot{g}(t))_M d(t)) .$$

As $X(g(t)d(t)) = g(t)X(d(t))$, we conclude

$$X(d(t)) = \dot{d}(t) + (g^{-1}(t)\dot{g}(t))_M d(t) . \tag{4.17}$$

So, similarly to the unconstrained case [159], we can split the reconstruction process in two steps,

1. find a curve $\xi(t)$ in $\mathfrak{g}_\mu$ such that $\xi(t)_M(d(t)) = X(d(t)) - \dot{d}(t)$,

2. find a curve $g(t)$ in G_μ such that $\dot{g}(t) = g(t)\xi(t)$, $g(0) = e$.

A standard procedure in the reconstruction of the dynamics [159] is the selection of a connection on an appropriate principal fiber bundle: for instance, if we choose an arbitrary connection Υ on the principal G_μ-bundle $M' \longrightarrow M_\mu$, Υ enables us to horizontally lift the integral curves of the reduced system from M_μ to M'. Therefore, we can take $d(t)$ as the horizontal lift of $c_\mu(t)$ with $d(0) = m$, that is, $\pi_\mu(d(t)) = c_\mu(t)$ and $\Upsilon(\dot{d}(t)) = 0$.

Making use of the connection Υ, we can replace (i) above by

(i') $\xi(t) = \Upsilon(\xi(t)_M(d(t))) = \Upsilon(X(d(t)) - \dot{d}(t)) = \Upsilon(X(d(t)))$.

In the remainder of this section, we show that for mechanical Lagrangian systems, the selection of the connection can be done in a natural way. Indeed, we first show that if the bundle $\varsigma_\mu : Q \longrightarrow Q/G_\mu$ has a connection, this induces a connection on $\rho : M' \longrightarrow M_\mu$. Then, we see how for mechanical systems there is always a natural connection on $\varsigma_\mu : Q \longrightarrow Q/G_\mu$.

Denote by $\mu' = \mu|_{\mathfrak{g}_\mu} \in \mathfrak{g}_\mu^*$, the restriction of μ to $\mathfrak{g}_\mu$, and consider the mapping $J_\mu : TQ \longrightarrow \mathfrak{g}_\mu^*$ defined by $J_\mu(v_q) = J(v_q)|_{\mathfrak{g}_\mu}$. We have that $\varsigma_\mu : Q \longrightarrow Q/G_\mu$ is a principal G_μ-bundle. Let $\gamma \in \Lambda^1(Q, \mathfrak{g}_\mu)$ be a connection form on it. We recall now the tangent bundle version of the cotangent bundle reduction theorem of Satzer, Marsden and Kummer (see [1, 128, 159]).

Theorem 4.3.2. *Assume that there is a vector field $Y_\mu \in \mathfrak{X}(Q)$ such that $Y_\mu(Q) \subset (J_\mu)^{-1}(\mu')$ and Y_μ is G_μ-invariant, that is $Y_\mu(gq) = gY_\mu(q)$, for $g \in G_\mu$. Then, there is an embedding $\varphi_\mu : (TQ)_\mu \longrightarrow T(Q/G_\mu)$ whose range is a vector subbundle with base Q/G_μ. This embedding is onto if and only if $\mathfrak{g} = \mathfrak{g}_\mu$.*

Proof. The proof simply consists in making the necessary translations with respect to the proof in the cotangent bundle picture. The vector field Y_μ induces, by equivariance, the vector field $\tilde{Y}_\mu$ on Q/G_μ: $\tilde{Y}_\mu \circ \varsigma_\mu = \varsigma_{\mu*} Y_\mu$. Define the projection $\tau_\mu : (TQ)_\mu \longrightarrow Q/G_\mu$ by $\tau_\mu([v_q]) = [q]$, so that $\varsigma_\mu \circ \tau = \tau_\mu \circ \pi_\mu$, where we recall that $\pi_\mu : J^{-1}(\mu) \longrightarrow (TQ)_\mu$ is the canonical projection. Let $t_\mu : J_\mu^{-1}(\mu') \longrightarrow J_\mu^{-1}(0)$ be given by $t_\mu(v_q) = v_q - Y_\mu(q)$ and let $\varphi_\mu : (TQ)_\mu \longrightarrow T(Q/G_\mu)$ be the map induced by the relation $\varphi_\mu \circ \pi_\mu = \varsigma_{\mu*} \circ t_\mu$, defined on the set $J^{-1}(\mu)$. Then φ_μ is an embedding and it is easy to see that it is onto iff $\mathfrak{g} = \mathfrak{g}_\mu$ by comparing $J_\mu^{-1}(0)$ and $J^{-1}(0)$. $\square$

The commutative diagram in Figure 4.2 will help us handle the theorem.

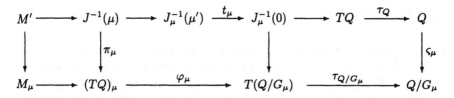

Fig. 4.2. Illustration of the result in Theorem 4.3.2

The vector field Y_μ postulated in the hypothesis of Theorem 4.3.2 can be chosen to be consistent with the principal connection $\gamma \in \Lambda^1(Q, \mathfrak{g}_\mu)$, by means of

$$Y_\mu(q) = \mathcal{F}L^{-1}(\mu' \circ \gamma(q)), \quad q \in Q, \tag{4.18}$$

i.e. Y_μ is the Legendre transform of the μ'-component of the connection γ.

The connection $\gamma \in \Lambda^1(Q, \mathfrak{g}_\mu)$ induces a connection $\Upsilon \in \Lambda^1(M', \mathfrak{g}_\mu)$ by pullback, $\Upsilon = (\tau_Q \cdot t_\mu)^* \gamma$ so that $\Upsilon_{v_q}(U_{v_q}) = \gamma_q(T\tau_Q \cdot U_{v_q})$ for all $U_{v_q} \in T_{v_q}M'$. Taking this into account, we can rewrite (i') above as

(i') $\xi(t) = \Upsilon(X(d(t))) = \gamma(T\tau_Q \cdot X(d(t))) = \gamma(d(t))$.

In the following, we specialize our discussion to the case of Lagrangian systems of natural type.

Natural Lagrangian systems In case we have a Lagrangian of mechanical type, $L = T - V$, where T is the kinetic energy of a Riemannian metric g on Q and V is a potential energy, we know that the nonholonomic Lagrangian system fulfills the compatibility condition (cf. Section 3.4.1). Making use of the special geometry of mechanical systems, we can naturally define a connection on the principal fiber bundle $\varsigma_\mu : Q \longrightarrow Q/G_\mu$. We discuss this next.

Let $\mathcal{V}_{\varsigma_\mu} = \ker T\varsigma_\mu$ and consider $\mathcal{H} = \mathcal{V}_{\varsigma_\mu}^\perp$, the orthogonal complement of $\mathcal{V}_{\varsigma_\mu}$ with respect to the metric g. We define the *mechanical connection*, γ_{mech}, as the connection on $Q \longrightarrow Q/G_\mu$ whose horizontal subspace is $\mathcal{H}$.

As a consequence of this choice, we can rewrite (i') as,

(i') $\xi(t) = \gamma_{mech}(q(t))(d(t))$,

with $q(t) = \tau_Q(d(t))$. If we define for each $q \in Q$ the μ-*locked inertia tensor* (see [158]), $I_\mu(q) : \mathfrak{g}_\mu \longrightarrow \mathfrak{g}_\mu^*$, by $\langle I_\mu(q)\zeta, \eta \rangle = \langle \zeta_Q(q), \eta_Q(q) \rangle$, we can verify that $\gamma_{mech}(v_q) = I_\mu^{-1}(q)J(v_q)$. We then rewrite (i') as,

(i') $\xi(t) = \gamma_{mech}(q(t))(d(t)) = I_\mu^{-1}(q(t))(\mu)$.

Results somehow related to this one can be found in [29].

4.4 The general case

IN this section, we consider the intermediate case between the extreme cases of purely kinematic and horizontal symmetries. Assume that at each $x \in M$, $\{0\} \subsetneq V_x \cap F_x \subsetneq V_x$. Following our discussion in Section 4.1 (see the computations we made in deriving equation (4.2)), we see that the momentum mapping J is no longer a conserved quantity for the constrained dynamics. However, a careful investigation of the structure of nonholonomic systems with symmetry leads us to obtain an equation describing the evolution along the integral curves of the components of the momentum mapping compatible with the constraints. The first result is this sense was given in [29] for nonholonomic mechanical systems and extended in [51] to general constrained systems (see [222] for a recent approach).

Let us consider for each $x \in M$ the following space

$$\mathfrak{g}^x = \{\xi \in \mathfrak{g} \mid \xi_M(x) \in F_x\} \, .$$

Recall that ξ_M is just the restriction of ξ_{TQ} to the G-invariant submanifold M. Since F is a vector bundle, we have that $\mathfrak{g}^x$ is a vector subspace of $\mathfrak{g}$. Putting

$$\mathfrak{g}^F = \coprod_{x \in M} \mathfrak{g}^x \, ,$$

where we use the symbol "$\coprod$" to denote the disjoint union of the vector spaces, we obtain a ("generalized") vector bundle over M, with canonical projection $\mathfrak{g}^F \longrightarrow M : \xi \in \mathfrak{g}^x \longmapsto x$. In general, this bundle need not have constant rank. However, for the subsequent discussion we make the simplifying assumption that $\mathfrak{g}^F$ is a genuine vector bundle over M, the fibers of which have constant dimension (independent of the base point). This assumption is valid for a large variety of examples.

Now, consider the "restriction" of the natural momentum mapping $J = J^L : TQ \longrightarrow \mathfrak{g}^*$ (cf. Section 2.9) to the symmetry directions that are compatible with the constraints, which are precisely given by the bundle $\mathfrak{g}^F$. Define a smooth section $J^{(c)} : M \longrightarrow (\mathfrak{g}^F)^*$ of the dual bundle $(\mathfrak{g}^F)^*$ as follows,

$$J^{(c)}(x) : \mathfrak{g}^x \longrightarrow \mathbb{R} \, , \ J^{(c)}(x)(\xi) = \langle J(x), \xi \rangle \, .$$

The mapping $J^{(c)}$ will be called the *constrained momentum mapping* [29, 51].

Remark 4.4.1. In the works [29, 50], the reduction theory is developed in terms of the vector bundle $\mathfrak{g}^M \longrightarrow Q$, defined by,

$$\mathfrak{g}^q = \{\xi \in \mathfrak{g} \mid \xi_{TQ}(v_q) \in F_{v_q} \text{ for all } v_q \in M \cap T_qQ\} \, .$$

The *nonholonomic momentum mapping* $J^{nh} : TQ \longrightarrow (\mathfrak{g}^M)^*$ is then defined by

$$\langle J^{nh}(v_q), \xi \rangle = \langle J(v_q), \xi \rangle \,,$$

for all $v_q \in TQ$ and $\xi \in \mathfrak{g}^q$. In fact, J^{nh} restricts naturally to M, $J^{nh}_{|M} : M \longrightarrow (\mathfrak{g}^M)^*$. For simplicity, we will usually denote this mapping by J^{nh}, instead of $J^{nh}_{|M}$. We will only restore the notational distinction when the confusion is possible.

The vector bundle we are considering here was introduced in [51] in a more general constrained setting. The relation between both bundles $\mathfrak{g}^M \longrightarrow Q$ and $\mathfrak{g}^F \longrightarrow M$ is given by the following result. By definition, we have that

$$\mathfrak{g}^q = \bigcap_{v_q \in M \cap T_q Q} \mathfrak{g}^{v_q}$$

for all $q \in Q$. The fibers $\mathfrak{g}^q$ and $\mathfrak{g}^{v_q}$ of $\mathfrak{g}^M$ and $\mathfrak{g}^F$, resp., do not coincide in general. Indeed, let us take $\xi \in \mathfrak{g}^{v_q}$ and $w_q \in M \cap T_q Q$. We want to see if $\xi \in \mathfrak{g}^{w_q}$, i.e. $\xi_{TQ}(w_q) \in F_{w_q}$. Applying the musical mapping $\flat_L$, this is equivalent to $\flat_L(\xi_{TQ}(w_q)) = (dJ_\xi)_{w_q} \in \flat_L(F_{w_q}) = (F_{w_q}^\perp)^\circ$. Now, $F^\perp$ is locally generated by the Hamiltonian vector fields $Z_1, \ldots, Z_m$ (cf. Section 3.4.1). Consequently, we would have $(dJ_\xi)_{w_q} Z_i(w_q) = 0$, $1 \leq i \leq m$. But

$$(dJ_\xi)_{w_q} Z_i(w_q) = \omega_L(\xi_{TQ}, Z_i)(w_q)$$
$$= -S^*(d\phi_i)_{w_q}(\xi_{TQ}(w_q)) = -(d\phi_i)_{w_q}(\xi_Q^v(w_q)) \,.$$

In coordinates, if we write $\xi_Q(q) = f^A(q) \frac{\partial}{\partial q^A}$, the right-hand side becomes

$$(d\phi_i)_{w_q}(\xi_Q^v(w_q)) = \frac{\partial \phi_i}{\partial \dot{q}^A}(q, \dot{q}) f^A(q) \,,$$

which in general will not be zero (even though it is zero at v_q). However, if the constraints are linear or affine, the term

$$\frac{\partial \phi_i}{\partial \dot{q}^A}(q, \dot{q})$$

only depends on the base point $q \in Q$, and $\xi \in \mathfrak{g}^{v_q}$ implies $\xi \in \mathfrak{g}^{w_q}$ for all $w_q \in M \cap T_q Q$. Therefore $\mathfrak{g}^q = \mathfrak{g}^{w_q}$ for all $w_q \in M \cap T_q Q$. In such a case, we will identify the momentum mappings $J^{(c)}$ and J^{nh}.

Given a smooth section $\overline{\xi}$ of the vector bundle $\mathfrak{g}^F$, $\overline{\xi} : M \longrightarrow \mathfrak{g}^F$, we can then define a smooth function $J^{(c)}_{\overline{\xi}}$ on M according to

$$J^{(c)}_{\overline{\xi}} = \langle J^{(c)}, \overline{\xi} \rangle \,.$$

The section $\overline{\xi}$ induces a vector field Ξ on M by putting

$$\Xi(x) = (\overline{\xi}(x))_M(x) \,, \ \forall x \in M \,.$$

Theorem 4.4.2. *[51] Let X be the solution of (3.12). For any smooth section $\bar{\xi}$ of $\mathfrak{g}^F$ we then have*

$$X(J_{\bar{\xi}}^{(c)}) = (\mathcal{L}_{\Xi}\theta_L)(X) \,. \tag{4.19}$$

Proof. We have that

$$X(J_{\bar{\xi}}^{(c)}) = dJ_{\bar{\xi}}^{(c)}(X) = i_X di_{\Xi} j^* \theta_L = i_X i_{\Xi} j^* \omega_L + i_X \mathcal{L}_{\Xi} j^* \theta_L \,.$$

Since $j^* \omega_L(\Xi, X) = -(i_X \omega_L)(\Xi)$ and $\Xi \in F$ this further becomes

$$X(J_{\bar{\xi}}^{(c)}) = i_X \mathcal{L}_{\Xi} j^* \theta_L = -i_{[\Xi,X]} j^* \theta_L + \mathcal{L}_{\Xi} i_X j^* \theta_L = (\mathcal{L}_{\Xi}\theta_L)(X) \,.$$

$\square$

Note that for the above result we do not have to require X to be G-invariant. Equation (4.19) is called *the momentum equation* for the given constrained system.

Remark 4.4.3. An important observation is the following: for nonholonomic Lagrangian systems, recall that the solution X of (3.12) is a SODE vector field. As a consequence, $(\mathcal{L}_{\Xi}\theta_L)(X)$ can be obtained without having to compute explicitly the dynamics X, since $\mathcal{L}_{\Xi}\theta_L$ is a semi-basic one-form.

Remark 4.4.4. In the case of linear constraints (recall Remark 4.4.1) we precisely recover the result established by Bloch *et al* [29]. Indeed, a global section $\bar{\xi}$ of the vector bundle $\mathfrak{g}^M \longrightarrow Q$ induces a vector field $\tilde{\Xi}$ on Q as follows

$$\tilde{\Xi}(q) = (\bar{\xi}(q))_Q(q) \in T_q Q \,,$$

for all $q \in Q$. Then, it can be seen after some computations [50] that the nonholonomic momentum equation (4.19) can be rewritten as

$$X(J_{\bar{\xi}}^{nh}) = \tilde{\Xi}^c(L) \,. \tag{4.20}$$

Remark 4.4.5. Let $\bar{\xi}$ be a constant section of $\mathfrak{g}^F$, i.e. $\bar{\xi}(x) = \xi^0 \in \mathfrak{g}$ for all $x \in M$. We may then identify the corresponding vector field Ξ with the infinitesimal generator ξ_M^0 and, clearly, $J_{\bar{\xi}}^{(c)} = (J_{\xi^0})_{|M}$. Moreover, by construction, ξ_M^0 is a horizontal symmetry. The momentum equation (4.19) then leads to

$$X(J_{\bar{\xi}}^{(c)}) = X(J_{\xi^0})_{|M} = 0 \,,$$

i.e. we have obtained a conserved quantity of X associated with the horizontal symmetry ξ_M^0. This is again a manifestation of Noether's theorem for constrained systems.

4.4.1 Reduction

In this section, we are going to develop a reduction procedure in the general case via the constrained momentum mapping. We will assume throughout the discussion that the constraints are affine, so that we will make use of the nonholonomic momentum mapping $J^{nh} : TQ \longrightarrow (\mathfrak{g}^M)^*$.

As we have remarked above, the main difficulty (and precisely the important point) for nonholonomic systems is that the momentum is not a conserved quantity. So, instead of fixing a value $\mu \in \mathfrak{g}^*$ for the momentum as in the traditional approach of symplectic reduction [1, 166, 178], we will take a C^∞-section $\mu : Q \longrightarrow (\mathfrak{g}^M)^*$ of the dual vector bundle $(\mathfrak{g}^M)^*$, with canonical projection $\pi^* : (\mathfrak{g}^M)^* \longrightarrow Q$. Now, consider the level set

$$(J^{nh})^{-1}(\mu) = \{v_q \in M \mid J^{nh}(v_q) = \mu(q)\}$$

In general, $(J^{nh})^{-1}(\mu)$ will not be a submanifold of M. We will denote the inclusion by $j : (J^{nh})^{-1}(\mu) \hookrightarrow M$.

Assume that the vector bundle $\mathfrak{g}^M \longrightarrow Q$ has constant rank r, and choose $\overline{\xi}_1, \ldots, \overline{\xi}_r$, r linearly independent sections. Consider r functions on M, $f_i : M \longrightarrow \mathbb{R}$, defined by $f_i = \langle \mu, \overline{\xi}_i \rangle \circ \tau_Q - J^{nh}_{\overline{\xi}_i}$. For each i, we denote $TQ_{\overline{\xi}_i} = f_i^{-1}(0)$. Then, it is not hard to see that

$$(J^{nh})^{-1}(\mu) = \bigcap_{i=1}^{r} TQ_{\overline{\xi}_i} .$$

In the following proposition, we characterize the desired sections μ and give certain conditions to assure the existence of a differentiable structure on $(J^{nh})^{-1}(\mu)$.

Proposition 4.4.6. *If 0 is a weakly regular value of f_i for $1 \leq i \leq r$, then $TQ_{\overline{\xi}_i}$ is a submanifold of M. If, in addition, the intersection $\bigcap_{i=1}^{r} TQ_{\overline{\xi}_i}$ is clean, then $(J^{nh})^{-1}(\mu)$ is a submanifold of M, and X is tangent to it if and only if*

$$X(\langle \mu, \overline{\xi}_i \rangle \circ \tau_Q) = \Xi_i^c(L) , \tag{4.21}$$

for all $1 \leq i \leq r$.

Proof. Given the above discussion, it only remains to prove the equivalence. Assume that the section μ fulfills equation (4.21), i.e. $X(\langle \mu, \overline{\xi}_i \rangle \circ \tau_Q) = \Xi_i^c(L)$, for all $1 \leq i \leq r$. Then, using the nonholonomic momentum equation, we have that $X(f_i) = 0$. So X is tangent to the level submanifold $TQ_{\overline{\xi}_i}$ for each i. As $T((J^{nh})^{-1}(\mu)) = \bigcap_{i=1}^{r} T(TQ_{\overline{\xi}_i})$, it follows that $X \in T((J^{nh})^{-1}(\mu))$. The converse is obvious. $\qquad\square$

In the sequel, we will assume the hypothesis of Proposition 4.4.6.

Lemma 4.4.7. *We have*

$$T^\perp((J_{|M}^{nh})^{-1}(\mu)) = T^\perp M + \text{span}\left\{X_{\tilde{f}_1}, \ldots, X_{\tilde{f}_r}\right\}.$$

Proof. We will distinguish now between $J^{nh} : TQ \to (\mathfrak{g}^M)^*$ and $J_{|M}^{nh} : M \to (\mathfrak{g}^M)^*$. We have that $(J_{|M}^{nh})^{-1}(\mu) = (J^{nh})^{-1}(\mu) \cap M$. Consider $\tilde{f}_i$, the natural extension of f_i to TQ, $\tilde{f}_i = \langle \mu, \bar{\xi}_i \rangle \circ \tau_Q - J_{\bar{\xi}_i}^{nh}$. For each i, denote $\widetilde{TQ}_{\bar{\xi}_i} = \tilde{f}_i^{-1}(0)$. It is clear that $TQ_{\bar{\xi}_i} = \widetilde{TQ}_{\bar{\xi}_i} \cap M$. We also have that $(J^{nh})^{-1}(\mu) = \bigcap_{i=1}^r \widetilde{TQ}_{\bar{\xi}_i}$. A simple counting of dimensions shows that $\dim T((J^{nh})^{-1}(\mu)) \geq \dim TQ - r$. Consequently, we have that $\dim T^\perp((J^{nh})^{-1}(\mu)) \leq r$. On the other hand, it easy to check that $X_{\tilde{f}_i} \in T^\perp((J^{nh})^{-1}(\mu))$, $1 \leq i \leq r$. Then, we have proved that

$$T^\perp((J^{nh})^{-1}(\mu)) = \text{span}\left\{X_{\tilde{f}_1}, \ldots, X_{\tilde{f}_r}\right\}.$$

Finally,

$$T^\perp((J_{|M}^{nh})^{-1}(\mu)) = T^\perp M + T^\perp((J^{nh})^{-1}(\mu)) = T^\perp M + \text{span}\left\{X_{\tilde{f}_1}, \ldots, X_{\tilde{f}_r}\right\}.$$

$\square$

In order to perform the reduction, we need a G-action on the vector bundle $(\mathfrak{g}^M)^*$, playing the role of the coadjoint action of G on $\mathfrak{g}^*$. The following lemma enables us to go further in that direction

Lemma 4.4.8. *For $g \in G$ and $\xi \in \mathfrak{g}^M$, put $Ad^M(g, \xi) = Ad_g(\xi)$. Then, for $\xi \in \mathfrak{g}^q$, we have that $Ad^M(g, \xi) \in \mathfrak{g}^{gq}$, and $Ad^M : G \times \mathfrak{g}^M \longrightarrow \mathfrak{g}^M$ is a well-defined "action" on the vector bundle $\mathfrak{g}^M$.*

Proof. The unique thing that remains to be proved is that Ad^M is well-defined, because the properties $Ad_e^M = \text{Id}$ and $Ad_{gh}^M = Ad_g^M \circ Ad_h^M$ follow directly from the fact that Ad is a G-action. Thus let us take $g \in G$ and $\xi \in \mathfrak{g}^q$, which is to say that $\xi_{TQ}(v_q) \in F_{v_q}$ for all $v_q \in T_qQ \cap M$. As the vector bundle F is G-invariant, we have that $(Ad_g(\xi))_{TQ}(g \cdot v_q) = (\Phi_g)_*(\xi_{TQ}(v_q))$ belongs to $F_{g \cdot v_q}$, for all $v_q \in T_qQ \cap M$, namely, $(Ad_g(\xi))_{TQ}(w_q) \in F_{w_q}$, for all $w_q \in T_qQ \cap M$. Consequently, $Ad_g(\xi) \in \mathfrak{g}^{gq}$ and Ad^M is well-defined. $\square$

In a similar way, we can consider the G-"action" on $(\mathfrak{g}^M)^*$ defined by

$$CoAd^M : G \times (\mathfrak{g}^M)^* \longrightarrow (\mathfrak{g}^M)^*$$
$$(g, \eta) \longmapsto CoAd^M(g, \eta) = CoAd_g(\eta)$$

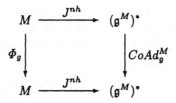

Fig. 4.3. G-equivariance of the nonholonomic momentum mapping.

Note that the nonholonomic momentum mapping $J^{nh} : M \longrightarrow (\mathfrak{g}^M)^*$ is G-equivariant with respect to this action, that is to say, $CoAd_g(J^{nh}(v_q)) = J^{nh}(g \cdot v_q)$, for all $g \in G$, and the diagram in Figure 4.3 is commutative.

The last ingredient we need to define is the "isotropy group" of the action $CoAd^M$ corresponding to a section $\mu : Q \longrightarrow (\mathfrak{g}^M)^*$. This is defined as

$$G_\mu = \{g \in G \,|\, CoAd_g^M \circ \mu = \mu \circ \Phi_g\} \,,$$

where we mean by $CoAd_g^M \circ \mu = \mu \circ \Phi_g$ that $CoAd_g^M(\mu(q)) = \mu(gq)$ for all $q \in Q$. Note that G_μ is a Lie subgroup of G.

Now, we can define a G_μ-action on the manifold $(J^{nh})^{-1}(\mu)$ in the following way

$$\begin{aligned}
\Theta : \ G_\mu \times (J^{nh})^{-1}(\mu) &\longrightarrow (J^{nh})^{-1}(\mu) \\
(g, v_q) &\longmapsto \Theta(g, v_q) = g \cdot v_q
\end{aligned}$$

The definition of the group G_μ and the equivariance of $J^{nh} : M \longrightarrow (\mathfrak{g}^M)^*$ imply that this action is well defined, as we check in the following

Lemma 4.4.9. *The mapping Θ is well defined.*

Proof. Take $g \in G_\mu$ and $v_q \in (J^{nh})^{-1}(\mu)$. By equivariance, we have $J^{nh}(\Theta(g, v_q)) = CoAd_g^M(J^{nh}(v_q)) = CoAd_g^M(\mu(q))$. Finally, by definition of G_μ, it follows that $\Theta(g, v_q) \in (J^{nh})^{-1}(\mu)$. $\qquad\square$

We can consider the action Θ as the restriction to $(J^{nh})^{-1}(\mu)$ of a G_μ-action on M, $\Theta_M : G_\mu \times M \longrightarrow M$. Both Θ and Θ_M will be free and proper actions, because they inherit these properties from the original action $\hat{\Psi} : G \times TQ \longrightarrow TQ$. Then, the orbit spaces M/G_μ and $\overline{(J^{nh})^{-1}(\mu)} = (J^{nh})^{-1}(\mu)/G_\mu$ are differentiable manifolds, and we have two principal G_μ-bundles $\pi : M \longrightarrow M/G_\mu$ and $\pi_{|(J^{nh})^{-1}(\mu)} : (J^{nh})^{-1}(\mu) \longrightarrow \overline{(J^{nh})^{-1}(\mu)}$, respectively.

An approach based on symplectic reduction In this section, we make use of ideas from the symplectic reduction scheme developed by Bates and Śniatycki [18] (cf. Section 4.1) for the reduction of nonholonomic systems via the nonholonomic momentum mapping.

Now, we define a (generalized) vector subbundle U_μ of $TM_{|(J^{nh})^{-1}(\mu)}$, whose fiber at $x \in (J^{nh})^{-1}(\mu)$ is given by

$$(U_\mu)_x = \{v \in F_x \cap T_x(J^{nh})^{-1}(\mu) \ / \ \omega_L(v,\tilde\xi) = 0, \text{ for all } \tilde\xi \in (\mathcal{V}_\mu)_x \cap F_x\} .$$
(4.22)

In general, U_μ need not be of constant rank. For the further discussion, however, we will assume that U_μ is a genuine vector bundle over $(J^{nh})^{-1}(\mu)$. Note that $U_\mu = F \cap T((J^{nh})^{-1}(\mu)) \cap (\mathcal{V}_\mu \cap F)^\perp$, where $(\mathcal{V}_\mu \cap F)^\perp$ is the ω_L-complement of $\mathcal{V}_\mu \cap F$ in $TTQ_{|(J^{nh})^{-1}(\mu)}$. The bundle U_μ is G_μ-invariant and, hence, it projects onto a subbundle $\overline{U}_\mu$ of $T(\overline{M}_\mu)_{|\overline{(J^{nh})^{-1}(\mu)}}$.

Let us now denote by ω_μ the restriction of ω_L to U_μ. Clearly, ω_μ is also G_μ-invariant and by the very definition of the vector bundle U_μ, the 2-form ω_μ pushes down to a 2-form $\overline{\omega}_\mu$ on $\overline{U}_\mu$. Similarly, the restriction of dE_L to U_μ, denoted by $d_\mu E_L$, pushes down to a 1-form $d_\mu \overline{E}_L$ on $\overline{U}_\mu$, which is simply the restriction of $d\overline{E}_L$ to $\overline{U}_\mu$. Note that neither $\overline{\omega}_\mu$ nor $d_\mu \overline{E}_L$ are differential forms on $\overline{(J^{nh})^{-1}(\mu)}$; they are exterior forms on a vector bundle over $\overline{(J^{nh})^{-1}(\mu)}$, with smooth dependence on the base point.

Proposition 4.4.10. *Let X be the solution of* (3.12)*. Then, its projection $\overline{X}_\mu$ onto $\overline{(J^{nh})^{-1}(\mu)}$ is a section of $\overline{U}_\mu$ satisfying the equation*

$$i_{\overline{X}_\mu} \overline{\omega}_\mu = d_\mu \overline{E}_L .$$
(4.23)

Proof. Similar as for Proposition 4.1.4.   □

Remark 4.4.11. It should be noticed that, in general, the 2-form $\overline{\omega}_\mu$ may be degenerate. So, the reduced dynamics is not uniquely determined by equation (4.23).

Almost-Poisson reduction The idea of this approach is to project the nonholonomic bracket (cf. Section 3.4.1) to the reduced space $\overline{(J^{nh})^{-1}(\mu)}$ via the G_μ-action $\Theta : G_\mu \times (J^{nh})^{-1}(\mu) \longrightarrow (J^{nh})^{-1}(\mu)$. For that purpose we will make use of the almost-Poisson reduction scheme discussed in Section 2.8.1.

With the notation of that section, we have that $N = (J^{nh})^{-1}(\mu)$. It seems to be quite reasonable to take as the control bundle E at each point v_q of $(J^{nh})^{-1}(\mu)$ just the tangent at v_q to the G_μ-orbit of v_q, i.e. $E_{v_q} = T_{v_q}(G_\mu \cdot v_q)$. It is easy to see that (N, E) is a reductive structure, with $S = \overline{(J^{nh})^{-1}(\mu)}$.

We now investigate whether (M, N, E) is a reducible triple. We have that

$$E^o = \text{span}\left\{d\chi \mid \chi \in C^\infty_{G_\mu}(M)\right\},$$

where $C^\infty_{G_\mu}(M)$ denotes the G_μ-invariant functions on M. Then,

$$\sharp_M(E^o) = \text{span}\left\{X^M_\chi \mid \chi \in C^\infty_{G_\mu}(M)\right\}.$$

Note that X^M_χ denotes the Hamiltonian vector field associated with the function $\chi : M \longrightarrow \mathbb{R}$ by the musical mapping $\sharp_M$ induced by the almost-Poisson bivector field Λ_M. But,

$$\begin{aligned}
X^M_\chi(v) &= \{v, \chi\}_M = \omega_L(\bar{P}(X_{\tilde{v}}), \bar{P}(X_{\tilde{\chi}})) \circ j_M \\
&= \omega_L(X_{\tilde{v}}, \bar{P}(X_{\tilde{\chi}})) \circ j_M = \bar{P}(X_{\tilde{\chi}})(v),
\end{aligned}$$

for all $v \in C^\infty(M)$, where $\tilde{\chi}$ denotes an arbitrary extension of χ to TQ. So $X^M_\chi = \bar{P}(X_{\tilde{\chi}})$ and

$$\sharp_M(E^o) = \text{span}\left\{\bar{P}(X_{\tilde{\chi}}) \mid \chi \in C^\infty_{G_\mu}(M)\right\}.$$

In addition, since $E + N = T(G_\mu \cdot) + T((J^{nh})^{-1}(\mu)) = T((J^{nh})^{-1}(\mu)) = N$, we have, according to Theorem 2.8.5, that (M, N, E) is a reducible triple if and only if

$$\sharp_M(E^o) \subseteq T((J^{nh})^{-1}(\mu)) \iff \bar{P}(X_{\tilde{\chi}})(f_i) \circ j = 0, \ 1 \le i \le r, \ \forall \chi \in C^\infty_{G_\mu}(M)$$
$$\iff \{f_i, \chi\}_M \circ j = 0, \ 1 \le i \le r, \ \forall \chi \in C^\infty_{G_\mu}(M) \tag{4.24}$$

In the purely kinematic case, as we have discussed in Section 4.2, the nonholonomic momentum mapping is trivial, and therefore the conditions (4.24) hold trivially (in fact, $(J^{nh})^{-1}(\mu) = M$). In the horizontal case, we would have $\mathfrak{g}^M = \mathfrak{g} \times Q$, so $r = \dim G$. Taking a constant section $\mu(q) = (\mu, q)$ and a basis of the Lie algebra $\mathfrak{g}$, $\xi_1, \ldots, \xi_r$, we could write $f_i = \langle \mu, \xi_i \rangle - J_{\xi_i}$, $1 \le i \le r$. Then $\{f_i, \chi\}_M \circ j = -\bar{P}(X_{f_i})(\chi) \circ j = (\xi_i)_M(\chi) \circ j$. However, the conditions (4.24) will not be fulfilled in general, because $C^\infty_{G_\mu}(M) \ne C^\infty_G(M)$.

Almost-Poisson mappings The obstruction we have found above in the horizontal case to reduce the nonholonomic bracket $\{\cdot, \cdot\}_M$ to $\overline{(J^{nh})^{-1}(\mu)}$ via $((J^{nh})^{-1}(\mu)), T(G_\mu \cdot))$ leads us to develop another scheme of reduction which takes into account the whole Lie group G. For that purpose, let us define the following mapping

$$k : \overline{(J^{nh})^{-1}(\mu)} \xrightarrow{k_\mu} M/G_\mu \xrightarrow{p} M/G = \overline{M}.$$

On $\overline{M}$, we have the natural almost-Poisson structure induced by (M, Λ_M). The idea of this section is to study under which conditions there exists an almost-Poisson structure on $\overline{(J^{nh})^{-1}(\mu)}$ in such a way that k is an almost-Poisson mapping. In this case, then for each pair of functions $\bar{\lambda}, \bar{\sigma} : \overline{M} \longrightarrow \mathbb{R}$ we would have that

$$\{\lambda_\mu, \sigma_\mu\}_\mu = \{\bar{\lambda}, \bar{\sigma}\}_{\overline{M}} \circ k\,,$$

with $\bar{\lambda} \circ k = \lambda_\mu$ and $\bar{\sigma} \circ k = \sigma_\mu$.

In fact, taking $\bar{\lambda}_1, \bar{\lambda}_2 : \overline{M} \longrightarrow \mathbb{R}$ with $\bar{\lambda}_1 \circ k = \bar{\lambda}_2 \circ k = \lambda_\mu$, we would have

$$\{\bar{\lambda}_1, \bar{\sigma}\}_{\overline{M}} \circ k = \{\bar{\lambda}_2, \bar{\sigma}\}_{\overline{M}} \circ k\,, \ \forall \bar{\sigma} \in C^\infty(\overline{M})\,. \tag{4.25}$$

In case of k being injective, this equality would be a necessary and sufficient condition to obtain an almost-Poisson bracket $\{\cdot, \cdot\}_\mu$ on $\overline{(J^{nh})^{-1}(\mu)}$, making k an almost-Poisson morphism. Moreover, in this case, $\{\cdot, \cdot\}_\mu$ will be unique satisfying that property.

We will discuss if equality (4.25) is fulfilled. Equivalently, given $\bar{\lambda} : \overline{M} \longrightarrow \mathbb{R}$ with $\bar{\lambda} \circ k = 0$, we want to verify if

$$\{\bar{\lambda}, \bar{\sigma}\}_{\overline{M}} \circ k = 0\,, \ \forall \bar{\sigma} \in C^\infty(\overline{M})\,.$$

Consider the following commutative diagram

$$
\begin{array}{ccccc}
(J^{nh})^{-1}(\mu) & \xrightarrow{\ j\ } & M & \xrightarrow{\ id_M\ } & M \\[2pt]
\Big\downarrow{\scriptstyle \pi_{(J^{nh})^{-1}(\mu)}} & & \Big\downarrow{\scriptstyle \pi} & & \Big\downarrow{\scriptstyle \rho_{|M}} \\[2pt]
\overline{(J^{nh})^{-1}(\mu)} & \xrightarrow{\ k_\mu\ } & M/G_\mu & \xrightarrow{\ p\ } & \overline{M}
\end{array}
$$

Then, we have that $\{\rho_{|M}^* \bar{\lambda}, \rho_{|M}^* \bar{\sigma}\}_M \circ j = \{\bar{\lambda}, \bar{\sigma}\}_{\overline{M}} \circ \rho_{|M} \circ j = \{\bar{\lambda}, \bar{\sigma}\}_{\overline{M}} \circ k \circ \pi_{(J^{nh})^{-1}(\mu)}$. In addition, $\rho_{|M}^* \bar{\lambda} \circ j = \bar{\lambda} \circ k \circ \pi_{(J^{nh})^{-1}(\mu)} = 0$.

It is clear that

$$\{\rho_{|M}^* \bar{\lambda}, \rho_{|M}^* \bar{\sigma}\}_M \circ j = 0 \Longleftrightarrow \{\bar{\lambda}, \bar{\sigma}\}_{\overline{M}} \circ k = 0\,.$$

Therefore, now our question is rephrased as follows: given $\lambda \in C_G^\infty(M)$ with $\lambda \circ j = 0$, we would like to verify if

$$\{\lambda, \sigma\}_M \circ j = 0\,, \ \forall \sigma \in C_G^\infty(M)\,.$$

By definition, we have that $\{\lambda, \sigma\}_M = \omega_L(\bar{\mathcal{P}}(X_{\tilde{\lambda}}), \bar{\mathcal{P}}(X_{\tilde{\sigma}})) \circ j_M$, where $\tilde{\lambda}, \tilde{\sigma}$ are arbitrary extensions of λ, σ to TQ, $\tilde{\lambda} \circ j_M = \lambda$, $\tilde{\sigma} \circ j_M = \sigma$. Without loss of generality, we can suppose them to be G-invariant.

Now, $(j_M \circ j)^* \tilde{\lambda} = j^* \lambda = 0$. Therefore we deduce

$$0 = (j_M \circ j)^* d\tilde{\lambda} = (j_M \circ j)^* i_{X_{\tilde{\lambda}}} \omega_L .$$

If we could assure that $\bar{\mathcal{P}}(X_{\tilde{\sigma}}) \in T((J^{nh})^{-1}(\mu))$, then we would have

$$\begin{aligned}
\{\lambda, \sigma\}_M \circ j &= \omega_L(X_{\tilde{\lambda}}, \bar{\mathcal{P}}(X_{\tilde{\sigma}})) \circ (j_M \circ j) \\
&= \omega_L(X_{\tilde{\lambda}}, (j_M \circ j)_* \bar{\mathcal{P}}(X_{\tilde{\sigma}})) \circ (j_M \circ j) \\
&= (j_M \circ j)^* i_{X_{\tilde{\lambda}}} \omega_L(\bar{\mathcal{P}}(X_{\tilde{\sigma}})) = 0 .
\end{aligned}$$

Therefore, if we guarantee that $\bar{\mathcal{P}}(X_{\tilde{\sigma}}) \in T((J^{nh})^{-1}(\mu))$, $\forall \tilde{\sigma} \in C^\infty_G(TQ)$, then (4.25) holds. We characterize when this occurs in the following

Proposition 4.4.12. *Let σ be a G-invariant function on M, and $\tilde{\sigma}$ one G-invariant extension of σ to TQ. Then,*

$$\bar{\mathcal{P}}(X_{\tilde{\sigma}}) \in T((J^{nh})^{-1}(\mu)) \iff \{\sigma, f_i\}_M \circ j = 0, \ 1 \le i \le r. \quad (4.26)$$

Proof. Take $\sigma \in C^\infty_G(M)$. We have that

$$\bar{\mathcal{P}}(X_{\tilde{\sigma}}) \in T((J^{nh})^{-1}(\mu)) \iff \omega_L(\bar{\mathcal{P}}(X_{\tilde{\sigma}}), Z) \circ j_M \circ j = 0 ,$$

for all $Z \in T^\perp((J^{nh})^{-1}(\mu))$. By Lemma 4.4.7, we know $T^\perp((J^{nh})^{-1}(\mu)) = T^\perp M + \text{span}\left\{ X_{\tilde{f}_1}, \ldots, X_{\tilde{f}_r} \right\}$. As $F \cap TM \subset TM$, then $T^\perp M \subset (F \cap TM)^\perp$. Thus we have that $\omega_L(\bar{\mathcal{P}}(X_{\tilde{\sigma}}), Z) = 0$ for every $Z \in T^\perp M$. Then, for all $1 \le i \le r$,

$$\bar{\mathcal{P}}(X_{\tilde{\sigma}}) \in T((J^{nh})^{-1}(\mu)) \iff \omega_L(\bar{\mathcal{P}}(X_{\tilde{\sigma}}), X_{f_i}) \circ j_M \circ j = \{\sigma, f_i\}_M \circ j = 0 .$$

$\square$

Consequently, in case we have

$$\{\sigma, f_i\}_M \circ j = 0, \ 1 \le i \le r, \ \forall \sigma \in C^\infty_G(M), \quad (4.27)$$

we have proved that equality (4.25) holds good. Conditions (4.27) will not be fulfilled in general. In the following section, we will see that in the case of horizontal symmetries, k is injective and conditions (4.27) are satisfied, and therefore, there is a well-defined (unique) almost-Poisson structure on $\overline{(J^{nh})^{-1}(\mu)}$, so that $k : \overline{(J^{nh})^{-1}(\mu)} \longrightarrow \overline{M}$ is an almost-Poisson morphism.

Concerning the dynamics, if k is injective, then $k_*(\overline{X}_\mu) = \overline{X}$. The restriction of the energy E_L to $(J^{nh})^{-1}(\mu)$ is G_μ-invariant, so it induces a function on $\overline{(J^{nh})^{-1}(\mu)}$, $(E_L)_\mu : \overline{(J^{nh})^{-1}(\mu)} \longrightarrow \mathbb{R}$. One can easily check that $(\overline{E}_L)_{|\overline{M}} \circ k = (E_L)_\mu$. If, in addition, (4.25) holds, we have that k is an almost-Poisson mapping, or equivalently,

$$k_*(X^\mu_{\overline{\lambda} \circ k}) = X^M_{\overline{\lambda}} \circ k, \ \forall \overline{\lambda} \in C^\infty(\overline{M}) .$$

In particular, taking $(\overline{E}_L)_{|\overline{M}}$, we have that

$$
\begin{aligned}
X^\mu_{(E_L)_\mu}(\lambda_\mu) &= X^\mu_{(E_L)_\mu}(\bar\lambda \circ k) = k_*(X^\mu_{(E_L)_\mu})(\bar\lambda) \\
&= X^{\overline{M}}_{(\overline{E}_L)_{|\overline{M}}}(\bar\lambda) \circ k = \overline{X}(\bar\lambda) \circ k = k_*(\overline{X}_\mu)(\bar\lambda) = \overline{X}_\mu(\lambda_\mu),
\end{aligned}
$$

for all $\lambda_\mu \in C^\infty(\overline{(J^{nh})^{-1}(\mu)})$. Therefore, $X^\mu_{(E_L)_\mu} = \overline{X}_\mu$. Then, we can conclude that the evolution of any function $\lambda_\mu \in C^\infty(\overline{(J^{nh})^{-1}(\mu)})$ along the integral curves of $\overline{X}_\mu$ on $\overline{(J^{nh})^{-1}(\mu)}$ is given by

$$
\dot\lambda_\mu = \overline{X}_\mu(\lambda_\mu) = \{\lambda_\mu, (E_L)_\mu\}_\mu . \tag{4.28}
$$

Almost-Poisson mappings: the horizontal case Assume that the nonholonomic Lagrangian system with symmetry under consideration falls into the horizontal case (cf. Section 4.3). Next, we show that $k : \overline{(J^{nh})^{-1}(\mu)} \longrightarrow \overline{M}$ is injective and that the conditions (4.27) are satisfied.

Proposition 4.4.13. *Let $k : \overline{(J^{nh})^{-1}(\mu)} = M_\mu \longrightarrow \overline{M}$ be the composition of $k_\mu : M_\mu \longrightarrow M/G_\mu$ and $p : M/G_\mu \longrightarrow \overline{M}$. Then we can define on M_μ a unique almost-Poisson structure so that k is an almost-Poisson mapping.*

Proof. It is an easy exercise to prove that k is injective in the case of horizontal symmetries. From the analysis of the previous section, we know that it is sufficient to verify the conditions (4.27). Now, taking $\xi_1, \ldots, \xi_r$ a base of the Lie algebra $\mathfrak{g}$, we have that $f_i = \langle \mu, \xi_i \rangle - J_{\xi_i}$, $1 \le i \le r$. Given $\sigma \in C^\infty_G(M)$, we deduce that

$$
\{\sigma, f_i\}_M \circ j = (\xi_i)_{TQ}(\tilde\sigma) \circ j_M \circ j = (\xi_i)_M(\sigma) \circ j = 0,
$$

due to the G-invariance of σ. $\square$

On the other hand, we have that the symplectic distribution $F \cap TM$ induces a symplectic distribution $F_\mu \cap TM_\mu$ in $T(TQ)_\mu = T(TQ)_\mu$, that is to say

$$
T(TQ)_{\mu|M_\mu} = (F_\mu \cap TM_\mu) \oplus (F_\mu \cap TM_\mu)^{\perp_\mu},
$$

with induced projectors for each $\bar v_q \in M_\mu$

$$
\begin{aligned}
\bar P_\mu &: T_{\bar v_q}(TQ)_\mu \longrightarrow (F_\mu)_{\bar v_q} \cap T_{\bar v_q} M_\mu, \\
\bar Q_\mu &: T_{\bar v_q}(TQ)_\mu \longrightarrow ((F_\mu)_{\bar v_q} \cap T_{\bar v_q} M_\mu)^{\perp_\mu}.
\end{aligned}
$$

The above decomposition induces an almost-Poisson bracket $\{\cdot, \cdot\}_{M_\mu}$ on M_μ. Given $\lambda_\mu, \sigma_\mu : M_\mu \longrightarrow \mathbb{R}$, take $\tilde\lambda_\mu, \tilde\sigma_\mu$ arbitrary extensions to $(TQ)_\mu$, $\tilde\lambda_\mu \circ$

$j_{M_\mu} = \lambda_\mu$, $\tilde{\sigma}_\mu \circ j_{M_\mu} = \sigma_\mu$, with $j_{M_\mu} : M_\mu \hookrightarrow (TQ)_\mu$ the canonical inclusion, and define

$$\{\lambda_\mu, \sigma_\mu\}_{M_\mu} = (\omega_L)_\mu(\bar{P}_\mu(X^\mu_{\tilde{\lambda}_\mu}), \bar{P}_\mu(X^\mu_{\tilde{\sigma}_\mu})) \circ j_{M_\mu}.$$

Indeed, we have that both brackets coincide, $\{\cdot,\cdot\}_{M_\mu} = \{\cdot,\cdot\}_\mu$, as we prove in the following

Theorem 4.4.14. *Consider the almost-Poisson manifolds $(M_\mu, \{\cdot,\cdot\}_{M_\mu})$ and $(\overline{M}, \{\cdot,\cdot\}_{\overline{M}})$. Then $k : M_\mu \longrightarrow \overline{M}$ is an almost-Poisson mapping.*

Proof. The proof consists of a careful exercise of equalities. The following two commutative diagrams will be helpful,

$$
\begin{array}{ccc}
(J^{nh})^{-1}(\mu) = M' & \xrightarrow{\ j\ } & M \\
{\scriptstyle i}\downarrow & & \downarrow{\scriptstyle j_M} \\
J^{-1}(\mu) & \xrightarrow{\ j_\mu\ } & TQ
\end{array}
\qquad
\begin{array}{ccc}
M' & \xrightarrow{\ i\ } & J^{-1}(\mu) \\
{\scriptstyle \pi_{M'}}\downarrow & & \downarrow{\scriptstyle \pi_\mu} \\
M_\mu & \xrightarrow{\ j_{M_\mu}\ } & (TQ)_\mu
\end{array}
$$

Then, given $\lambda_\mu, \sigma_\mu : M_\mu \longrightarrow \mathbb{R}$, we have

$$
\begin{aligned}
\{\lambda_\mu, \sigma_\mu\}_\mu \circ \pi_{M'} &= \{\bar{\lambda}, \bar{\sigma}\}_{\overline{M}} \circ k \circ \pi_{M'} = \{\lambda, \sigma\}_M \circ j \\
&= \omega_L(\bar{P}(X_{\tilde{\lambda}}), \bar{P}(X_{\tilde{\sigma}})) \circ j_M \circ j = (j_M \circ j)^* \omega_L(\bar{P}(X_{\tilde{\lambda}}), \bar{P}(X_{\tilde{\sigma}})) \\
&= (\pi_\mu \circ i)^* (\omega_L)_\mu(\bar{P}(X_{\tilde{\lambda}}), \bar{P}(X_{\tilde{\sigma}})) \\
&= (\omega_L)_\mu(\bar{P}_\mu(X^\mu_{\tilde{\lambda}_\mu}), \bar{P}_\mu(X^\mu_{\tilde{\sigma}_\mu})) \circ j_{M_\mu} \circ \pi_{M'} \\
&= \{\lambda_\mu, \sigma_\mu\}_{M_\mu} \circ \pi_{M'}.
\end{aligned}
$$

$\square$

Remark 4.4.15. It should be noticed that from the general discussion above, it is concluded that for nonholonomic Lagrangian systems which fit in the horizontal case, Theorem 4.4.14 is the utmost one can say. That is, meanwhile conditions (4.27) are always fulfilled, conditions (4.24) are no longer satisfied in general. This means, in particular, that the almost-Poisson bracket $\{\cdot,\cdot\}_{M_\mu}$ is not the reduced bracket of $\{\cdot,\cdot\}_M$, as it was stated in [50] (Theorem 8.2). However, following (4.28), we know that for all $\lambda_\mu \in C^\infty(M_\mu)$, its evolution along the integral curves of the dynamics is given by

$$\dot{\lambda}_\mu = X_\mu(\lambda_\mu) = \{\lambda_\mu, (E_L)_\mu\}_{M_\mu}.$$

The nonholonomic free particle revisited We return now to the example of the nonholonomic free particle discussed in Section 4.2.1.

Recall that considering the action Ψ of $G = \mathbb{R}^2$ on the configuration space, the example falls into the general case. Let $\{e_1, e_2\}$ be the standard basis of $\mathbb{R}^2$ and $\{e^1, e^2\}$ its dual basis. We define a section of the vector bundle $(\mathbb{R}^2)^M$, $\bar{\xi} : M \longrightarrow (\mathbb{R}^2)^M$ by

$$\bar{\xi} : (x, y, z, \dot{x}, \dot{y}) \longmapsto e_1 + y e_2 .$$

Its corresponding nonholonomic momentum function is $J^{nh}_{\bar{\xi}} = \dot{x} + y\dot{z}$. From the section $\bar{\xi}$, we can construct the vector field Ξ,

$$\Xi = \frac{\partial}{\partial x} + y \frac{\partial}{\partial z} .$$

Therefore the momentum equation would be,

$$\frac{d}{dt} (\dot{x} + y\dot{z}) = \dot{z}\dot{y} .$$

Using the constraint $\phi = 0$, we may rewrite this equation as

$$\ddot{x} + \frac{y}{1+y^2} \dot{x}\dot{y} = 0 . \tag{4.29}$$

In [17], Bates *et al.* have obtained a constant of the motion for this problem, in addition to the energy, related with the symmetry group and the constraint. We are going to see now how the finding of this constant fits nice in the geometrical setting we have exposed here.

Following Theorem 3.4.1, it is interesting to realize that,

1. $\bar{\mathcal{P}}(\Xi)(E_L) = \Xi(E_L) = 0$, because $\Xi \in F \cap TM$ and E_L is G-invariant,
2. $\bar{\mathcal{P}}(X_\phi)(E_L) = \omega_L(X_{E_L}, \bar{\mathcal{P}}(X_\phi)) = -X(\phi) = 0$, because $X \in F \cap TM$.

Therefore, if we find functions f, g on TQ such that the vector field $Z = f\Xi + gX_\phi$ is Hamiltonian, say $Z = X_\varphi$, from Theorem 3.4.1 we could conclude that φ is a constant of the motion due to the symmetry and the constraint. In general, the condition "Z is Hamiltonian" will lead us to a quite complex first-order system of partial derivative equations. However, in this case, it is not difficult to prove (just a few computations) that $f = \frac{1}{\sqrt{1+y^2}}$ and $g = -\frac{y}{\sqrt{1+y^2}}$ will do. Consequently, we obtain the conservation law,

$$\varphi = \dot{x}\sqrt{1+y^2} .$$

Then we choose the following section of $(\mathbb{R}^2)^{M^*} \longrightarrow Q$

$$\mu : Q \longrightarrow (\mathbb{R}^2)^{M^*}$$
$$q \longmapsto \mu(q) : ((\mathbb{R}^2)^q)^* \to \mathbb{R}$$
$$e_1 + y e_2 \mapsto c\sqrt{1 + y^2},$$

where $q = (x, y, z)$. We have that $f : M \longrightarrow \mathbb{R}$, $f = \langle \mu, \bar{\xi} \rangle \circ \tau_Q - J_{\bar{\xi}}^{nh}$ is given by

$$f = c\sqrt{1 + y^2} - \dot{x}(1 + y^2).$$

The hypothesis of Proposition 4.4.6 are fulfilled. A direct computation shows that the section μ satisfies equation (4.21). Then $(J^{nh})^{-1}(\mu)$ is a submanifold of M. In fact,

$$(J^{nh})^{-1}(\mu) = \{(x, y, z, \dot{x}, \dot{y}) \mid \dot{x} = c\sqrt{1 + y^2}\} = \{(x, y, z, \dot{y})\}.$$

As the Lie group $G = \mathbb{R}^2$ is Abelian, the coadjoint action is trivial. Then it is easily seen that the isotropy group G_μ of the action $CoAd^M$ is $G_\mu = G$. So we have the action

$$\Theta : G_\mu \times (J^{nh})^{-1}(\mu) \longrightarrow \mathbb{R}$$
$$((r, s), (x, y, z, \dot{y})) \longmapsto (x + r, y, z + s, \dot{y}).$$

Consequently, $\overline{(J^{nh})^{-1}(\mu)} = \{y, \dot{y}\}$. We obtain that

$$X_f = -\frac{\partial}{\partial x} - y \frac{\partial}{\partial z} - \left(\frac{cy}{\sqrt{1 + y^2}} - \dot{z}\right)\frac{\partial}{\partial \dot{y}} \in F \cap TM.$$

Therefore, for all $\sigma \in C_G^\infty(M)$, we have

$$\{\sigma, f\}_M \circ j = X_f^M(\sigma) \circ j = \mathcal{P}(X_f)(\sigma) \circ j$$
$$= X_f(\sigma) \circ j = \frac{\partial \sigma}{\partial \dot{y}}\left(\frac{cy}{\sqrt{1 + y^2}} - y\dot{x}\right) = 0.$$

Moreover, the mapping k is injective,

$$k : \overline{(J^{nh})^{-1}(\mu)} \longrightarrow \overline{M}$$
$$(y, \dot{y}) \longmapsto (y, c\sqrt{1 + y^2}, \dot{y}).$$

Then, we know from the above discussion that there is a well-defined almost-Poisson structure on $\overline{(J^{nh})^{-1}(\mu)}$ which is given by

$$\{y, \dot{y}\}_\mu = 1.$$

As conditions (4.24) and conditions (4.27) are exactly the same (due to $G_\mu = G$), we have that $\{\cdot, \cdot\}_\mu$ is the reduced bracket of $\{\cdot, \cdot\}_M$. Indeed, $\{\cdot, \cdot\}_\mu$ is integrable, that is, it is a Poisson structure.

4.5 A special subcase: kinematic plus horizontal

$\mathbf{I}$N this section we treat a special subcase of the general one, in which the reduction procedure can be decomposed in a two-step procedure.

Consider a nonholonomic system with symmetry such that the bundle $\mathfrak{g}^F$ is trivial, i.e. $\mathfrak{g}^x = \mathfrak{g}_0$, $\forall x \in M$. Then, we have the following result.

Lemma 4.5.1. $\mathfrak{g}_0$ *is an ideal of $\mathfrak{g}$ which is invariant with respect to the adjoint representation.*

Proof. The invariance follows from Lemma 4.4.8, which implies that $Ad_g \mathfrak{g}_0 = \mathfrak{g}_0$, for all $g \in G$. This implies that $\mathfrak{g}_0$ is an ideal of $\mathfrak{g}$. $\qquad\square$

Next we consider G_0, the normal connected subgroup of G with Lie algebra $\mathfrak{g}_0$ and $\Psi_0 : G_0 \times Q \longrightarrow Q$, the restricted action to G_0. For the corresponding lifted action, it is clear that $\mathcal{V}_{0|M} \subset F \cap TM$, so we are in the case of horizontal symmetries treated in Section 4.3. Now we are going to proceed in the way described there to reduce the dynamics.

Let $\mu \in \mathfrak{g}_0^*$ be a weakly regular value of $J = J^L$. From the given assumptions on the action $\hat{\Psi}$, we have that G_μ^0, its isotropy group in G_0, acts freely and properly on the level set $J^{-1}(\mu)$. Under these conditions, $((TQ)_\mu = J^{-1}(\mu)/G_\mu^0, \omega_\mu)$ is a symplectic manifold. We also suppose that M and $J^{-1}(\mu)$ have a clean intersection, $M' = M \cap J^{-1}(\mu)$, which is a G_μ^0-invariant submanifold of $J^{-1}(\mu)$. We then consider $M_\mu = M'/G_\mu^0$. We can define a distribution F' on TQ along M' by putting $F'_{x'} = T_{x'}(J^{-1}(\mu)) \cap F_{x'}$, $\forall x' \in M'$ and in addition assume that F' has constant rank. Again, F' is G_μ^0-invariant and it projects onto a subbundle F_μ of $T(TQ)_\mu$ along M_μ. Finally, with the function $E_{L\mu}$ induced by the restriction of the energy function E_L to $J^{-1}(\mu)$, we have all the ingredients to apply Theorem 4.3.1 and obtain the following reduced constrained problem on $((TQ)_\mu, \omega_\mu)$,

$$\begin{cases} (i_{X_\mu}\omega_\mu - dE_{L\mu})_{|M_\mu} \in F_\mu^o \, , \\ X_\mu \in TM_\mu \, . \end{cases} \tag{4.30}$$

So far, we have reduced the constrained problem by the horizontal symmetries and have obtained again a constrained problem. In the following, we will investigate what happens with the symmetries we have not taken into account.

For this purpose, we consider the action $\Phi : G_\mu \cdot G_0/G_0 \times (TQ)_\mu \longrightarrow (TQ)_\mu$ defined by $\Phi(\overline{g}, \overline{p}) = \overline{\Phi(g, p)}$. Note that this action is well defined because we are not treating with all the remaining symmetries G/G_0, but only with the adequate ones to $(TQ)_\mu$. Indeed, we prove the next result.

Lemma 4.5.2. *The mapping Φ is well defined.*

Proof. We must verify that given $\bar{g}, \bar{h} \in G_\mu \cdot G_0/G_0$ and $\bar{p}, \bar{q} \in (TQ)_\mu$ so that $\bar{g} = \bar{h}$ and $\bar{p} = \bar{q}$, we have $\Phi(\bar{g}, \bar{p}) = \Phi(\bar{h}, \bar{q})$. Since $G_\mu \cdot G_0/G_0 \cong G_\mu/G_\mu \cap G_0 = G_\mu/G_\mu^0$, we can consider $\bar{g}, \bar{h}$ as elements of this latter group, so we have that $h^{-1}g \in G_\mu \cap G_0$. We also have that there exists $i \in G_\mu^0$ such that $p = iq$. Then $gp = giq = gih^{-1}hq$. Moreover, $gih^{-1} = (ih^{-1}g)^{g^{-1}} \in G_0$, because i and $h^{-1}g$ are in G_0, and this group is normal in G. Clearly $gih^{-1} \in G_\mu$, so finally we have that $gih^{-1} \in G_\mu^0$. We have obtained $\overline{gp} = \overline{hq}$, i.e. $\Phi(\bar{g}, \bar{p}) = \Phi(\bar{h}, \bar{q})$. $\qquad\square$

In a similar way, we can check easily that Φ is a symplectic action on $(TQ)_\mu$ and that M_μ, F_μ and H_μ are all G_μ/G_μ^0-invariant. We denote by $\rho_\mu : (TQ)_\mu \longrightarrow \overline{(TQ)_\mu}$ the canonical projection for Φ and $\mathcal{V}_\mu = \ker T\rho_\mu$.

Our aim is to prove that, under the assumption $TM_\mu = (F_\mu \cap TM_\mu) + \mathcal{V}_{\mu|M_\mu}$, the constrained problem with symmetries on $((TQ)_\mu, \omega_\mu)$ fits in the purely kinematic case (recall Remark 4.2.2). For this purpose, we identify now the fundamental vector fields of the action Φ.

Lemma 4.5.3. *Let $\zeta + \mathfrak{g}_\mu \cap \mathfrak{g}_0$ be an element of $\mathfrak{g}_\mu/\mathfrak{g}_\mu \cap \mathfrak{g}_0$, the Lie algebra of G_μ/G_μ^0. Then*

$$(\zeta + \mathfrak{g}_\mu \cap \mathfrak{g}_0)_{(TQ)_\mu}(\bar{p}) = T\pi_\mu \zeta_{J^{-1}(\mu)}(p), \ \forall p \in J^{-1}(\mu),$$

where $\pi_\mu : J^{-1}(\mu) \longrightarrow (TQ)_\mu$ is the projection mapping associated with the action of G_μ^0 on $J^{-1}(\mu)$ and $\zeta_{J^{-1}(\mu)}$ is the fundamental vector field corresponding to the action of G_μ on $J^{-1}(\mu)$.

Proof. We have

$$(\zeta + \mathfrak{g}_\mu \cap \mathfrak{g}_0)_{(TQ)_\mu}(\bar{p}) = \left(\frac{d}{dt}\right)_{|t=0} \Phi(\exp(t\zeta + \mathfrak{g}_\mu \cap \mathfrak{g}_0), \bar{p})$$

$$= \left(\frac{d}{dt}\right)_{|t=0} \Phi(\overline{\exp_\mu t\zeta}, \bar{p}) = \left(\frac{d}{dt}\right)_{|t=0} \overline{(\exp_\mu t\zeta \cdot p)} = T\pi_\mu \zeta_{J^{-1}(\mu)}(p).$$

$$\qquad\square$$

Now, we are at disposal of proving the former statement.

Proposition 4.5.4. *If $TM_\mu = (F_\mu \cap TM_\mu) + \mathcal{V}_{\mu|M_\mu}$, the reduced constrained system (4.30), considered with the action Φ on $(\widetilde{TQ})_\mu$, fits in the vertical case.*

Proof. We must prove that $(\mathcal{V}_\mu)_{\overline{x}} \cap (F_\mu)_{\overline{x}} = \{0\}$, $\forall \overline{x} \in M_\mu$. Suppose that $(\zeta + \mathfrak{g}_\mu \cap \mathfrak{g}_0)_{(TQ)_\mu}(\overline{x}) \in (F_\mu)_{\overline{x}}$ for some $\overline{x} \in M_\mu$. Recall that $F_{\mu \overline{x}} = T\pi_\mu F'_x$. Then, we have that there exists $Y \in F'_x$ such that $T\pi_\mu(Y) = (\zeta + \mathfrak{g}_\mu \cap \mathfrak{g}_0)_{(TQ)_\mu}(\overline{x}) = T\pi_\mu(\zeta_{J^{-1}(\mu)}(x))$ which, in turn, implies there exists $\xi \in \mathfrak{g}_\mu^0 = \mathfrak{g}_\mu \cap \mathfrak{g}_0$ such that $\zeta_{J^{-1}(\mu)}(x) = Y + \xi_{J^{-1}(\mu)}(x)$. Therefore, $(\zeta - \xi)_{J^{-1}(\mu)}(x) = Y$, which gives $\zeta - \xi \in \mathfrak{g}^x = \mathfrak{g}_0$. Obviously, $\zeta - \xi \in \mathfrak{g}_\mu$. Then, $\zeta + \mathfrak{g}_\mu \cap \mathfrak{g}_0 = \xi + \mathfrak{g}_\mu \cap \mathfrak{g}_0 = 0 + \mathfrak{g}_\mu \cap \mathfrak{g}_0$. $\square$

Next, we proceed as in Section 4.2.1. We obtain a principal connection Υ on the principal (G_μ/G_μ^0)-bundle $\rho_{\mu|M_\mu} : M_\mu \longrightarrow \overline{M}_\mu$, with horizontal subspace $\mathcal{U}_{\overline{x}} = (F_\mu)_{\overline{x}} \cap T_{\overline{x}} M_\mu$ at each point $\overline{x} \in M_\mu$.

Let θ_μ be the 1-form defined by $\pi_\mu^* \theta_\mu = j_\mu^* \theta_L$. Obviously, θ_μ is G_μ/G_μ^0-invariant. Let $\theta'_\mu = j_{M_\mu}^* \theta_\mu$, where $j_{M_\mu} : M_\mu \hookrightarrow (TQ)_\mu$ is the canonical inclusion. Then Proposition 4.2.1 can be applied to the reduced constrained problem (4.30) to give,

$$i_{\overline{X}_\mu} \overline{\omega} = d\overline{H}_\mu + \overline{\alpha_{X_\mu}}, \qquad (4.31)$$

where $\overline{\alpha_{X_\mu}}$ is the projection of α_{X_μ}, with $\alpha_{X_\mu} = i_{X_\mu}(\mathbf{h}^* d\theta'_\mu - d\mathbf{h}^* \theta'_\mu)$, and $\overline{\omega} = d(\overline{\theta'}_\mu)_h$, with $(\overline{\theta'}_\mu)_h$ the projection of $\mathbf{h}^* \theta_\mu$.

Remark 4.5.5. In general, the condition "$\mathfrak{g}^x$ does not depend on $x \in M$" seems to be quite restrictive. In [220], Śniatycki defined $\mathfrak{g}' \subset \mathfrak{g}$ by

$$\mathfrak{g}' = \{\xi' \in \mathfrak{g} \mid \exists \text{ a constant section } \overline{\xi} \text{ of } \mathfrak{g}^F \text{ with } \overline{\xi}(x) = \xi', \forall x \in M\} .$$

In other words, $\mathfrak{g}'$ consists of those elements of $\mathfrak{g}$ such that its corresponding infinitesimal generator of the induced action on M is a horizontal symmetry. If $\mathfrak{g}^x$ does not depend on $x \in M$, it is clear that $\mathfrak{g}_0 = \mathfrak{g}'$.

Śniatycki argues that $\mathfrak{g}'$ is an ideal of $\mathfrak{g}$ and then he considers the normal connected subgroup G' of G with Lie algebra $\mathfrak{g}'$. The reduction process is parallel to the one done here until we reach Proposition 4.5.4, which will not be true in general.

As we have seen for the reduction, the reconstruction of the dynamics would be a two-step process. First, to implement a purely kinematic-type reconstruction and secondly, an horizontal-type one.

4.5.1 The nonholonomic free particle modified

We are going to treat next the example of the nonholonomic free particle, but with a different constraint. As before, we have a particle moving in space, subject to the constraint

$$\phi = \dot{z} - x\dot{x} = 0.$$

The Lagrangian function and the constraint submanifold are

$$L = \frac{1}{2}\left(\dot{x}^2 + \dot{y}^2 + \dot{z}^2\right), \quad M = \mathrm{span}\left\{\frac{\partial}{\partial x} + x\frac{\partial}{\partial z}, \frac{\partial}{\partial y}\right\}.$$

Consider the Lie group $G = \mathbb{R}^2$ and its action on Q,

$$\Psi: \quad G \times Q \quad \longrightarrow Q$$
$$((r, s), (x, y, z)) \longmapsto (x, y + r, z + s).$$

It is a simple verification to see that L and M are G-invariant. The infinitesimal generators of the lifted action of φ to TQ are $\{\frac{\partial}{\partial y}, \frac{\partial}{\partial z}\}$.

Choose local coordinates $m = (x, y, z, \dot{x}, \dot{y})$ on M. We find that,

$$F_{|M} = \mathrm{span}\left\{\frac{\partial}{\partial x} + x\frac{\partial}{\partial z}, \frac{\partial}{\partial y}, \frac{\partial}{\partial \dot{x}}, \frac{\partial}{\partial \dot{y}}, \frac{\partial}{\partial \dot{z}}\right\}, \quad V_m \cap F_m = \mathrm{span}\left\{\frac{\partial}{\partial y}\right\}.$$

Note that the fiber $(\mathbb{R}^2)^m$ does not depend on the base point $m \in M$. Then, the bundle $(\mathbb{R}^2)^F$ is trivial and we are just in the special subcase of the general case treated in this section. With the notations we have been using, $\mathfrak{g}_0 = \mathbb{R} \times \{0\}$ and $G_0 = \mathbb{R} \times \{0\}$. Let $\{e_1, e_2\}$ be the standard basis of $\mathbb{R}^2$ and $\{e^1, e^2\}$ its dual basis. We know that Ψ_0 is Hamiltonian, with momentum mapping,

$$J: \quad T\mathbb{R}^3 \quad \longrightarrow \mathbb{R}^*$$
$$(x, y, z, \dot{x}, \dot{y}, \dot{z}) \longmapsto \dot{y}e^1$$

Let $\mu = ae^1 \in \mathbb{R}^*$. We have that $G_\mu^0 = \mathbb{R}$ and $J^{-1}(\mu) = \{(x, y, x, \dot{x}, \dot{z})\}$. Therefore,

$$(T\mathbb{R}^3)_\mu = \{(x, z, \dot{x}, \dot{z})\}, \quad (\omega_L)_\mu = dx \wedge d\dot{x} + dz \wedge d\dot{z}.$$

We note that M and $J^{-1}(\mu)$ have a clean intersection $M' = \{(x, y, z, \dot{x})\}$ so that

$$M_\mu = \{(x, z, \dot{x})\}.$$

After some computations, we find that

$$F_\mu = \mathrm{span}\left\{\frac{\partial}{\partial x} + x\frac{\partial}{\partial z}, \frac{\partial}{\partial \dot{x}}, \frac{\partial}{\partial \dot{z}}\right\},$$

$$F_\mu \cap TM_\mu = \mathrm{span}\left\{\frac{\partial}{\partial \dot{x}} + x\frac{\partial}{\partial \dot{z}}, \frac{\partial}{\partial x} + x\frac{\partial}{\partial z} + \dot{x}\frac{\partial}{\partial \dot{z}}\right\}.$$

Finally, we obtain $(E_L)_\mu = \frac{1}{2}(\dot{x}^2 + \dot{z}^2 + a^2)$. With all these ingredients, we pose the following constrained problem (4.30) on $((T\mathbb{R}^3, (\omega_L)_\mu))$,

$$\begin{cases} (i_{(X)_\mu}(\omega_L)_\mu - d(E_L)_\mu)_{|M_\mu} \in F_\mu^o \, , \\ (X)_\mu \in TM_\mu \, . \end{cases} \tag{4.32}$$

Now, we are going to investigate what happens with the symmetries we have not used yet. We have that $G_\mu = \mathbb{R}^2$ and consequently $(G_\mu + G_0)/G_0 \cong \mathbb{R}$. Consider the action

$$\begin{aligned} \varPhi : (G_\mu + G_0)/G_0 \times (T\mathbb{R}^3)_\mu &\longrightarrow (T\mathbb{R}^3)_\mu \\ (s, (x, z, \dot{x}, \dot{z})) &\longmapsto (x, z + s, \dot{x}, \dot{z}) \, . \end{aligned}$$

The canonical projection ρ_μ is given by

$$\begin{aligned} \rho_\mu : (T\mathbb{R}^3)_\mu &\longrightarrow \overline{(T\mathbb{R}^3)}_\mu \\ (x, z, \dot{x}, \dot{z}) &\longmapsto (x, \dot{x}, \dot{z}) \, , \end{aligned}$$

and its restriction to M_μ is

$$\begin{aligned} \rho_{\mu|M_\mu} : M_\mu &\longrightarrow \overline{M}_\mu \\ (x, z, \dot{x}) &\longmapsto (x, \dot{x}) \, . \end{aligned}$$

The vertical bundle of $\varPhi$ is $\mathcal{V}_\mu = \mathrm{span}\left\{\frac{\partial}{\partial z}\right\}$. For each $m_\mu \in M_\mu$ we have that $(\mathcal{V}_\mu)_{m_\mu} \cap (F_\mu)_{m_\mu} = \{0\}$. Moreover, $TM_\mu = F_{\mu|M_\mu} \cap TM_\mu + \mathcal{V}_{\mu|M_\mu}$. Therefore, the constrained system (4.32) on $((T\mathbb{R}^3)_\mu, (\omega_L)_\mu)$ fits in the purely kinematic case, that is, we obtain a principal connection $\varUpsilon$ on the principal $\mathbb{R}$-bundle $\rho_{\mu|M_\mu} : M_\mu \longrightarrow \overline{M}_\mu$, with horizontal subspace $U_{m_\mu} = (F_\mu)_{m_\mu} \cap T_{m_\mu} M_\mu$ at each point $m_\mu \in M_\mu$. The connection one-form is given by

$$\varUpsilon = (dz)e \, ,$$

where $\{e\}$ is the canonical basis of the Lie algebra $(\mathfrak{g}_\mu + \mathfrak{g}_0)/\mathfrak{g}_0 \cong \mathbb{R}$. Define

$$\theta_\mu = -\dot{x}dx - \dot{z}dz \, .$$

We check that $\theta'_\mu = j^*_{M_\mu}\theta_\mu = -\dot{x}(dx + xdz)$. Next, we calculate the one-form α_{X_μ} on M_μ defined by the prescription $\alpha_{X_\mu} = i_{X_\mu}(\mathbf{h}^*d\theta'_\mu - d\mathbf{h}^*\theta'_\mu)$. First, we have that

$$\mathbf{h}^*d\theta'_\mu = d\mathbf{h}^*\theta'_\mu = (1 + x^2)d\dot{x} \wedge dx \, ,$$

and consequently $\alpha_{X_\mu} = 0$. Projecting onto $\overline{M}_\mu$, we obtain that

$$\overline{\omega} = (1 + x^2)dx \wedge d\dot{x} \, , \quad \overline{(E_L)}_\mu = \frac{1}{2}(\dot{x}^2(1 + x^2) + a^2) \, .$$

Now, following (4.31), we can write from the constrained problem (4.32), the reduced unconstrained system

$$i_{\overline{X}_\mu}\overline{\omega} = d\overline{(E_L)}_\mu \, . \tag{4.33}$$

From a straightforward computation we have that the solution $\overline{X}_\mu$ of equation (4.33) is the vector field

$$\overline{X}_\mu = \dot{x}\frac{\partial}{\partial x} - \frac{x\dot{x}^2}{1 + x^2}\frac{\partial}{\partial \dot{x}} \, .$$

5 Chaplygin systems

IN this chapter, we focus our attention on Chaplygin systems. Typical problems in Mechanics, such as the vertical and the inclined rolling disk, the nonholonomic free particle and the two wheeled carriage, can be interpreted as generalized Chaplygin systems in the sense defined in Chapter 4. Systems of this type also occur in many problems of robotic locomotion [117] and motions of micro-organisms at low Reynolds number [216]. In the previous chapter we have taken a symplectic approach to the dynamics of Chaplygin systems. Here, instead, we expose a nice geometric description in terms of affine connections for generalized Chaplygin systems of mechanical type. We also explore the relation between both approaches. This investigation will lead us to answer negatively a question by Koiller [120] concerning the existence of a canonical invariant measure for the reduced dynamics of generalized Chaplygin systems.

The chapter is organized as follows. In Section 5.1 we present the description of the reduction and reconstruction of the dynamics of generalized Chaplygin systems under the affine connection formalism. In Section 5.2, we motivate the study of the relation with the symplectic approach with two examples. This relation is analyzed in Section 5.3. These developments enable us to study several integrability aspects in Section 5.4 and present a simple counter example to Koiller's question.

5.1 Generalized Chaplygin systems

RECALL from Chapter 4 the structure of a so-called *generalized Chaplygin system* [29, 120, 137]. The configuration manifold Q of a Chaplygin system is a principal G-bundle $\pi : Q \longrightarrow Q/G$, and the constraint submanifold $\mathcal{D}$ is given by the horizontal distribution $\mathcal{H}$ of a principal connection γ on π. Furthermore, the system is described by a regular Lagrangian $L : TQ \longrightarrow \mathbb{R}$, which is G-invariant for the lifted action of G on TQ. In this chapter, we shall restrict our attention to systems of mechanical type for which $L = T - V$, where $T : TQ \longrightarrow \mathbb{R}$ is the kinetic energy, corresponding to a Riemannian metric g on Q, and $V : Q \longrightarrow \mathbb{R}$ is the potential energy.

In addition, we suppose that both the potential energy and the metric g are G-invariant so that

$$\mathcal{L}_{\xi_Q} V = 0, \quad \mathcal{L}_{\xi_Q} g = 0,$$

for all $\xi \in \mathfrak{g}$. In particular, all fundamental vector fields ξ_Q are Killing vector fields.

5.1.1 Reduction in the affine connection formalism

In this section, we restrict our attention to Lagrangians of "pure kinetic energy type", i.e. we assume $V = 0$. The reason for doing this is twofold. First, it makes the geometric picture more clear and tractable, in that the equations of motion for the nonholonomic mechanical system can then be seen as the geodesic equations of an affine connection. Secondly, the extention to systems with a nontrivial potential energy function is rather straightforward but, at least for those aspects of nonholonomic dynamics that are of interest here, it does not really tell us anything new.

From the discussion in Section 3.4.2 we know that the equations of motion of the system can be written as

$$\overline{\nabla}_{\dot{c}(t)} \dot{c}(t) = 0, \quad \dot{c}(0) \in \mathcal{D},$$

where $\overline{\nabla}$ is the nonholonomic affine connection. The following is a complementary view to the reduction process of generalized (or non-Abelian) Chaplygin systems as described in [120].

Let us define a metric $\tilde{g}$ on the base manifold Q/G as follows

$$\tilde{g}_x(u_x, v_x) = g_q(U_q, V_q), \quad x \in Q/G, \ u_x, v_x \in T_x(Q/G),$$

where $q \in \pi^{-1}(x)$ and U_q, V_q are horizontal vectors which project under π onto u_x and v_x, respectively. From the G-invariance of g we deduce that the right-hand side is independent of the chosen point q in the fiber $\pi^{-1}(x)$ and, hence, $\tilde{g}$ is well defined.

Proposition 5.1.1. *We have that for all $X, Y \in \mathfrak{X}(Q/G)$ and $\xi \in \mathfrak{g}$*

$$\mathcal{L}_{\xi_Q}(\overline{\nabla}_{X^h} Y^h) = 0.$$

Proof. Since ξ_Q is a Killing vector field, it follows from Proposition VI. 2.2 in [119] that

$$\mathcal{L}_{\xi_Q} \nabla^g_{X^h} Y^h = \nabla^g_{[\xi_Q, X^h]} Y^h + \nabla^g_{X^h} [\xi_Q, Y^h] = 0, \tag{5.1}$$

because Y^h and X^h are projectable. Therefore, we only need to prove that

$$\mathcal{L}_{\xi_Q}\left[(\nabla^g_{X^h}\mathcal{Q})\,Y^h\right] = 0\,.$$

This condition is equivalent to

$$\mathcal{L}_{\xi_Q}\left[\mathcal{Q}(\nabla^g_{X^h}Y^h)\right] = 0\,. \tag{5.2}$$

But, as $\mathcal{D}$ and $\mathcal{D}^\perp$ are G-invariant, we have that $\mathcal{L}_{\xi_Q}\mathcal{Q} = 0$. This, together with (5.1), imply (5.2). $\qquad\square$

Now, we define an affine connection on Q/G as follows: for $X, Y \in \mathfrak{X}(Q/G)$, put

$$\tilde{\nabla}_X Y = \pi_*(\overline{\nabla}_{X^h}Y^h)\,.$$

This is well-defined since, by Proposition 5.1.1, the vector field $\overline{\nabla}_{X^h}Y^h$ is projectable, and one easily verifies that $\tilde{\nabla}$ satisfies the properties of an affine connection. Then, we obtain the following important result.

Proposition 5.1.2 ([120]). *The geodesics of $\overline{\nabla}$, with initial condition in $\mathcal{D}$, project onto the geodesics of $\tilde{\nabla}$.*

Proof. Key fact for the proof is that $\mathcal{D}$ is geodesically invariant with respect to the nonholonomic affine connection $\overline{\nabla}$. $\qquad\square$

Consequently, we have found that the equations of motion of the given generalized Chaplygin system reduce to the geodesic equations of the induced affine connection $\tilde{\nabla}$ on Q/G.

Consider the following (0,3)-tensor field on Q,

$$K_q(U_q, V_q, W_q) = g_q(\mathbf{h}_\gamma U_q, (\Omega^\gamma(V_q, W_q))_Q(q))\,,$$

where $\mathbf{h}_\gamma$ is the horizontal projector and Ω^γ is the curvature of the connection γ. Observe that K is horizontal, i.e. it vanishes if one of its arguments is a vertical vector, and it is skew-symmetric in its last two arguments. Moreover, one can see that

$$K_{gq}(T\Psi_g(U_q), T\Psi_g(V_q), T\Psi_g(W_q)) = K_q(U_q, V_q, W_q)\,,$$

for all $g \in G$ and $q \in Q$. Consequently, K induces a (0,3)-tensor on the base manifold Q/G

$$\tilde{K}_x(u_x, v_x, w_x) = K_q(U_q, V_q, W_q)\,,$$

where $\pi(q) = x$ and U_q, V_q, W_q are tangent vectors in q projecting onto u_x, v_x, w_x, respectively. K (resp. $\tilde{K}$) is called the *metric connection tensor* on Q (resp. Q/G).

In [120], it was shown that the application of the so-called Hamel's approach to Mechanics, leads to two additional affine connections on Q/G, whose geodesics are also solutions of the reduced nonholonomic problem. These connections are given by

$$(\nabla_1^H)_X Y = \nabla_X^{\tilde{g}} Y + B(X,Y), \quad (\nabla_2^H)_X Y = \nabla_X^{\tilde{g}} Y + B(Y,X), \qquad (5.3)$$

where B is the (1,2)-tensor field defined by $\beta(B(X,Y)) = \tilde{K}(X,Y,\sharp_{\tilde{g}}\beta)$, for any $\beta \in \Lambda^1(Q/G)$, $X, Y \in \mathfrak{X}(Q/G)$. So, B represents the contorsion of ∇_1^H.

The following explicit formula for the connection $\tilde{\nabla}$ was then derived in [120] (up to a minor misprint),

$$\tilde{\nabla}_X Y = \nabla_X^{\tilde{g}} Y + \frac{1}{2}\left(B(X,Y) + B(Y,X) - C(X,Y)\right),$$

where C is the (1,2)-tensor field implicitly defined by

$$\beta(C(X,Y)) = K(\sharp_{\tilde{g}}\beta, X, Y),$$

for arbitrary $\beta \in \Lambda^1(Q/G)$, $X, Y \in \mathfrak{X}(Q/G)$. As noted in [120], the average of Hamel's connections, i.e. $\nabla^{H/2} = \frac{1}{2}(\nabla_1^H + \nabla_2^H)$, in general differs from $\tilde{\nabla}$, because of the skew-symmetric term $C(X,Y)$.

It is interesting to observe that from Proposition 3.4.2 one can deduce

$$\tilde{\nabla}\tilde{g} = 0,$$

that is, $\tilde{\nabla}$ is a metric connection. From Proposition 2.5.3 and the definition of B, it is readily seen that ∇_1^H is also a metric connection. In general, however, the connections ∇_2^H and $\nabla^{H/2}$ will not be metric. In fact, it is straightforward to prove the following result.

Proposition 5.1.3. *The following properties are equivalent,*

1. ∇_2^H *is metric;*

2. $\nabla^{H/2}$ *is metric;*

3. *the tensor field B is skew-symmetric;*

4. $\nabla^{H/2}$ *is the Levi-Civita connection of $\tilde{g}$.*

Later we will see that these properties are also equivalent to the vanishing of the 1-form $\overline{\alpha_X}$ and, hence, to the Hamiltonian nature of the reduced system (cf. Corollary 5.3.2).

The torsion of $\tilde{\nabla}$ is given by $\tilde{T}(X, Y) = \pi_* \overline{T}(X^h, Y^h)$, with $\overline{T}$ the torsion of $\overline{\nabla}$. Then, we see from the above that the metric connection $\tilde{\nabla}$ is the Levi-Civita connection associated with $\tilde{g}$ iff the torsion of $\overline{\nabla}$ takes values in the vertical tangent bundle to π for each pair of vectors in $\mathcal{D}$.

Finally, the following result shows that equality of $\tilde{\nabla}$ and $\nabla^{H/2}$ is a rather strong condition.

Proposition 5.1.4.

$$\tilde{\nabla} = \nabla^{H/2} \iff \tilde{\nabla} = \nabla^{H/2} = \nabla^{\tilde{g}} .$$

Proof. If $\tilde{\nabla} = \nabla^{H/2}$, then $\nabla^{H/2}$ is metric. By Proposition 5.1.3, this implies that $\nabla^{H/2}$ coincides with $\nabla^{\tilde{g}}$. The reverse implication is trivial.   □

5.1.2 Reconstruction

In view of Proposition 5.1.2 above, we see that, in the present setting, the reconstruction of the solution curves in Q of the given constrained system, from those of the reduced system on Q/G, consists of a horizontal lift operation with respect to the connection γ. We state next this simple fact,

Proposition 5.1.5. *Let $\tilde{c}(t)$ be a geodesic of $\tilde{\nabla}$ and choose $c(0) \in Q$ such that $\pi(c(0)) = \tilde{c}(0)$. Then, the geodesic starting at $c(0)$, with initial velocity $\dot{c}(0) \in \mathcal{D}_{c(0)}$, is precisely the horizontal lift of $\tilde{c}(t)$ with respect to the principal connection γ.*

5.2 Two motivating examples

IN this section we present two examples of Chaplygin systems. We formulate each of them under the symplectic and the affine connection approaches. That will serve us as a motivation for the study in Section 5.3 of the relation between both of them.

5.2.1 Mobile robot with fixed orientation

Consider the motion of a mobile robot whose body maintains a fixed orientation with respect to its environment [117] (see Figure 5.1). One such robot is the B12 Mobile Robot Base, manufactured by Real World Interface, Inc. The robot has three wheels, with radius R, which turn simultaneously about independent axes, and perform a rolling without sliding over a horizontal

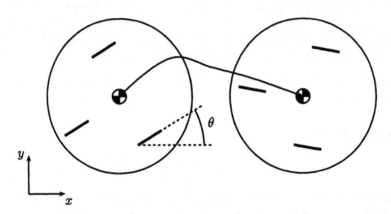

Fig. 5.1. A mobile robot with fixed orientation

floor. Let $(x, y) \in \mathbb{R}^2$ denote the position of the center of mass of the robot (in a Cartesian reference frame), $\theta \in \mathbb{S}^1$ the steering angle of the wheels and $\psi \in \mathbb{S}^1$ the rotation angle of the wheels in their rolling motion over the floor. The configuration space can then be modeled by $Q = \mathbb{S}^1 \times \mathbb{S}^1 \times \mathbb{R}^2$.

The Lagrangian function L is the kinetic energy function corresponding to the metric $g = m dx \otimes dx + m dy \otimes dy + J d\theta \otimes d\theta + 3 J_w d\psi \otimes d\psi$, where m is the mass of the robot, J its moment of inertia and J_w the axial moment of inertia of each wheel. The constraints are induced by the conditions that the wheels roll without sliding, in the direction in which they "point", and that the instantaneous contact points of the wheels with the floor have no velocity component orthogonal to that direction (cf. [117])

$$\dot{x} \sin\theta - \dot{y} \cos\theta = 0, \quad \dot{x} \cos\theta + \dot{y} \sin\theta - R\dot{\psi} = 0. \tag{5.4}$$

The constraint distribution $\mathcal{D}$ is then spanned by

$$\left\{ \frac{\partial}{\partial \theta}, \frac{\partial}{\partial \psi} + R\left(\cos\theta \frac{\partial}{\partial x} + \sin\theta \frac{\partial}{\partial y}\right) \right\}.$$

If we consider the Abelian action of $G = \mathbb{R}^2$ on Q by translations

$$\begin{aligned}
\Psi : \quad G \times Q &\longrightarrow Q \\
((a, b), (\theta, \psi, x, y)) &\longmapsto (\theta, \psi, a + x, b + y),
\end{aligned}$$

we see that the constraint distribution $\mathcal{D}$ can be interpreted as the horizontal subspace of the principal connection $\gamma = (dx - R\cos\theta d\psi)e_1 + (dy - R\sin\theta d\psi)e_2$, where $\{e_1, e_2\}$ is the canonical basis of $\mathbb{R}^2$ (identified with the Lie algebra of G).

The metric induced on Q/G here becomes

$$\tilde{g} = J d\theta \otimes d\theta + (3J_\omega + mR^2) d\psi \otimes d\psi .$$

The reduced Lagrangian L^* is the kinetic energy function corresponding to the metric $\tilde{g}$. Moreover, one easily verifies that the gyroscopic 1-form $\overline{\alpha_X}$ identically vanishes and, hence, the symplectic reduction (4.11) yields

$$i_{\overline{X}} \omega_{L^*} = dE_{L^*} ,$$

i.e. the reduced system is an unconstrained, purely Lagrangian system.

On the other hand, one can easily check that, in this example, the metric connection tensor $\tilde{K}$ also vanishes. Consequently, with the notations of the previous section, $\tilde{\nabla} = \nabla^{H/2}$ and by Proposition 5.1.4 we then have that $\tilde{\nabla} = \nabla^{H/2} = \nabla^{\tilde{g}}$.

5.2.2 Two-wheeled planar mobile robot

Consider the motion of a two-wheeled planar mobile robot (or "two-wheeled carriage") which is able to move in the direction in which it points and, in addition, can spin about a vertical axis [117, 120, 137, 188]. Let P be the intersection point of the horizontal symmetry axis of the robot and the horizontal line connecting the centers of the two wheels. The position and orientation of the robot is determined, with respect to a fixed Cartesian reference frame by $(x, y, \theta) \in SE(2)$, where $\theta \in \mathbb{S}^1$ is the heading angle and the coordinates $(x, y) \in \mathbb{R}^2$ locate the point P (see Figure 5.2). Let $\psi_1, \psi_2 \in \mathbb{S}^1$ denote the rotation angles of the wheels which are assumed to be controlled independently and roll without slipping on the floor. The configuration space of this system is $Q = \mathbb{S}^1 \times \mathbb{S}^1 \times SE(2)$.

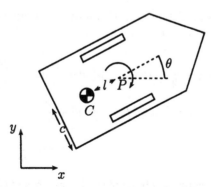

Fig. 5.2. A two-wheeled planar mobile robot

The Lagrangian function is the kinetic energy corresponding to the metric

$$g = m dx \otimes dx + m dy \otimes dy + m_0 l \cos \theta (dy \otimes d\theta + d\theta \otimes dy)$$
$$- m_0 l \sin \theta (dx \otimes d\theta + d\theta \otimes dx) + J d\theta \otimes d\theta + J_2 d\psi_1 \otimes d\psi_1 + J_2 d\psi_2 \otimes d\psi_2 \,,$$

where $m = m_0 + 2m_1$, m_0 is the mass of the robot without the wheels, J its moment of inertia with respect to the vertical axis, m_1 the mass of each wheel, J_w the axial moments of inertia of the wheels, and l the distance between the center of mass C of the robot and the point P.

The constraints, induced by the conditions that there is no lateral sliding of the robot and that the motion of the wheels also consists of a rolling without sliding, are

$$\dot{x} \sin \theta - \dot{y} \cos \theta = 0 \,,$$
$$\dot{x} \cos \theta + \dot{y} \sin \theta + c\dot{\theta} + R\dot{\psi}_1 = 0 \,,$$
$$\dot{x} \cos \theta + \dot{y} \sin \theta - c\dot{\theta} + R\dot{\psi}_2 = 0 \,,$$

where R is the radius of the wheels and $2c$ the lateral length of the robot. The constraint distribution $\mathcal{D}$ is then spanned by

$$\left\{ \frac{\partial}{\partial \psi_1} - \frac{R}{2} \left(\cos \theta \frac{\partial}{\partial x} + \sin \theta \frac{\partial}{\partial y} + \frac{1}{c} \frac{\partial}{\partial \theta} \right) \,, \right.$$
$$\left. \frac{\partial}{\partial \psi_2} - \frac{R}{2} \left(\cos \theta \frac{\partial}{\partial x} + \sin \theta \frac{\partial}{\partial y} - \frac{1}{c} \frac{\partial}{\partial \theta} \right) \right\} \,.$$

If we consider the action of $G = SE(2)$ on Q, $\Psi : G \times Q \longrightarrow Q$ defined by

$$(a, b, \alpha), (\psi_1, \psi_2, x, y, \theta)) \mapsto$$
$$(\psi_1, \psi_2, a + x \cos \alpha - y \sin \alpha, b + x \sin \alpha + y \cos \alpha, \alpha + \theta) \,,$$

we see that the constraint distribution $\mathcal{D}$ can be interpreted as the horizontal subspace of the principal connection

$$\gamma = \left(dx + \frac{R}{2} \cos \theta d\psi_1 + \frac{R}{2} \cos \theta d\psi_2 + y(d\theta + \frac{R}{2c} d\psi_1 - \frac{R}{2c} d\psi_2) \right) e_1$$
$$+ \left(dy + \frac{R}{2} \sin \theta d\psi_1 + \frac{R}{2} \sin \theta d\psi_2 - x(d\theta + \frac{R}{2c} d\psi_1 - \frac{R}{2c} d\psi_2) \right) e_2$$
$$+ (d\theta + \frac{R}{2c} d\psi_1 - \frac{R}{2c} d\psi_2) e_3 \,,$$

where $\{e_1, e_2, e_3\}$ is the canonical basis of the Lie algebra of G, with associated fundamental vector fields

$$(e_1)_Q = \frac{\partial}{\partial x} \,, \quad (e_2)_Q = \frac{\partial}{\partial y} \,, \quad (e_3)_Q = \frac{\partial}{\partial \theta} - y \frac{\partial}{\partial x} + x \frac{\partial}{\partial y} \,.$$

The curvature of γ is

$$\Omega = \frac{R^2}{2c}(\sin\theta\, e_1 - \cos\theta\, e_2)d\psi_1 \wedge d\psi_2\,.$$

The induced metric on Q/G is given by

$$\tilde{g} = (J_2 + m\frac{R^2}{4} + J\frac{R^2}{4c^2})d\psi_1 \otimes d\psi_1 + (J_2 + m\frac{R^2}{4} + J\frac{R^2}{4c^2})d\psi_2 \otimes d\psi_2$$
$$+ (m\frac{R^2}{4} - J\frac{R^2}{4c^2})(d\psi_1 \otimes d\psi_2 + d\psi_2 \otimes d\psi_1)\,.$$

The Lagrangian L^* is the kinetic energy function induced by $\tilde{g}$. The gyroscopic 1-form $\overline{\alpha_X}$ here becomes

$$\overline{\alpha_X} = \frac{m_0 l R^3}{4c^2}(\dot{\psi}_2 - \dot{\psi}_1)(\dot{\psi}_1 d\psi_2 - \dot{\psi}_2 d\psi_1)\,.$$

Then, the symplectic reduction (4.11) yields

$$i_{\overline{X}}\omega_{L^*} = dE_{L^*} + \overline{\alpha_X}\,.$$

On the other hand, the metric connection tensor $\tilde{K}$ is given by

$$\tilde{K} = \frac{m_0 l R^3}{4c^2}(d\psi_1 \otimes d\psi_1 \wedge d\psi_2 - d\psi_2 \otimes d\psi_1 \wedge d\psi_2)\,.$$

It is easily seen that, in this case, the tensor field B is not skew-symmetric and so, by Proposition 5.1.3, $\nabla^{H/2} \neq \nabla^{\tilde{g}}$. In addition, the Christoffel symbols of the metric connection $\tilde{\nabla}$ are

$$\begin{array}{ll}
\Gamma^{\psi_1}_{\psi_1\psi_1} = K_1\,, & \Gamma^{\psi_2}_{\psi_1\psi_1} = K_2\,, \\
\Gamma^{\psi_1}_{\psi_1\psi_2} = -K_2\,, & \Gamma^{\psi_2}_{\psi_1\psi_2} = -K_1\,, \\
\Gamma^{\psi_1}_{\psi_2\psi_1} = -K_1\,, & \Gamma^{\psi_2}_{\psi_2\psi_1} = -K_2\,, \\
\Gamma^{\psi_1}_{\psi_2\psi_2} = K_2\,, & \Gamma^{\psi_2}_{\psi_2\psi_2} = K_1\,,
\end{array}$$

where

$$K_1 = m_0 R^5 l\frac{J_2 - mc^2}{4c^2(2Jc^2 + j_2 R^2)(2J + mc^2)}\,,$$
$$K_2 = R^3 l\frac{4Jc^2 + (J_2 + mc^2)R^2 m_0}{4c^2(2Jc^2 + j_2 R^2)(2J + mc^2)}\,.$$

Clearly, the torsion $\tilde{T}$ does not vanish, and so $\tilde{\nabla} \neq \nabla^{\tilde{g}}$.

5.3 Relation between both approaches

THE above examples show us the following intriguing fact. In the case of the mobile robot with fixed orientation, the 1-form $\overline{\alpha_X}$ identically vanishes, and thus the reduced problem has no external gyroscopic force in its unconstrained symplectic formulation. Consequently, since there is no potential, the solutions of the reduced system are geodesics of the Levi-Civita connection $\nabla^{\tilde{g}}$. Indeed, we verified that $\tilde{\nabla} = \nabla^{H/2} = \nabla^{\tilde{g}}$. However, in the case of the two-wheeled mobile robot we obtained $\overline{\alpha_X} \neq 0$ and $\tilde{\nabla} \neq \nabla^{\tilde{g}} \neq \nabla^{H/2}$. Apparently, there exists a relation between the properties of the contorsions of the connections considered in Section 5.1.1 and the vanishing (or not) of the gyroscopic 1-form.

Using the definition of α_X in (4.6), one can check that the following relation holds [49, 51]: for any $Y \in \mathfrak{X}(\mathcal{D})$,

$$\alpha_X(Y) = \mathbf{v}_\Upsilon(Y)(\theta_L(X)) + \theta_L(R(X,Y)) + \theta_L(\mathbf{h}_\Upsilon[X, \mathbf{v}_\Upsilon(Y)]),$$

where R is the tensor field of type (1,2) on $\mathcal{D}$ given by

$$R = \frac{1}{2}[\mathbf{h}_\Upsilon, \mathbf{h}_\Upsilon],$$

with $[\cdot, \cdot]$ denoting the Nijenhuis bracket of type (1,1)-tensor fields. The relation between R and Ω^Υ, the curvature tensor of the principal connection Υ, is given by $R(U,V)(v) = (\Omega^\Upsilon(U_v, V_v))_{TQ}(v)$ for any $U, V \in \mathfrak{X}(\mathcal{D})$ and $v \in \mathcal{D}$.

In particular, if we take a horizontal vector field $Y \in \mathcal{U}$, we deduce from the above that $\alpha_X(Y) = \theta_L(R(X,Y))$. Herewith, the action of the gyroscopic 1-form $\overline{\alpha_X}$ on a vector field $Z \in \mathfrak{X}(T(Q/G))$, evaluated at a point $w_{\bar{q}} \in T(Q/G)(\cong \overline{\mathcal{D}})$, becomes

$$\overline{\alpha_X}(Z)(w_{\bar{q}}) = \alpha_X(Z^h)(v_q) = (\theta_L)_{v_q}(R(X, Z^h)(v_q))$$
$$= (\theta_L)_{v_q}\left((\Omega^\Upsilon(X_{v_q}, Z^h_{v_q}))_{TQ}(v_q)\right)$$
$$= (\theta_L)_{v_q}\left((\Omega^\Upsilon(\tau_{Q_*}X_{v_q}, \tau_{Q_*}Z^h_{v_q}))_{TQ}(v_q)\right),$$

for an arbitrary $v_q \in \mathcal{D}$ such that $\rho(v_q) = w_{\bar{q}}$, and where the last equality has been derived using (4.10). In these expressions, Z^h is the horizontal lift of Z with respect to Υ. Recalling that the Poincaré-Cartan 1-form θ_L and the Legendre mapping (induced by the given Lagrangian) $\mathcal{F}L : TQ \longrightarrow T^*Q$ are related by $(\theta_L)_{v_q}(u) = \langle \mathcal{F}L(v_q), \tau_{Q_*}(u)\rangle$, for any $u \in T_{v_q}TQ$, and taking into account that X is a SODE, we further obtain

$$\overline{\alpha_X}(Z)(w_{\bar{q}}) = \langle \mathcal{F}L(v_q), (\Omega^\Upsilon(\tau_{Q_*}X_{v_q}, \tau_{Q_*}Z^h_{v_q}))_Q(q)\rangle$$
$$= g_q\left(v_q, (\Omega^\Upsilon(v_q, \tau_{Q_*}Z^h_{v_q}))_Q(q)\right)$$
$$= g_q\left(v_q, (\Omega^\Upsilon(v_q, (\tau_{Q/G_*}Z_{w_{\bar{q}}})^h_q))_Q(q)\right). \qquad (5.5)$$

Note that in the last expression, the horizontal lift of $\tau_{Q/G_*}Z_{w_{\bar{q}}}$ is the one with respect to γ.

An important observation is that the expression (5.5) immediately shows that the gyroscopic 1-form $\overline{\alpha_X}$ is semi-basic with respect to the canonical fibration $\tau_{Q/G} : T(Q/G) \longrightarrow Q/G$. Indeed, assume $(\tau_{Q/G})_* \circ Z = 0$, then it readily follows that $\overline{\alpha_X}(Z) = 0$.

Elaborating (5.5) a bit further, using the metric connection tensor $\tilde{K}$ and the contorsion B introduced in Section 5.1.1, we find

$$\overline{\alpha_X}(Z)(w_{\bar{q}}) = g_q\left(v_q, \Omega(v_q, (\tau_{Q/G_*}Z_{w_{\bar{q}}})_q^h))_Q(q)\right)$$
$$= \tilde{K}_{\bar{q}}(w_{\bar{q}}, w_{\bar{q}}, \tau_{Q/G_*}Z_{w_{\bar{q}}}) = \tilde{g}_{\bar{q}}(B(w_{\bar{q}}, w_{\bar{q}}), \tau_{Q/G_*}Z_{w_{\bar{q}}}).$$

This proves the next result, which was already implicit in the work of Koiller [120].

Proposition 5.3.1. *An explicit relation between the gyroscopic 1-form and the contorsion tensor field B, defined in Section 5.1.1, is given by*

$$(\overline{\alpha_X})_{w_{\bar{q}}}(u) = \tilde{g}_{\bar{q}}(B(w_{\bar{q}}, w_{\bar{q}}), \tau_{Q/G_*}u),$$

for all $u \in T_{w_{\bar{q}}}T(Q/G))$, $w_{\bar{q}} \in T(Q/G)$.

From this we can deduce, taking into account Proposition 5.1.3,

Corollary 5.3.2. *The following statements are equivalent,*

1. $\overline{\alpha_X}$ *vanishes identically;*

2. B *is skew-symmetric;*

3. $\nabla^{H/2} = \nabla^{\tilde{g}}$.

Remark 5.3.3. One can think of simple examples in which B is skew-symmetric but nonzero. Consequently, if $\overline{\alpha_X}$ vanishes, this does not imply $\tilde{\nabla} = \nabla^{\tilde{g}}$, although in such a case both connections do have the same geodesics (recall the discussion in Section 2.5.1).

By means of Proposition 5.3.1, one can also recover the gyroscopic character of $\overline{\alpha_X}$, established already in Proposition 4.2.7. For that purpose, let us define the following 2-form on $T(Q/G)$,

$$\Xi(Y, Z)(w_{\bar{q}}) = \tilde{g}_{\bar{q}}(B(w_{\bar{q}}, \tau_{Q/G_*}Y_{w_{\bar{q}}}), \tau_{Q/G_*}Z_{w_{\bar{q}}}). \tag{5.6}$$

One readily verifies that Ξ is indeed bilinear and by Proposition 2.5.3, $\Xi(Y, Y) = 0$. It is then easy to check that

$$\overline{\alpha_X} = i_{\overline{X}}\Xi\,.$$

In local coordinates q^a $(a = 1, \dots, n - k)$ on Q/G, we have that

$$\Xi = \sum_{a<b} \dot{q}^e B^c_{ea} \tilde{g}_{bc} dq^a \wedge dq^b\,, \qquad \overline{\alpha_X} = \sum_{a,e,c} \dot{q}^a \dot{q}^e B^c_{ea} \tilde{g}_{bc} dq^b\,. \qquad (5.7)$$

A careful calculation, very similar to the one performed to prove Proposition 5.3.1, reveals that for generalized Chaplygin systems of mechanical type, the 2-form $\overline{\Sigma}$ of Proposition 4.2.7 and the above 2-form Ξ coincide.

5.4 Integrability aspects and the existence of an invariant measure

5.4.1 Koiller's question

In [120], the author wonders whether there might exist an invariant measure for the reduced equations of a generalized Chaplygin system. In the following, we shall deal with this problem.

The existence of a measure which is invariant under the flow of a given dynamical system is a strong property. Indeed, using an integrating factor it is possible to derive from it (locally) an integral of the motion. This fact plays an important role in discussions concerning the integrability of the system under consideration, as illustrated by the following theorem.

Theorem 5.4.1 ([9]). *Suppose that the system $\dot{x} = X(x)$, $x \in N$, with N an n-dimensional smooth manifold, admits an invariant measure and has $n - 2$ first integrals $F_1, \dots, F_{n-2}$. Suppose also that $F_1, \dots, F_{n-2}$ are independent on the invariant set $N_c = \{x \in N : F_s(x) = c_s, 1 \leq s \leq n - 2\}$. Then*

– *the solutions of the differential equation lying on N_c can be found by quadratures.*

Moreover, if L_c is a compact connected component of the level set N_c and if X does not vanish on L_c, then

– *L_c is a smooth manifold diffeomorphic to a two-torus;*
– *one can find angular coordinates φ_1, $\varphi_2 \bmod (2\pi)$ on L_c in terms of which the differential equations take the simple form*

$$\dot{\varphi}_1 = \frac{\omega_1}{\Phi(\varphi_1, \varphi_2)}\,, \quad \dot{\varphi}_2 = \frac{\omega_2}{\Phi(\varphi_1, \varphi_2)}\,,$$

where ω_1, ω_2 are constant and Φ is a smooth positive function which is 2π-periodic in φ_1, φ_2.

By the Riesz representation theorem, we know that each volume form on an orientable manifold induces a unique measure on the Borel σ-algebra [1]. Therefore, with a view on tackling the integrability problem of generalized Chaplygin systems, it is worth looking for invariant volume forms under the flow of the reduced dynamics $\overline{X}$. This is what we intend to do in the sequel.

In this section, we deal with natural Chaplygin systems which may have non-trivial potential energy V. The reduced equations of motion are (cf. equation (4.11))

$$i_{\overline{X}}\omega_{L^*} = dE_{L^*} + \overline{\alpha_X} , \qquad (5.8)$$

where $L^* = \frac{1}{2}\tilde{g} - \tilde{V}$, with $\tilde{g}$ and $\tilde{V}$ the metric and the potential function on Q/G induced by, respectively, g and V. Remember that the energy E_{L^*} is a constant of the motion. The local expression for the reduced dynamics takes the form (cf. (4.13))

$$\overline{X} = \dot{q}^a \frac{\partial}{\partial q^a} - \left(\tilde{g}^{ab}(\alpha_b + \frac{\partial \tilde{V}}{\partial q^b}) + \dot{q}^b \dot{q}^c \tilde{\Gamma}^a_{bc} \right) \frac{\partial}{\partial \dot{q}^a} ,$$

where $\tilde{\Gamma}^a_{bc}$ are the Christoffel symbols of the Levi-Civita connection $\nabla^{\tilde{g}}$.

The gyroscopic systems usually encountered in the Mechanics and Control literature [252, 258] differ in a crucial way from the ones we obtain through the reduction of a generalized Chaplygin system. In fact, the common situation in Mechanics is that of a system, with configuration space P, described by an equation of the form

$$i_\Gamma \omega_{\mathbb{L}} = dE_{\mathbb{L}} + \alpha ,$$

where $\mathbb{L} : TP \longrightarrow \mathbb{R}$ is a (regular) Lagrangian and where the gyroscopic force is represented by a 1-form $\alpha = i_\Gamma(\tau_P^* \Pi)$, with Π a closed 2-form on P. These systems are then Hamiltonian with respect to the symplectic 2-form $\omega = \omega_{\mathbb{L}} - \tau_P^* \Pi$, and thus they admit an invariant measure, determined by the volume form $\omega^n = \omega_{\mathbb{L}}^n$.

In some sense, our reduced system (5.8) exhibits the opposite behavior. Indeed, the 2-form Ξ, defined by (5.6), is semi-basic but in general it is not basic, i.e. it is not the pull-back of a 2-form on the base Q/G. This can be readily deduced from its local expression (5.7). Moreover, the following property is easily proved.

Proposition 5.4.2. *The 2-form Ξ is closed if and only if it is identically zero.*

Note, in passing, that a similar property also applies to the gyroscopic 1-form $\overline{\alpha_X}$. The semi-basic character of Ξ ensures, however, that the 2-form

$\omega_{L^\bullet} - \varXi$ is still nondegenerate and, consequently, we have that the equation (5.8) can be rewritten in the form

$$i_{\overline{X}}\omega = dE_{L^\bullet},\tag{5.9}$$

with $\omega = \omega_{L^\bullet} - \varXi$ an almost symplectic form (i.e. a nondegenerate, but not necessarily closed 2-form).

In [226], S.V. Stanchenko has studied Chaplygin systems of mechanical type with an Abelian Lie group in terms of differential forms, in a way which shows many links to the symplectic approach described in Section 4.2.1. In our setting, his results can be generalized to the non-Abelian case for any kind of generalized Chaplygin system.

Assume, following [226], that there exists a function $F \in C^\infty(T(Q/G))$ such that

$$dF \wedge \theta_{L^\bullet} = \varXi.\tag{5.10}$$

Putting $N = \exp F$, we have that

$$d(N\omega) = d(N\omega_{L^\bullet} - N\varXi) = d(N\omega_{L^\bullet} - dN \wedge \theta_{L^\bullet}) = 0.$$

Since (5.9) can still be written as

$$i_{\overline{X}/N}(N\omega) = dE_{L^\bullet},$$

we deduce that $\mathcal{L}_{\overline{X}/N}(N\omega) = 0$. Consequently,

$$0 = \mathcal{L}_{\overline{X}/N}(N\omega)^n = \mathcal{L}_{\overline{X}}N^{n-1}\omega^n,$$

and we see that $N^{n-1}\omega^n$ is an invariant volume form. This proves the following result.

Theorem 5.4.3 ([226]). *Condition* (5.10) *is sufficient for the existence of an invariant measure for the reduced Chaplygin equations* (5.9).

Remark 5.4.4. Stanchenko observes that if F satisfies (5.10), the semi-basic character of both $\theta_{L^\bullet}$ and $\varXi$ imply that F is necessarily the pullback of a function on Q/G.

It turns out that condition (5.10) can be relaxed to some extent: it suffices to require the almost symplectic 2-form ω to be *globally conformal symplectic*, that is, that there exists a function $F \in C^\infty(T(Q/G))$ such that

$$dF \wedge \omega = -d\omega.\tag{5.11}$$

Theorem 5.4.3 still holds in this case, with (5.11) replacing (5.10). The previous remark also remains valid: the function F is necessarily the pullback of a function on Q/G. Note that (5.10) obviously implies (5.11).

However, even the weaker condition (5.11) is not necessary in general for the existence of an invariant volume form on $T(Q/G)$. To derive a necessary condition, let us suppose that μ is an invariant volume form for the dynamics $\overline{X}$ on $T(Q/G)$. We then necessarily have that $\mu = k\omega^n$, for some nowhere vanishing function $k \in C^\infty(T(Q/G))$. Restricting ourselves to a connected component of Q/G if need be, we may always assume k is strictly positive. It follows that

$$
\begin{aligned}
0 = \mathcal{L}_{\overline{X}}\mu &= \overline{X}(k)\,\omega^n + k\,\mathcal{L}_{\overline{X}}\omega^n \\
&= \overline{X}(k)\,\omega^n + nk\,\mathcal{L}_{\overline{X}}\omega \wedge \omega^{n-1} = \overline{X}(k)\,\omega^n - nk\,i_{\overline{X}}d\,\Xi \wedge \omega^{n-1}\,.
\end{aligned}
$$

The $2n$-form $i_{\overline{X}}d\,\Xi \wedge \omega^{n-1}$ determines a function $h \in C^\infty(T(Q/G))$ by

$$
i_{\overline{X}}d\,\Xi \wedge \omega^{n-1} = \frac{h}{n}\,\omega^n\,. \tag{5.12}
$$

Therefore, we have

$$
\overline{X}(k) = kh \quad \text{or, equivalently,} \quad \overline{X}(\ln k) = h\,. \tag{5.13}
$$

This essentially yields the same characterization as the one derived in [226]. Now, conversely, assume there exists a function k satisfying (5.13), with h defined by (5.12). Going through the above computations in reverse order then shows that the $2n$-form $k\omega^n$ is an invariant volume form of $\overline{X}$. We may therefore conclude that the existence of a globally defined function k for which (5.13) holds is not only a necessary but also a sufficient condition for the existence of an invariant volume form. It is interesting to note that in [226], Stanchenko has proved that in case the reduced Lagrangian is of kinetic energy type, $L^* = \frac{1}{2}\tilde{g}$, and if there exists a solution k of (5.13) which is basic, i.e. which is the pullback of a function on the base space Q/G, then the volume form $\mu = k\omega^n$ remains an invariant of the reduced dynamics if a potential energy function $\tilde{V} \in C^\infty(Q/G)$ is included in the Lagrangian L^* (coming from a G-invariant potential added to the given Chaplygin system).

Obviously, however, equation (5.13) is not a very handy criterium to deal with in practice. We will now see that, at least for a subclass of generalized Chaplygin systems, it may be replaced by a more easily manageable condition.

From (5.12), we can deduce a local expression for h. After some computations, we get

$$
h = \sum_{a,b} \tilde{g}^{ab} \frac{\partial \alpha_b}{\partial \dot{q}^a}\,,
$$

and, using (5.7), this further becomes

$$h = \sum_{f,b} \tilde{g}^{fb} \dot{q}^e \tilde{g}_{bc}(B^c_{ef} + B^c_{fe}) = \sum_{c,e} \dot{q}^e (B^c_{ec} + B^c_{ce}) \,.$$

Note that $S^* dh$ is the pullback of a basic 1-form, i.e. $S^* dh = \tau^*_{Q/G}\beta$, where the local expression for β reads

$$\beta = h_e(q)dq^e = \sum_c (B^c_{ec} + B^c_{ce})dq^e \,,$$

with $h_e = \partial h / \partial \dot{q}^e$. Let us assume now that there exists a basic function k for which (5.13) holds. We then have that $S^*(d\overline{X}(\ln k)) = d(\ln k)$. Therefore, taking the differential of both hand-sides of (5.13) and applying S^* to the resulting equality, we obtain

$$d(\ln k) = \beta \,,$$

or, in local coordinates,

$$\frac{\partial(\ln k)}{\partial q^e} = h_e(q) = \sum_c (B^c_{ec} + B^c_{ce}), \ e = 1,\ldots,n \,.$$

We thus see that, if (5.13) admits a solution k which is basic, then the 1-form β is exact.

It turns out that, for systems for which $\tilde{V} = 0$ (i.e. L^* is a pure kinetic energy Lagrangian), the previous result even has a more definitive character. Indeed, let $\tilde{V} = 0$ and suppose there exists a function $k \in C^\infty(T(Q/G))$ (not necessarily basic) satisfying (5.13). Then, we have that

$$S^* d\overline{X}(\ln k) = S^* dh \,.$$

In local coordinates, this becomes

$$\frac{\partial \overline{X}(\ln k)}{\partial \dot{q}^e} = h_e(q), \ e = 1,\ldots,n \,.$$

But $\overline{X}(\ln k) = \dot{q}^a \dfrac{\partial(\ln k)}{\partial q^a} + \overline{X}^a \dfrac{\partial(\ln k)}{\partial \dot{q}^a}$, where $\overline{X}^a = -(\tilde{g}^{ab}\alpha_b + \dot{q}^b \dot{q}^c \tilde{\Gamma}^a_{bc})$, and so we have

$$\frac{\partial \overline{X}(\ln k)}{\partial \dot{q}^e} = \frac{\partial(\ln k)}{\partial q^e} + \dot{q}^a \frac{\partial^2(\ln k)}{\partial q^a \partial \dot{q}^e} + \frac{\partial \overline{X}^a}{\partial \dot{q}^e}\frac{\partial(\ln k)}{\partial \dot{q}^a} + \overline{X}^a \frac{\partial^2(\ln k)}{\partial \dot{q}^a \partial \dot{q}^e} \,.$$

In points 0_q of the zero section of $T(Q/G)$ this reduces to

$$\frac{\partial \overline{X}(\ln k)}{\partial \dot{q}^e}\bigg|_{0_q} = \frac{\partial(\ln k)}{\partial q^e}\bigg|_{0_q} \,.$$

If we now define the basic function $\tilde{k} = k \circ s$, where $s : Q/G \to T(Q/G), q \mapsto (q, 0)$ determines the zero section, we derive from the above that

$$\frac{\partial(\ln \tilde{k})}{\partial q^e}(q) = \frac{\partial(\ln k)}{\partial q^e}(q, 0) = h_e(q), \ e = 1, \ldots, n,$$

and, hence, it follows again that the 1-form β is exact.

Conversely, if the 1-form β is exact, say $\beta = df$ for some $f \in C^\infty(Q/G)$, and putting $k = \exp(\tau_{Q/G}{}^* f)$, one easily verifies that k satisfies (5.13). This obviously also holds in the presence of a potential (i.e. if $\tilde{V} \neq 0$).

Summarizing the above, we have proved the following interesting result.

Theorem 5.4.5. *For a generalized Chaplygin system with a Lagrangian of kinetic energy type, i.e. $L = \frac{1}{2}g$, there exists an invariant volume form for the reduced dynamics $\overline{X}$ on $T(Q/G)$ iff the basic 1-form β, defined by $S^* dh = \tau_{Q/G}^* \beta$ (with h given by (5.12)), is exact. The 'if' part also holds if L is of the form $L = \frac{1}{2}g - V$, with V a G-invariant potential.*

Therefore, if we find a particular example of a system with a kinetic energy type Lagrangian for which β is not exact, we shall have proved that the answer to Koiller's question about the existence of an invariant volume form for all generalized Chaplygin systems, is negative. In particular, for such a counter example it suffices to show that the corresponding β is not closed.

5.4.2 A counter example

Let us consider the following modified version of the classical example of the nonholonomic free particle. Let a particle be moving in space, $Q = \mathbb{R}^3$, subject to the nonholonomic constraint

$$\phi = \dot{z} - yx\dot{x} = 0 .$$

The Lagrangian function is the kinetic energy corresponding to the standard metric $g = dx \otimes dx + dy \otimes dy + dz \otimes dz$. Therefore,

$$L = \frac{1}{2}\left(\dot{x}^2 + \dot{y}^2 + \dot{z}^2\right) .$$

The constraint submanifold is defined by the distribution

$$\mathcal{D} = \text{span}\left\{\frac{\partial}{\partial x} + yx\frac{\partial}{\partial z}, \frac{\partial}{\partial y}\right\} .$$

Consider the Lie group $G = \mathbb{R}$ with its trivial action by translations on Q,

$$\Phi: \quad G \times Q \quad \longrightarrow Q$$
$$(s, (x, y, z)) \longmapsto (x, y, z + s).$$

Note that $\mathcal{D}$ is the horizontal subspace of a connection γ on the principal fiber bundle $Q \longrightarrow Q/G$, where $\gamma = dz - yxdx$. Therefore, this system belongs to the class of generalized Chaplygin systems.

The curvature of γ is given by

$$\Omega^\gamma = xdx \wedge dy.$$

The induced metric $\tilde{g}$ on $Q/G \cong \mathbb{R}^2$ is

$$\tilde{g} = (1 + x^2 y^2)dx \otimes dx + dy \otimes dy.$$

The metric connection tensor $\tilde{K}$ is determined by

$$\tilde{K}(\frac{\partial}{\partial x}, \frac{\partial}{\partial x}, \frac{\partial}{\partial y}) = x^2 y, \quad \tilde{K}(\frac{\partial}{\partial y}, \frac{\partial}{\partial x}, \frac{\partial}{\partial y}) = 0.$$

Then, the contorsion of the affine connection ∇_1^H here reads

$$B = x^2 y dx \otimes dx \otimes \frac{\partial}{\partial y} - \frac{x^2 y}{1 + x^2 y^2} dx \otimes dy \otimes \frac{\partial}{\partial x}.$$

Finally, the 1-form β associated with the reduced Chaplygin system is

$$\beta = -\frac{x^2 y}{1 + x^2 y^2} dy,$$

which is clearly not closed. Hence, according to Theorem 5.4.5, there is no invariant volume form for the reduced dynamics.

Note that in this example the distribution $\mathcal{D}$ has 'length' 1 at all points of $\mathbb{R}^3$ not belonging to the plane $x = 0$, since $\mathcal{D} + [\mathcal{D}, \mathcal{D}]$ spans the full tangent space at all points for which $x \neq 0$.

Remark 5.4.6. In the example of the nonholonomic free particle (cf. Section 4.2.1), the constraint distribution does have length 1 everywhere. After performing the appropriate computations, we obtain

$$\beta = -\frac{y}{1 + y^2} dy,$$

which is clearly exact: $\beta = df$, with $f = \frac{1}{2}\ln\left(\frac{1}{1 + y^2}\right)$. Then, $k = \frac{1}{\sqrt{1 + y^2}}$ and the invariant measure defined by $k\omega^n$ leads, using Euler's integrating factor technique, to the constant of the motion

$$\varphi = \dot{x}\sqrt{1 + y^2}$$

which was also obtained by a different method in Section 4.4.1.

6 A class of hybrid nonholonomic systems

IN this chapter, we deal with systems subject to nonholonomic linear constraints that can be "redundant" at some points. Otherwise said, we treat systems for which, depending on the specific configuration, the set of constraints to be satisfied are different, but in such a way that they all are "glued" in a smooth manner. This precisely corresponds to the situation in which the constraints are given by a generalized differentiable codistribution $\mathcal{D}$ on Q, as defined in Chapter 2. There has been other works, among which we mention [65], which have addressed similar situations, but they have been focused mainly on solving the trajectory planning problem. Here, instead, we pay attention to the problem of determining what occurs to the system when it changes from one set of constraints to another. To be exact, we identify the points in the configuration space where a discrete dynamics, instead of the continuous one, acts on the system and we describe its behavior. Recent developments in this line include [87, 199, 251].

The chapter is organized as follows. In Section 6.1 we introduce the class of systems under consideration. Section 6.2 briefly reviews the classical treatment on impulsive forces, together with some modern results. This will be helpful in the subsequent discussion. Section 6.3 contains the main results of the chapter. We show that the points in which the discrete dynamics governs the motion of the system precisely correspond to the *singular* points, that is, those where the constraints change. We discuss the behavior of the discrete dynamics and obtain a jump rule to compute the changes in momentum. Finally, in Section 6.4 we present two examples to illustrate the theory.

6.1 Mechanical systems subject to constraints of variable rank

LET us consider a mechanical system with Lagrangian function $L : TQ \to \mathbb{R}$, $L(v) = \frac{1}{2}g(v, v) - (V \circ \tau_Q)(v)$, where g is a Riemannian metric on Q and V is a function on the configuration space Q (the potential). Suppose, in addition, that the system is subject to a set of constraints given by a generalized differentiable codistribution $\mathcal{D}$ on Q, i.e. we assume that $\pi_Q(\mathcal{D}) = Q$.

The motions of the system are forced to take place satisfying the constraints imposed by $\mathcal{D}$.

We know from Section 2.2 that the codistribution $\mathcal{D}$ induces a decomposition of Q into regular and singular points. We write

$$Q = R \cup S.$$

Let us fix R_c, a connected component of R. We can consider the restriction of the codistribution to R_c, $\mathcal{D}_c = \mathcal{D}_{|R_c} : R_c \subset Q \longrightarrow T^*Q$. Obviously, we have that $\mathcal{D}_c$ is a regular codistribution, that is, it has constant rank. Then, let us denote by $\mathcal{D}_c^o : R_c \longrightarrow TQ$ the annihilator of $\mathcal{D}_c$. Now, we can consider the dynamical problem with regular Lagrangian L, subject to the regular codistribution $\mathcal{D}_c^o$ and apply the theory for nonholonomic Lagrangian systems exposed in the preceding chapters.

Consequently, our problem is solved on each connected component of R. The situation changes radically if the motion reaches a singular point. The rank of the constraint codistribution can vary suddenly and the classical derivation of the equations of motion for nonholonomic Lagrangian systems is no longer valid. Our objective is to explore the response of the system when such a situation arises.

Remark 6.1.1. We note that the notion of singular point considered here is different from the one considered in [65]. In that paper, the authors treat the case of generalized constraints given by a globally defined set of 1-forms, $\omega_1, \ldots, \omega_l$. Then, they consider the l-form

$$\Omega = \omega_1 \wedge \cdots \wedge \omega_l.$$

The singular set consists of the points for which $\Omega(q) = 0$, that is, the points, q, such that $\{\omega_1(q), \ldots, \omega_l(q)\}$ are linearly dependent. Applying this notion to the example of a generalized codistribution we gave in Section 2.2 (Example 2.2.2), $Q = \mathbb{R}^2$, $\mathcal{D}_{(x,y)} = \text{span}\{\phi(x)(dx - dy)\}$, with

$$\phi(x) = \begin{cases} 0 & x \leq 0, \\ e^{-\frac{1}{x^2}} & x > 0, \end{cases}$$

the set of singular points would be the half-plane $\{x \leq 0\}$.

It will turn out to be that the behavior of the constraint forces acting on mechanical systems subject to generalized constraints has many resemblances with the impulsive case. In order to make the discussion more clear, we review in the following section the classical treatment of impulsive forces and then turn our attention to the case of generalized constraints.

6.2 Impulsive forces

IN this section, we present a review of the classical treatment of mechanical systems with impulsive forces [6, 108, 188, 198, 207, 243]. This field has traditionally been studied by a rich variety of methods (analytical, numerical and experimental), being a meeting point among physicists, mechanical engineers and mathematicians (see [39] for an overview on the subject). Recently, such systems have been brought into the context of Geometric Mechanics [100, 101, 102, 130]. We will give here a brief review of the classical approach. These ideas will be useful in understanding the behavior of the constraint forces acting on mechanical systems subject to generalized constraints. Both situations are not the same, but have many points in common, as we will see in the following.

Consider a system of n particles in $\mathbb{R}^3$ such that the particle r has mass m_r. Introducing coordinates $(q^{3r-2}, q^{3r-1}, q^{3r})$ for the particle r, we denote by Q the configuration manifold $\mathbb{R}^{3n}$ and by $F_r = (F^{3r-2}, F^{3r-1}, F^{3r})$ the resultant of all forces acting on the rth particle.

The motion of the particle r in an interval $[t, t']$ is determined by the system of integral equations

$$m_r(\dot{q}^k(t') - \dot{q}^k(t)) = \int_t^{t'} F^k(\tau)d\tau \,, \tag{6.1}$$

where $3r - 2 \leq k \leq 3r$ and k is an integer. The integrals of the right-hand side are the components of the *impulse* of the force F_r. Equation (6.1) establishes the relation between the impulse and the momentum change, i.e. 'impulse is equal to momentum change'. Equation (6.1) is a generalized writing of Newton's second law, stated in integral form in order to allow us to consider the case of velocities with finite jump discontinuities. This is precisely the case of impulsive forces, which generate a finite non-zero impulse at some time instants.

If F is impulsive there exists an instant t_0 such that

$$\lim_{t \to t_0} \int_{t_0}^t F(\tau)d\tau = P \neq 0 \,. \tag{6.2}$$

Equation (6.2) implies that the impulsive force has an infinite magnitude at the point t_0, but we are assuming that its impulse P is well defined and bounded. The expression

$$P \cdot \delta(t_0) = \lim_{t \to t_0} F(t)$$

can be mathematically seen as a Dirac delta function concentrated at t_0.

Now, we will derive the equations for impulsive motion following the discussion in [207]. In the sequel, the velocity vector of the rth particle, $(\dot{q}^{3r-2}, \dot{q}^{3r-1}, \dot{q}^{3r})$, will be denoted by $\dot{q}^r$. Then, the system of integral equations (6.1) can be written as

$$m_r(\dot{q}^r(t_0 + \epsilon) - \dot{q}^r(t_0 - \epsilon)) = \int_{t_0-\epsilon}^{t_0+\epsilon} F_r(\tau)d\tau .$$

Multiplying by the virtual displacements at the point $q(t_0)$, we obtain

$$(p_r(t_0 + \epsilon) - p_r(t_0 - \epsilon)) \cdot \delta q^r = \int_{t_0-\epsilon}^{t_0+\epsilon} F_r(\tau)d\tau \cdot \delta q^r .$$

For the entire system, one has

$$\sum_{r=1}^{n} \left\{ p_r(t_0 + \epsilon) - p_r(t_0 - \epsilon) - \int_{t_0-\epsilon}^{t_0+\epsilon} F_r'(\tau)d\tau \right\} \cdot \delta q^r$$
$$= \sum_{r=1}^{n} \int_{t_0-\epsilon}^{t_0+\epsilon} F_r''(\tau)d\tau \cdot \delta q^r , \quad (6.3)$$

where F_r' and F_r'' are, respectively, the resultant of the *given* forces and of the *constraint reaction* forces acting on the rth particle at time τ.

Now, take a local chart (q^A), $1 \leq A \leq 3n$ on a neighborhood U of $q(t_0)$ and consider the identification $T_q Q \equiv \mathbb{R}^{3n}$, which maps each $v_q \in T_q Q$ to (v^A), such that

$$v_q = v^A \left(\frac{\partial}{\partial q^A} \right)_q ,$$

for each $q \in U$. Let us suppose that the constraints are given on U by the 1-forms $\omega_i = \mu_{iA} dq^A$, $1 \leq i \leq m$. Then, we have that $\mu_{iA}(q(t)) = \mu_{iA}(q(t_0)) + O(t - t_0)$ along the trajectory $q(t)$. As the virtual displacements at the point $q(t)$ satisfy by definition

$$\sum \mu_{iA}(q(t))(\delta q(t))^A = 0 , \quad 1 \leq i \leq m ,$$

we conclude that $\sum \left(\mu_{iA}(q(t_0))(\delta q(t))^A + O(t - t_0) \right) = 0$. Therefore, we have that

$$\delta q^r(t) = \delta q^r(t_0) + O(\epsilon) , \quad t \in [t_0 - \epsilon, t_0 + \epsilon] ,$$

that is, the virtual displacements at $q(t)$ can be approximated by the virtual displacements at $q(t_0)$. As a consequence, in the right-hand side of (6.3) we can write

$$\int_{t_0-\epsilon}^{t_0+\epsilon} F_r''(\tau)d\tau \cdot \delta q^r = \int_{t_0-\epsilon}^{t_0+\epsilon} F_r''(\tau) \cdot \delta q^r \, d\tau = \int_{t_0-\epsilon}^{t_0+\epsilon} F_r''(\tau) \cdot \delta q^r(\tau)d\tau + O(\epsilon).$$

The first term after the last equality is the virtual work done by the constraint forces along the trajectory, and this work is zero since we are considering ideal constraints. The second one goes to zero as ϵ tends to zero.

In the presence of given impulsive forces acting on m particles, say, at time t_0, we have

$$\lim_{t \to t_0} \int_{t_0}^{t} F_{r'}(\tau)d\tau = P_{r'} \neq 0 \, , \, 1 \leq r' \leq m \, .$$

Then, taking the limit $\epsilon \to 0$ in (6.3), we obtain the equation for impulsive motion [188, 207]

$$\sum_{r=1}^{n} \{p_r(t_0)_+ - p_r(t_0)_- - P_r\} \cdot \delta q^r = 0 \, . \tag{6.4}$$

An example in which equation (6.4) may be applied is when we strike with a cue a billiard ball which is initially at rest. In that case we are exerting an impulsive force that puts the ball into motion. But what happens when the ball collides with the edge of the billiard? What we see is that it bounces, i.e. it suffers again a discontinuous jump in its velocity. The constraint imposed by the wall of the billiard exerts an impulsive force on the ball. When the impulsive force is caused by constraints, such constraints are called *impulsive constraints*. There is a number of different situations in which they can appear. In the following, we examine them.

In the presence of linear constraints of type $\Psi = 0$, where $\Psi = b_k(q)\dot{q}^k$ (a situation which covers the case of unilateral holonomic constraints, such as the impact against a wall, and more general types of constraints, such as instantaneous nonholonomic constraints), the constraint force, $F = F_k \, dq^k$, is given by $F_k = \mu \cdot b_k$, where μ is a Lagrange multiplier. Then the constraint is impulsive if and only if

$$\lim_{t \to t_0} \int_{t_0}^{t} \mu \cdot b_k d\tau = P_k \neq 0 \, ,$$

for some k. The impulsive force may be caused by different circumstances: the function b_k is discontinuous at t_0, the Lagrange multiplier μ is discontinuous at t_0 or both.

The presence of such constraints does not invalidate equation (6.4). It merely means that the virtual displacements δq^r must satisfy certain additional conditions, which are just those imposed by the constraints. So, in the absence of impulsive external forces and in the presence of impulsive constraints, we would have

$$\sum_{r=1}^{n} \Delta p_r(t_0) \cdot \delta q^r = 0, \tag{6.5}$$

where $\Delta p_r(t_0) = p_r(t_0)_+ - p_r(t_0)_-$.

Remark 6.2.1. In general, equation (6.5) is not enough to determine the jump of the momentum. One usually needs additional physical hypothesis, related with elasticity, plasticity, etc. to obtain the post-impact momentum. In this respect, there are two classical approaches, the Newtonian approach and the Poisson approach [39, 229]. The Newtonian approach relates the normal component of the rebound velocity to the normal component of the incident velocity by means of an experimentally determined coefficient of restitution e, where $0 \le e \le 1$. The Poisson approach divides the impact into compression and decompression phases, and relates the impulse in the restitution phase to the impulse in the compression phase.

Remark 6.2.2. It could happen that impulsive constraints and impulsive forces were present at the same time. For example, in the collision between a rigid lamina and an immobile plane surface, we must take into account not only the normal component of the contact force, but also the friction force associated with the contact. It is not innocuous the way the friction is entered into the picture. In fact, the Newton and Poisson approaches have been revealed to be physically inconsistent in certain situations. On the one hand, Newton approach can show energy gains [116, 229]. On the other hand, Poisson's rule is not satisfactory since non-frictional dissipation does not vanish for perfectly elastic impacts [39, 229]. This surprising consequence of the impact laws is only present when the velocity along the impact surface (slip) stops or reverses during collision, due precisely to the friction. Stronge [229, 230] proposed a new energetically consistent hypothesis for rigid body collisions with slip and friction. It should be noticed that the three approaches are equivalent if slip does not stop during collision and in the perfectly inelastic case ($e = 0$).

Recently, a new Newton-style model of partly elastic impacts has been proposed [227] which, interestingly, always dissipates energy, unlike the classical formulation of the Newtonian approach discussed in [229].

In the frictionless case, one can prove the following

Theorem 6.2.3 (Carnot's theorem [102, 207]). *The energy change due to impulsive constraints is always a loss of energy.*

6.3 Generalized constraints

RECALL the situation in Section 6.1. We have a mechanical system subject to a set of constraints given by a generalized codistribution $\mathcal{D}$ on

Q. We want to explore the behavior of the system when its motion reaches a singular point of the codistribution, where indeed the number of constraint functions changes suddenly.

Consider a trajectory of the system, $q(t)$, which reaches a singular point at time t_0, i.e. $q(t_0) \in S$, such that $q((t_0 - \epsilon, t_0)) \subset R$ and $q((t_0, t_0 + \epsilon)) \subset R$ for sufficiently small $\epsilon > 0$. The motion along the trajectory $q(t)$ is governed by the following equation, which is, as in the impulsive case, an integral writing of Newton's second law, to consider possible finite jump discontinuities in the velocities (or the momenta). That is, on any interval $t \leq t' < \infty$

$$p_A(t') - p_A(t) = \int_t^{t'} F_A(\tau) d\tau \,, \tag{6.6}$$

at each component, where F is the resultant of all the forces action on the trajectory $q(t)$. In our case, the unique forces acting are the constraint reaction forces.

The nature of the force can become impulsive because of the change of rank of the codistribution $\mathcal{D}$. We summarize the situations that can be found in Table 6.1. On entering the singular set, the rank of the codistribution $\mathcal{D}$ at the singular point $q(t_0)$ can be the same as at the preceding points (Case 1) or can be lower (Cases 2 and 3). In these two latter situations, the constraints have collapsed at $q(t_0)$ and this induces a finite jump in the constraint force. As the magnitude of the force is not infinite, there is no abrupt change in the momentum. Consequently, in all cases, we find no momentum jumps on entering the singular set.

	$q(t_0 - \epsilon)$: pre-points	$q(t_0)$: singular point	$q(t_0 + \epsilon)$: post-points
Case 1	$\rho = r$	$\rho_0 = r_0 = r$	$\rho > r$
Case 2	$\rho = r$	$\rho_0 = r_0 < r$	$\rho = r_0$
Case 3	$\rho = r$	$\rho_0 = r_0 < r$	$\rho > r_0$

Table 6.1. Possible cases. The rank of $\mathcal{D}$ is denoted by ρ.

On leaving the singular set, the rank of $\mathcal{D}$ at the posterior points can be the same as at $q(t_0)$ (Case 2) or can be higher (Cases 1 and 3). In Case 2 nothing special occurs. In Cases 1 and 3, the trajectory must satisfy, immediately after the point $q(t_0)$, *additional constraints* which were not present before. It is in this sense that we affirm that the constraint force can become impulsive: if the motion which passes through the singular set and tries to enter the regular one again does not satisfy the new constraints, then it experiences a jump of its momentum, due to the presence of the constraint force. In this way, the new values of the momentum satisfy the constraints. But one has to be careful: the impulsive force will act just on leaving S, on the regular set.

Consequently, we must take into account the virtual displacements associated with the posterior regular points. The underlying idea of the mathematical derivation of the momentum jumps in Section 6.3.1 is the following: take an infinitesimal posterior point $q(t)$ to $q(t_0)$, forget for a moment the presence of the constraints on the path $q((t_0, t))$ and derive the momentum jump at $q(t)$ due to the appearance of the additional constraints. Afterwards, make a limit process $t \to t_0$, canceling out the interval (t_0, t) where we 'forgot' the constraints. In any case, we will make the convention that the jump happens at $q(t_0)$.

We illustrate the above discussion in the following example.

Example 6.3.1. Consider a particle in the plane subject to the constraints imposed by the generalized codistribution in Example 2.2.2. The Lagrangian function is

$$L = \frac{1}{2}m(\dot{x}^2 + \dot{y}^2).$$

On the half-plane $R_1 = \{x < 0\}$ the codistribution is zero and the motion is free. Consequently the trajectories are

$$x = \dot{x}_0 t + x_0, \quad y = \dot{y}_0 t + y_0.$$

If the particle starts its motion with initial conditions $x_0 = -1$, $y_0 = 1$, $\dot{x}_0 = 1$, $\dot{y}_0 = 0$, after a time 1, it reaches the singular set $S = \{x = 0\}$. If the motion crosses the y-axis, something abrupt occurs on entering the half-plane $R_2 = \{x > 0\}$, where the codistribution is no longer zero and, indeed, imposes the additional constraint $\dot{x} = \dot{y}$ (Case 1). We know that the integral manifolds of $\mathcal{D}$ on R_2 are half-lines of slope 1, so the particle suffers a finite jump in the velocity on going through the singular part in order to adapt its motion to the prescribed direction (Figure 6.1).

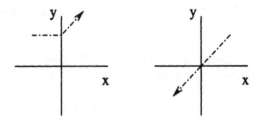

Fig. 6.1. Possible trajectories in Example 6.3.1

If, on the contrary, the particle starts on R_2, say with initial conditions $x_0 = 1$, $y_0 = 1$, $\dot{x}_0 = -1$, $\dot{y}_0 = -1$, after a certain time, it reaches the set

S. On crossing it, nothing special happens, because the particle finds less constraints to fulfill, indeed, there are no constraints (Case 2). Its motion on R_1 is free, on a straight line of slope 1 and with constant velocity equal to the one at the singular point of crossing (Figure 6.1).

6.3.1 Momentum jumps

Now, we derive a formula, strongly inspired by the theory of impulsive motion, for the momentum jumps which can occur due to the changes of rank of the codistribution $\mathcal{D}$ in Cases 1 and 3.

At $q(t_0)$ we define the following vector subspaces of $T^*_{q(t_0)}Q$

$$\mathcal{D}^-_{q(t_0)} = \{\alpha \in T^*_{q(t_0)}Q \mid \exists \tilde{\alpha} : (t_0 - \epsilon, t_0) \to T^*Q, \tilde{\alpha}(t) \in \mathcal{D}_{q(t)}, \lim_{t \to t_0^-} \tilde{\alpha}(t) = \alpha\},$$

$$\mathcal{D}^+_{q(t_0)} = \{\alpha \in T^*_{q(t_0)}Q \mid \exists \tilde{\alpha} : (t_0, t_0 + \epsilon) \to T^*Q, \tilde{\alpha}(t) \in \mathcal{D}_{q(t)}, \lim_{t \to t_0^+} \tilde{\alpha}(t) = \alpha\}.$$

From the definition of $\mathcal{D}^-_{q(t_0)}$ and $\mathcal{D}^+_{q(t_0)}$ we have that

$$(\mathcal{D}^-_{q(t_0)})^\perp = \lim_{t \to t_0^-} (\mathcal{D}_{q(t)})^\perp \quad \text{and} \quad (\mathcal{D}^+_{q(t_0)})^\perp = \lim_{t \to t_0^+} (\mathcal{D}_{q(t)})^\perp$$

where $^\perp$ denotes the orthogonal complement with respect to the cometric induced by g, and the limits $(\mathcal{D}^\perp)^-$ and $(\mathcal{D}^\perp)^+$ are defined as in the case of $\mathcal{D}^-$ and $\mathcal{D}^+$.

Since $\mathcal{D}$ is a differentiable codistribution, then $\mathcal{D}_{q(t_0)} \subseteq \mathcal{D}^-_{q(t_0)}$ and $\mathcal{D}_{q(t_0)} \subseteq \mathcal{D}^+_{q(t_0)}$. Along the interval $[t_0, t]$, we have

$$p_A(t) - p_A(t_0) = \int_{t_0}^t F_A(\tau)\, d\tau .$$

Multiplying by the virtual displacements at the point $q(t)$ and summing in A, we obtain

$$\sum_{A=1}^n (p_A(t) - p_A(t_0)) \cdot \delta q^A_{|q(t)} = \sum_{A=1}^n \int_{t_0}^t F_A(\tau) d\tau \cdot \delta q^A_{|q(t)}. \tag{6.7}$$

Since we are dealing with ideal constraints, the virtual work vanishes, i.e.

$$\sum_{A=1}^n \int_{t_0}^t F_A(\tau) \delta q^A_{|q(\tau)} \, d\tau = 0 .$$

If t is near t_0, then τ is close to t, and $q(\tau)$ is near $q(t)$, so δq remains both nearly constant and nearly equal to its value at time t throughout the time interval $(t_0, t]$, in the same way we exposed in Section 6.2. Therefore,

$$\sum_{A=1}^{n} \left(\int_{t_0}^{t} F_A(\tau)\, d\tau \right) \cdot \delta q^A_{|q(t)} = \sum_{A=1}^{n} \int_{t_0}^{t} F_A(\tau)\delta q^A_{|q(\tau)}\, d\tau + O(t - t_0) = O(t - t_0).$$

Consequently, equation (6.7) becomes

$$\sum_{A=1}^{n} (p_A(t) - p_A(t_0)) \cdot \delta q^A_{|q(t)} = O(t - t_0). \tag{6.8}$$

Taking limits we obtain $\lim_{t \to t_0^+} \left(\sum_{A=1}^{n}(p_A(t) - p_A(t_0)) \cdot \delta q^A_{|q(t)} \right) = 0$, which implies

$$\sum_{A=1}^{n}(p_A(t_0)_+ - p_A(t_0)) \lim_{t \to t_0^+} \delta q^A_{|q(t)} = 0, \tag{6.9}$$

or, in other words,

$$(p_A(t_0)_+ - p_A(t_0))\, dq^A \in \lim_{t \to t_0^+} \mathcal{D}_{q(t)} = \mathcal{D}^+_{q(t_0)}. \tag{6.10}$$

Conclusion: Following the above discussion, we will deduce the existence of jump of momenta depending on the relation between $\mathcal{D}^-_{q(t_0)}$ and $\mathcal{D}^+_{q(t_0)}$. The possible cases are shown in Table 6.2.

$\mathcal{D}^+_{q(t_0)} \subseteq \mathcal{D}^-_{q(t_0)}$	there is no jump of momenta
$\mathcal{D}^+_{q(t_0)} \not\subseteq \mathcal{D}^-_{q(t_0)}$	possibility of jump of momenta

Table 6.2. The two cases that may arise in studying the jump of momenta.

In the second case in Table 6.2, we have a jump of momenta if the 'pre-impact' momentum $p(t_0)_- = p(t_0)$ does not satisfy the constraints imposed by $\mathcal{D}^+_{q(t_0)}$, i.e.

$$p_A(t_0)_- dq^A \notin (\mathcal{D}^+_{q(t_0)})^\perp.$$

Our proposal for the equations which determine the jump is then

$$\begin{cases} (p_A(t_0)_+ - p_A(t_0)_-)\, dq^A \in \mathcal{D}^+_{q(t_0)} \\ p_A(t_0)_+\, dq^A \in (\mathcal{D}^+_{q(t_0)})^\perp. \end{cases}$$

The first equation has been derived above (cf. (6.10)) from the generalized writing of Newton's second law (6.1). The second equation simply encodes the fact that the 'post-impact' momentum must satisfy the new constraints imposed by $\mathcal{D}^+_{q(t_0)}$.

Remark 6.3.2. In Cases 1 and 3, the virtual displacements at $q(t_0)$ are radically different from the ones at the regular posterior points, because of the change of rank. From a dynamics point of view, these are the 'main' ones, since it is on the regular set where an additional constraint reaction force acts. As we have seen, the momentum jump happens on just leaving S, due to the presence of this additional constraint force on the regular set. Note that with the procedure we have just derived, we are taking into account precisely the virtual displacements at the regular posterior points, and not those of $q(t_0)$. If we took the virtual displacements at $q(t_0)$ and multiply by them in (6.7), we would obtain non-consistent jump conditions. This is easy to see, for instance, in Example 6.3.1.

An explicit derivation of the momentum jumps for Cases 1 and 3 would be as follows. Let m be the maximum between $\rho = r$, the rank at the regular preceding points, and $\rho = s$, the rank at the regular posterior points. Then there exists a neighborhood U of $q(t_0)$ and 1-forms $\omega_1, \ldots, \omega_m$ such that

$$\mathcal{D}_q = \text{span} \{\omega_1(q), \ldots, \omega_m(q)\} , \ \forall q \in U .$$

Let us suppose that $\omega_1, \ldots, \omega_s$ are linearly independent at the regular posterior points (if not, we reorder them). Obviously, at $q(t_0)$, these s 1-forms are linearly dependent. In the following, we will denote by ω_i the 1-form evaluated at $q(t)$, (t time immediately posterior to t_0) i.e. $\omega_i \equiv \omega_i(q(t))$, in order to simplify notation.

Since the Lagrangian is of the form $L = T - V$, where T is the kinetic energy of the Riemannian metric g, we have that

$$\omega_{jA}(q(t))\dot{q}^A(t) = \sum_{A,B} \omega_{jA} g^{AB} p_B(t) = 0 , \ j = 1, \ldots, s . \tag{6.11}$$

Using the metric g we have the decomposition $T_q^*Q = \mathcal{D}_q \oplus \mathcal{D}_q^\perp$, for each $q \in Q$. The two complementary projectors associated with this decomposition are

$$\mathcal{P}_q : T_q^*Q \longrightarrow \mathcal{D}_q^\perp , \quad \mathcal{Q}_q : T_q^*Q \longrightarrow \mathcal{D}_q .$$

Let C be the symmetric matrix with entries $C_{ij} = \omega_{iA} g^{AB} \omega_{jB}$, or $C = \omega g^{-1}\omega^T$ with the obvious notations. The projector $\mathcal{P}_q$ is given by $\mathcal{P}_q(\alpha_q) = \alpha_q - C^{ij}\alpha_q(Z_i)\omega_j$, for $\alpha_q \in T^*Q$, where

$$Z_i = g^{AB}\omega_{iB} \left(\frac{\partial}{\partial q^A}\right)_q ,$$

and C^{ij} are the entries of the inverse matrix of C.

By definition

$$p_A(t_0)_+ dq^A|_{q(t_0)} = \lim_{t \to t_0^+} (p_A(t) dq^A|_{q(t)}) .$$

From (6.11), $\mathcal{P}_{q(t)}(p_A(t) dq^A|_{q(t)}) = p_A(t) dq^A|_{q(t)}$ and then

$$p_A(t_0)_+ dq^A|_{q(t_0)} = \left(\lim_{t \to t_0^+} \mathcal{P}_{q(t)} \right) (p_A(t_0)_+ dq^A|_{q(t_0)}) \in (\mathcal{D}_{q(t_0)}^+)^\perp . \qquad (6.12)$$

Combining (6.10) and (6.12), we obtain

$$p_A(t_0)_+ dq^A|_{q(t_0)} = \left(\lim_{t \to t_0^+} \mathcal{P}_{q(t)} \right) [p_A(t_0)_- dq^A|_{q(t_0)}] .$$

In coordinates, this can be expressed as

$$p_A(t_0)_+ = p_A(t_0)_- - \lim_{t \to t_0^+} \left(\sum_{i,j,A,B} C^{ij} \omega_{jB} g^{BC} \omega_{iA} \right) \Big|_{q(t)} p_C(t_0)_- , \qquad (6.13)$$

for $A = 1, \ldots, n$. Equation (6.13) can be written in matrix form as follows

$$p(t_0)_+ = \left(Id - \lim_{t \to t_0^+} (\omega^T C^{-1} \omega g^{-1})_{|q(t)} \right) p(t_0)_- . \qquad (6.14)$$

With the derived jump rule, we are able to prove the following version of Carnot's theorem for generalized constraints.

Theorem 6.3.3 (Carnot's theorem for generalized constraints). *The kinetic energy will only decrease by the use of the jump rule (6.14).*

Proof. We have that

$$g\left(p(t_0)_+, p(t_0)_+\right) = g\left(\left(\lim_{t \to t_0^+} \mathcal{P}_{q(t)} \right) p(t_0)_-, p(t_0)_- - \left(\lim_{t \to t_0^+} \mathcal{Q}_{q(t)} \right) p(t_0)_- \right)$$

$$= g\left(\left(\lim_{t \to t_0^+} \mathcal{P}_{q(t)} \right) p(t_0)_-, p(t_0)_- \right)$$

$$= g\left(p(t_0)_-, p(t_0)_-\right) - g\left(\left(\lim_{t \to t_0^+} \mathcal{Q}_{q(t)} \right) p(t_0)_-, p(t_0)_- \right) .$$

Since

$$g\left(\left(\lim_{t \to t_0^+} \mathcal{Q}_{q(t)} \right) p(t_0)_-, p(t_0)_- \right) =$$

$$= g\left(\left(\lim_{t \to t_0^+} \mathcal{Q}_{q(t)} \right) p(t_0)_-, \left(\lim_{t \to t_0^+} \mathcal{Q}_{q(t)} \right) p(t_0)_- \right) \geq 0 ,$$

we can conclude that $\frac{1}{2} g(p(t_0)_+, p(t_0)_+) \leq \frac{1}{2} g(p(t_0)_-, p(t_0)_-)$. $\qquad \square$

In fact, the jump rule (6.14) has the following alternative interpretation. Let $p \in \mathcal{D}^+_{q(t_0)}$ and observe that

$$g(p - p(t_0)_-, p - p(t_0)_-) = g(p(t_0)_-, p(t_0)_-) + g(p, p - 2p(t_0)_-)$$

$$= g(p(t_0)_-, p(t_0)_-) + g\left(p, p - 2\left(\lim_{t \to t_0^+} \mathcal{P}_{q(t)}\right) p(t_0)_-\right).$$

Now, note that the covector

$$p = \left(\lim_{t \to t_0^+} \mathcal{P}_{q(t)}\right) p(t_0)_- \in \mathcal{D}^+_{q(t_0)}$$

is such that the expression $g(p - p(t_0)_-, p - p(t_0)_-)$ is minimized among all the covectors belonging to $\mathcal{D}^+_{q(t_0)}$. Therefore, the derived jump rule (6.14) can be stated as follows:

> the 'post-impact' momenta $p(t_0)_+$ is such that the kinetic energy corresponding to the difference of the 'pre-impact' and 'post-impact' momentum is minimized among all the covectors satisfying the constraints.

This is an appropriate version for generalized constraints of the well known jump rule for perfectly inelastic collisions [181]. This is even more clear in the holonomic case, as is shown in Section 6.3.2.

Remark 6.3.4. So far, we have been dealing with impulsive constraints. More generally, we can consider the presence of external impulsive forces associated with external inputs or controls. Then, equation (6.10) must be modified as follows

$$(p_A(t_0)_+ - p_A(t_0)_- - P'_A(t_0)) \, dq^A_{|q(t_0)} \in \lim_{t \to t_0^+} \mathcal{D}_{q(t)} = \mathcal{D}^+_{q(t_0)}, \qquad (6.15)$$

where $P'_A(t_0)$, $1 \le A \le n$, are the external impulses at time t_0. Observe that if $q(t_0)$ is a regular point then

$$p_A(t_0)_+ = p_A(t_0)_- + P'_A(t_0) - \sum_{i,j,A,B} C^{ij} \omega_{jB} g^{BC} \omega_{iA} P'_C(t_0) \qquad (6.16)$$

and, if $q(t_0)$ is a singular point, we have

$$p_A(t_0)_+ = p_A(t_0)_- + P'_A(t_0)$$

$$- \lim_{t \to t_0^+} \left(\sum_{i,j,A,B} C^{ij} \omega_{jB} g^{BC} \omega_{iA}\right)\Bigg|_{q(t)} (p_C(t_0)_- + P'_C(t_0)). \qquad (6.17)$$

6.3.2 The holonomic case

We show in this section a meaningful interpretation of the proposed jump rule (6.14) in case the codistribution $\mathcal{D}$ is partially integrable (cf. Section 2.2).

Let us consider a trajectory $q(t) \in Q$ which reaches a singular point $q(t_0) \in S$ and falls in either Case 1 or Case 3. Since $Q = \bar{R}$, we have that $q(t_0) \in \bar{L}$, where L is the leaf of $\mathcal{D}$ which contains the regular posterior points of the trajectory $q(t)$. On leaving $q(t_0)$, we have seen that the trajectory suffers a finite jump in its momentum in order to satisfy the constraints imposed by $\mathcal{D}$, which in this case implies that the trajectory after time t_0 belongs to the leaf L. Consequently, the jump can be interpreted as a perfectly inelastic collision against the 'wall' represented by the leaf L!

Let us see it revisiting Example 6.3.1.

Example 6.3.5. Consider again the situation in Example 6.3.1. If the motion of the particle starts on the left half-plane going towards the right one, then it is easy to see that $\mathcal{D}^-_{(0,y)} = \{0\}$ and $\mathcal{D}^+_{(0,y)} = \text{span}\{dx - dy\}$. As $\mathcal{D}^-_{(0,y)} \not\subseteq \mathcal{D}^-_{(0,y)}$, a jump of momenta is possible. In fact, if the 'pre-impact' velocity $(\dot{x}_0, \dot{y}_0)$ does not satisfy $\dot{x}_0 = \dot{y}_0$, the jump occurs and is determined by $\Delta v(t_0) \in \mathcal{D}^+_{(0,y)}$ and $\dot{x}(t_0^+) = \dot{y}(t_0^+)$. Consequently, we obtain

$$\dot{x}(t_0^+) = \frac{\dot{x}_0 + \dot{y}_0}{2}, \quad \dot{y}(t_0^+) = \frac{\dot{x}_0 + \dot{y}_0}{2}.$$

We would have obtained the same result if we had considered that our particle hits, in a perfectly inelastic collision, against the 'wall' represented by the half-line of slope 1 contained in $\{x > 0\}$ passing through the point $(0, y)$.

If the particle starts on the right half-plane towards the left one, the roles are reversed and $\mathcal{D}^-_{(0,y)} = \text{span}\{dx - dy\}$, $\mathcal{D}^+_{(0,y)} = \{0\}$. We have that $\mathcal{D}^+_{(0,y)} \subseteq \mathcal{D}^-_{(0,y)}$ and therefore, according to Table 6.2, there is no jump.

6.4 Examples

NEXT, we are going to develop two examples illustrating the above discussion. First, we treat a variation of the classical example of the rolling sphere [188, 207]. Secondly, we take one example from [65].

6.4.1 The rolling sphere

Consider the example of a homogeneous sphere rolling on a table which remains still (that is, the example treated in Section 3.2.2 with $\Omega(t) \equiv 0$). The configuration space is $Q = \mathbb{R}^2 \times SO(3)$: (x, y) denotes the position of the center of the sphere and (φ, θ, ψ) denote the Eulerian angles.

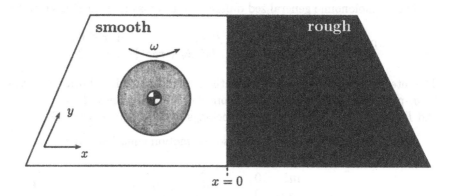

Fig. 6.2. The rolling sphere on a 'special' surface

Let us suppose that the plane is smooth if $x < 0$ and absolutely rough if $x > 0$ (see Figure 6.2). On the smooth part, we assume that the motion of the ball is free, that is, the sphere can slip. But if it reaches the rough half-plane, the sphere begins rolling without slipping, because of the presence of the constraints imposed by the roughness. We are interested in knowing the trajectories of the sphere and, in particular, the possible changes in its dynamics because of the crossing from one half-plane to the other.

The kinetic energy of the sphere is

$$T = \frac{1}{2} \left(\dot{x}^2 + \dot{y}^2 + k^2 (\omega_x^2 + \omega_y^2 + \omega_z^2) \right) ,$$

where ω_x, ω_y and ω_z are the angular velocities with respect to the inertial frame, given by

$$\omega_x = \dot{\theta} \cos \psi + \dot{\varphi} \sin \theta \sin \psi ,$$
$$\omega_y = \dot{\theta} \sin \psi - \dot{\varphi} \sin \theta \cos \psi ,$$
$$\omega_z = \dot{\varphi} \cos \theta + \dot{\psi} .$$

The potential energy is not considered here since it is constant.

The condition of rolling without sliding of the sphere when $x > 0$ implies that the point of contact of the sphere and the plane has zero velocity

$$\phi^1 = \dot{x} - r\omega_y = 0, \quad \phi^2 = \dot{y} + r\omega_x = 0,$$

where r is the radius of the sphere.

Following the classical procedure, we introduce quasi-coordinates 'q_1', 'q_2' and 'q_3' such that '$\dot{q}_1$'$= \omega_x$, '$\dot{q}_2$'$= \omega_y$ and '$\dot{q}_3$'$= \omega_z$ (see Section 3.2.2).

The nonholonomic generalized differentiable codistribution $\mathcal{D}$ is given by

$$\mathcal{D}_{(x,y,\phi,\theta,\psi)} = \begin{cases} \{0\}, & \text{if } x \leq 0, \\ \text{span}\,\{dx - rdq^2, dy + rdq^1\}, & \text{if } x > 0. \end{cases}$$

The intersection of the regular set of the generalized codistribution and the (x, y)-plane has two connected components, the half-planes $R_1 = \{x < 0\}$ and $R_2 = \{x > 0\}$. The line $\{x = 0\}$ belongs to the singular set of $\mathcal{D}$.

On R_1 the codistribution is zero, so the motion equations are

$$\begin{aligned} m\ddot{x} &= 0, \\ m\ddot{y} &= 0, \end{aligned} \qquad \begin{aligned} mk^2\dot{\omega}_x &= 0, \\ mk^2\dot{\omega}_y &= 0, \\ mk^2\dot{\omega}_z &= 0. \end{aligned} \tag{6.18}$$

On R_2 we have to take into account the constraints to obtain the following equations of motion

$$\begin{aligned} m\ddot{x} &= \lambda_1, \\ m\ddot{y} &= \lambda_2, \end{aligned} \qquad \begin{aligned} mk^2\dot{\omega}_x &= r\lambda_2, \\ mk^2\dot{\omega}_y &= -r\lambda_1, \\ mk^2\dot{\omega}_z &= 0, \end{aligned} \tag{6.19}$$

with the constraint equations $\dot{x} - r\omega_y = 0$ and $\dot{y} + r\omega_x = 0$. One can compute the Lagrange multipliers by the algebraic procedure described in Section 3.4.1.

Suppose that the sphere starts its motion at a point of R_1 with the following initial conditions at time $t = 0$: $x_0 < 0$, y_0, $\dot{x}_0 > 0$, $\dot{y}_0$, $(\omega_x)_0$, $(\omega_y)_0 > 0$ and $(\omega_z)_0$. Integrating equations (6.18) we have that if $x(t) < 0$

$$\begin{aligned} x(t) &= \dot{x}_0 t + x_0, \\ y(t) &= \dot{y}_0 t + y_0, \end{aligned} \qquad \begin{aligned} \omega_x(t) &= (\omega_x)_0, \\ \omega_y(t) &= (\omega_y)_0, \\ \omega_z(t) &= (\omega_z)_0. \end{aligned} \tag{6.20}$$

At time $\bar{t} = -x_0/\dot{x}_0$ the sphere finds the rough surface of the plane, where the codistribution is no longer zero and it is suddenly forced to roll without sliding (Case 1). Following the discussion in Section 6.3.1, we calculate the instantaneous change of velocity (momentum) at $x = 0$.

First of all we compute the matrix $\mathcal{C}$,

$$\mathcal{C} = \begin{pmatrix} 1 & 0 & 0 & -r & 0 \\ 0 & 1 & r & 0 & 0 \end{pmatrix} \begin{pmatrix} 1 & 0 & 0 & 0 & 0 \\ 0 & 1 & 0 & 0 & 0 \\ 0 & 0 & k^{-2} & 0 & 0 \\ 0 & 0 & 0 & k^{-2} & 0 \\ 0 & 0 & 0 & 0 & k^{-2} \end{pmatrix} \begin{pmatrix} 1 & 0 \\ 0 & 1 \\ 0 & r \\ -r & 0 \\ 0 & 0 \end{pmatrix}$$

$$= (1 + r^2 k^{-2}) \begin{pmatrix} 1 & 0 \\ 0 & 1 \end{pmatrix} .$$

Next, a direct computation shows that the projector $\mathcal{P}$ does not depend on the base point

$$\mathcal{P} = \begin{pmatrix} \frac{r^2}{r^2+k^2} & 0 & 0 & \frac{r}{r^2+k^2} & 0 \\ 0 & \frac{r^2}{r^2+k^2} & -\frac{r}{r^2+k^2} & 0 & 0 \\ 0 & \frac{-rk^2}{r^2+k^2} & \frac{k^2}{r^2+k^2} & 0 & 0 \\ \frac{rk^2}{r^2+k^2} & 0 & 0 & \frac{k^2}{r^2+k^2} & 0 \\ 0 & 0 & 0 & 0 & 1 \end{pmatrix} .$$

Therefore, we have

$$(p_x)_+ = \frac{r^2(p_x)_0 + r(p_2)_0}{r^2 + k^2} , \qquad (p_1)_+ = \frac{-rk^2(p_y)_0 + k^2(p_1)_0}{r^2 + k^2} ,$$

$$(p_y)_+ = \frac{r^2(p_y)_0 - r(p_1)_0}{r^2 + k^2} , \qquad (p_2)_+ = \frac{rk^2(p_x)_0 + k^2(p_2)_0}{r^2 + k^2} ,$$

$$(p_3)_+ = (p_3)_0 .$$

Now, using the relation between the momenta and the quasi-velocities

$$p_x = \dot{x} , p_y = \dot{y} , p_1 = k^2 \omega_x , p_2 = k^2 \omega_y , p_3 = k^2 \omega_z ,$$

we deduce that

$$\dot{x}_+ = \frac{r^2 \dot{x}_0 + rk^2(\omega_y)_0}{r^2 + k^2} , \qquad (\omega_x)_+ = \frac{-r\dot{y}_0 + k^2(\omega_x)_0}{r^2 + k^2} ,$$

$$\dot{y}_+ = \frac{r^2 \dot{y}_0 - rk^2(\omega_x)_0}{r^2 + k^2} , \qquad (\omega_y)_+ = \frac{r\dot{x}_0 + k^2(\omega_y)_0}{r^2 + k^2} , \qquad (6.21)$$

$$(\omega_z)_+ = (\omega_z)_0 .$$

Finally, integrating equations (6.19) at time $\bar{t} = -x_0/\dot{x}_0$ with initial conditions given by (6.21) we obtain that if $t > \bar{t}$,

$$x(t) = \frac{r^2 \dot{x}_0 + rk^2(\omega_y)_0}{r^2 + k^2}(t - \bar{t}),$$

$$y(t) = \frac{r^2 \dot{y}_0 - rk^2(\omega_x)_0}{r^2 + k^2}(t - \bar{t}) + \dot{y}_0 \bar{t} + y_0,$$

$$\omega_x(t) = \frac{-r\dot{y}_0 + k^2(\omega_x)_0}{r^2 + k^2},$$

$$\omega_y(t) = \frac{r\dot{x}_0 + k^2(\omega_y)_0}{r^2 + k^2},$$

$$\omega_z(t) = (\omega_z)_0.$$

$$(6.22)$$

6.4.2 Particle with constraint

This example is taken from [65]. Let us consider the motion of a particle of unit mass in $\mathbb{R}^3$ subject to the following constraint

$$\phi = (y^2 - x^2 - z)\dot{x} + (z - y^2 - xy)\dot{y} + x\dot{z} = 0.$$

In addition, let us assume that there is a central force system centered at the point $(0, 0, 1)$ with force field given by

$$F = -xdx - ydy + (1 - z)dz.$$

Then, the Lagrangian function of the particle is

$$L = T - V = \frac{1}{2}(\dot{x}^2 + \dot{y}^2 + \dot{z}^2 + x^2 + y^2 + z^2 - 2z),$$

and the constraint defines a generalized differentiable codistribution $\mathcal{D}$, whose singular set is $S = \{(x, y, z) : x = 0, z = y^2\}$.

On R, the regular set of $\mathcal{D}$, the dynamics can be computed following the standard symplectic procedure described in Section 3.4.1 to obtain $X = \Gamma_L + \lambda Z$, where

$$\Gamma_L = \dot{x}\frac{\partial}{\partial x} + \dot{y}\frac{\partial}{\partial y} + \dot{z}\frac{\partial}{\partial z} - x\frac{\partial}{\partial \dot{x}} - y\frac{\partial}{\partial \dot{y}} - (z - 1)\frac{\partial}{\partial \dot{z}},$$

$$Z = -\left((y^2 - x^2 - z)\frac{\partial}{\partial \dot{x}} + (z - y^2 - xy)\frac{\partial}{\partial \dot{y}} + x\frac{\partial}{\partial \dot{z}}\right),$$

and λ is given by

$$\lambda = -\frac{\Gamma_L(\phi)}{Z(\phi)} = \frac{-2x\dot{x}^2 + y\dot{y}\dot{x} - 2y\dot{y}^2 - x\dot{y}^2 + \dot{y}\dot{z} + x^3 + y^3 - yz + x}{(y^2 - x^2 - z)^2 + (z - y^2 - xy)^2 + x^2}.$$

$$(6.23)$$

Consequently, the motion equations on R are

$$m\ddot{x} + x = \lambda(y^2 - x^2 - z),$$
$$m\ddot{y} + y = \lambda(z - y^2 - xy), \tag{6.24}$$
$$m\ddot{z} + z - 1 = \lambda x,$$

together with the constraint equation $\phi = 0$.

From the discussions of [65], we know that in this case there is an integral surface, C, of the constraint ϕ, that is, a surface on which all motions satisfy the constraint. This surface is

$$C = \{(x, y, x) : z - x^2 - y^2 + xy = 0\}.$$

Note that $S \subset C$. Therefore, if a motion takes place on the cone-like surface C, it is confined to stay on this critical surface, unless it reaches a singular point. In this case, the space of allowable motions is suddenly increased (in fact, $T\mathbb{R}^3$), and the motion can 'escape' from C. In addition, this proves that the unique way to pass from one point of the exterior of the C to the interior, or vice versa, is through the singular set S.

In particular, we are interested in knowing

1. Is there any trajectory satisfying equations (6.24) which passes through the singular set?

2. if so, which are the possible momentum jumps due to the changes in the rank of the codistribution $\mathcal{D}$?

So far, we do not know an answer for the question of the existence of a motion of (6.24) crossing S. It seems that on approaching a singular point, the constraint force can become increasingly higher (cf. equation (6.23)). Consequently, this force possibly 'disarranges' the approaching of the motion to S. Numerical simulations are quite useless in this task, because of the special nature of the problem: the hard restriction given by the fact that a motion crossing the cone-like surface C must do it through the singular part S. Indeed, the numerical simulation performed in [65] crosses the surface C through points which are not in S, which is not consistent.

Concerning the second question, let us suppose that there is a trajectory of the dynamical system (6.24), $q(t) = (x(t), y(t), z(t))$, that passes through a singular point at time t_0, i.e. $x(t_0) = 0$ and $z(t_0) = y(t_0)^2$. The rank of the codistribution $\mathcal{D}$ at the immediately preceding and posterior points is 1, meanwhile at $q(t_0)$ it is 0 (Case 3). So, the change in the rank of $\mathcal{D}$ could induce a possible jump of the momentum.

A direct computation shows that the projector $\mathcal{P}$ depends explicitly on the base point $q \in Q$. Equivalently, we have that $\mathcal{D}_{q(t_0)}^+$ depends strongly on the trajectory $q(t)$. In fact, taking two curves $q_1(t)$, $q_2(t)$ passing through $q(t_0)$ at time t_0, and satisfying $x_1(t) \ll z_1(t) - y_1^2(t)$ and $z_2(t) - y_2^2(t) \ll x_2(t)$ when $t \to t_0^+$ respectively, one can easily see that $\mathcal{D}_{q_1(t_0)}^+ \neq \mathcal{D}_{q_2(t_0)}^+$ (the expression $f(t) \ll g(t)$ when $t \to t_0^+$ means that $\lim_{t \to t_0^+} f(t)/g(t) = 0$).

Consequently, we are not able to give an answer to question (ii) (in case the first one was true) unless we assume some additional information: for example, that the balance between $x(t)$ and $z(t) - y^2(t)$ is the same for $t \to t_0^-$ and $t \to t_0^+$. In such a case, $\mathcal{D}_{q(t_0)}^- = \mathcal{D}_{q(t_0)}^+$ and we would conclude that there is no jump. In mechanical phenomena of the type sliding-rolling, as the ones studied in Section 6.4.1 this kind of 'indeterminacy' will not occur in general. Recently, it has been brought to our attention ongoing research efforts on singular vector fields [87] which might be of help in the further analysis of this case.

7 Nonholonomic integrators

IN this chapter, we address the problem of constructing integrators for mechanical systems with nonholonomic constraints. This problem has been stated in a number of recent works [59, 256], including the presentation of open problems in symplectic integration given in [176]. The study of systems subject to holonomic constraints is a well established topic of research in the literature. For instance, the popular Verlet algorithm for unconstrained mechanical systems was adapted to handle holonomic constraints, resulting in the Shake algorithm [209] and the Rattle algorithm [5]. The case of general Hamiltonian systems subject to holonomic constraints has also been studied [106, 205, 206]. A different approach, based on the Dirac theory of constraints, may be found in [133]. Energy-momentum integrators derived from discrete directional derivatives and discrete versions of Hamiltonian mechanics have also been recently adapted to deal with holonomic constraints [89, 90].

Our approach to this topic follows the developments on variational integrators started by Veselov [183, 246, 247]. The underlying idea to variational integrators comes from the observation that much of the geometric properties of the continuous flow of the Euler-Lagrange equations can be explained by the fact that they are derived from a variational principle. Hence, instead of discretizing directly the equations of motion, one discretizes the Lagrangian itself, and derives the discrete equations from this discrete Lagrangian through a discrete principle. In this way, one expects to obtain discrete equations that also enjoy interesting geometric properties.

The chapter is organized as follows. In Section 7.1 we briefly introduce some basic notions and results related with mechanical integrators. Section 7.2 presents a brief review of variational integrators. In Section 7.3 we propose a discrete version of the Lagrange-d'Alembert principle. We show that, when the constraints are holonomic, this discrete principle leads us to recover variational integrators, so that nonholonomic integrators can be understood as its generalization. We discuss several ways to construct them in Section 7.4 and we study the geometric invariance properties of the proposed algorithm in Section 7.5. Section 7.6 contains two examples illustrating the performance of the nonholonomic integrator.

7.1 Symplectic integration

IN this section, we briefly introduce some common notions and results from the literature on geometric integration. The interested reader is referred to [158, 160, 210] for more thorough expositions.

An algorithm is a collection of maps $F_h : P \longrightarrow P$ depending smoothly on $h \in [0, h_0)$ and $z \in P$. Given a symplectic manifold (P, ω), an algorithm is said to be

- a *symplectic integrator* if each $F_h : P \longrightarrow P$ is a symplectic map.
- an *energy integrator* if $E \circ F_h = E$, with the algorithm being consistent with $X = X_E$.
- a *momentum integrator* if $J \circ F_h = J$, where $J : P \longrightarrow \mathfrak{g}^*$ is the momentum map for the action of a Lie group G.

Any algorithm having one or more of these properties is called a *mechanical integrator*.

The choice of a specific integrator depends on the specific problem under consideration. For instance, in molecular dynamics simulation and planetary motions, the preservation of the symplectic form (and the associated volume) is important for long time runs, otherwise one may obtain totally inconsistent solutions. On the other hand, the exact conservation of a momentum first integral is essential to problems in attitude control in satellite dynamics simulation, since this is the basic physical principle driving the reorientation of the system.

Unfortunately, in general one cannot preserve all three at the same time,

Theorem 7.1.1 ([86]). *Assume that an algorithm for a given Hamiltonian system X_H with a symmetry group G*

- *is energy preserving, symplectic, momentum preserving and G-equivariant, and*
- *the dynamics of X_H is nonintegrable on the reduced space (in the sense that any other conserved quantity is functionally dependent on H).*

Then, the algorithm already gives the exact solution of the given problem up to a time reparameterization.

Colloquially speaking, this result means that, if the time step is fixed, obtaining an energy-symplectic-momentum integrator is the same as exactly obtaining the continuous flow. Recently, it has been shown that the construction of energy-symplectic-momentum integrators is indeed possible if one allows varying time steps [109], instead of fixed ones. As a result of Theorem 7.1.1, mechanical integrators are divided into subclasses, such as symplectic-momentum and energy-momentum.

The ways employed so far in the discovery and development of mechanical integrators include: the search among existing algorithms to find ones with special algebraic properties that make them symplectic or energy-preserving (examples are the symplectic Runge-Kutta schemes [210], the Rattle and Shake algorithms [5, 134, 209], etc.); the composition of known mechanical integrators, preserving their geometric properties, to increase the order of accuracy [259]; methods like symplectic correctors or product formulas [8, 160], etc.

Here we shall focus our attention on the variational approach to construct mechanical integrators: *variational integrators* and extensions.

7.2 Variational integrators

MECHANICAL integrators based on the Veselov discretization technique [183, 246, 247] have been studied intensively in the last years and are by now well known [34, 109, 110, 161, 256]. We briefly review here the main ideas of this approach.

Let Q be a n-dimensional configuration manifold and $L_d : Q \times Q \longrightarrow \mathbb{R}$ a smooth map playing the role of a discrete Lagrangian. The *action sum* is the map $S : Q^{N+1} \longrightarrow \mathbb{R}$ defined by

$$S = \sum_{k=0}^{N-1} L_d(q_k, q_{k+1}) \,, \tag{7.1}$$

where $q_k \in Q$ for $k \in \{0, 1, \ldots, N\}$ and k is the discrete time. The discrete variational principle states that the evolution equations extremize the action sum, given fixed end points q_0, q_N. That is, we have

$$dS = \sum_{k=1}^{N-1} (D_1 L_d(q_k, q_{k+1}) + D_2 L_d(q_{k-1}, q_k)) \, dq_k \equiv 0 \,, \tag{7.2}$$

which leads to the discrete Euler-Lagrange (DEL) equations,

$$D_1 L_d(q_k, q_{k+1}) + D_2 L_d(q_{k-1}, q_k) = 0 \,. \tag{7.3}$$

Under appropriate regularity assumptions on the discrete Lagrangian L_d (namely, that the mapping $D_1 L_d(q, \cdot) : Q \longrightarrow T_q^* Q$ is invertible $\forall q \in Q$), the DEL equations define a map $\Phi : Q \times Q \longrightarrow Q \times Q$, $\Phi(q_{k-1}, q_k) = (q_k, q_{k+1})$ which describes the discrete time evolution of the system.

Now, define the *fiber derivative* or *discrete Legendre transform* corresponding to L_d by

$$\mathcal{F}L_d : Q \times Q \longrightarrow T^*Q$$
$$(q, q') \longmapsto (q', D_2 L_d(q, q')) ,$$

and the 2-form ω_{L_d} on $Q \times Q$ by pulling back the canonical 2-form $\omega_Q = -d\Theta_Q$ from T^*Q,

$$\omega_{L_d} = \mathcal{F}L_d^*(\omega_Q) .$$

The alternative discrete fiber derivative $\tilde{\mathcal{F}}L_d(q, q') = (q, -D_1 L_d(q, q'))$ may also be used and the results obtained will be essentially unchanged. A fundamental fact is that the algorithm Φ exactly preserves the symplectic form ω_{L_d}, that is, $\Phi^*\omega_{L_d} = \omega_{L_d}$ (see [256]). Indeed, one has that

$$\Phi^*\Omega_{L_d} = -\Phi^*\mathcal{F}L_d^* d\Theta_Q = -d(\mathcal{F}L_d \circ \Phi)^*\Theta_Q$$
$$= -d(\tilde{\mathcal{F}}L_d \circ \Phi)^*\Theta_Q \stackrel{DEL}{=} -d\mathcal{F}L_d^*\Theta_Q = \Omega_{L_d} . \qquad (7.4)$$

If we further assume that the discrete Lagrangian is invariant under the action of a Lie group G on Q, one can define the associated discrete momentum map, $J_d : Q \times Q \longrightarrow \mathfrak{g}^*$ (where $\mathfrak{g}^*$ denotes the dual of the Lie algebra $\mathfrak{g}$ of G), by

$$\langle J_d(q, q'), \xi \rangle = \langle D_2 L_d(q, q'), \xi_Q(q') \rangle .$$

Here, ξ_Q denotes the fundamental vector field corresponding to the element $\xi \in \mathfrak{g}$.

A second fundamental fact is that the discrete momentum is exactly preserved by the algorithm Φ [256]. This is seen in the following sequence of steps: by the invariance of the discrete Lagrangian L_d,

$$L(\exp(s\xi)q_k, \exp(s\xi)q_{k+1}) = L(q_k, q_{k+1}) .$$

Differentiating with respect to s and setting $s = 0$ yields

$$D_1 L_d(q_k, q_{k+1})\xi_Q(q_k) + D_2 L_d(q_k, q_{k+1})\xi_Q(q_{k+1}) = 0 . \qquad (7.5)$$

The DEL algorithm implies

$$D_1 L_d(q_k, q_{k+1})\xi_Q(q_k) + D_2 L_d(q_{k-1}, q_k)\xi_Q(q_k) = 0 . \qquad (7.6)$$

Subtracting equation (7.5) from equation (7.6), we find that

$$D_2 L_d(q_k, q_{k+1})\xi_Q(q_{k+1}) = D_2 L_d(q_{k-1}, q_k)\xi_Q(q_k) . \qquad (7.7)$$

Finally, the result follows from (7.7) since

$$(J_d)_\xi(q_k, q_{k+1}) - (J_d)_\xi(q_{k-1}, q_k) =$$
$$= D_2 L_d(q_k, q_{k+1}) \xi_Q(q_{k+1}) - D_2 L_d(q_{k-1}, q_k) \xi_Q(q_k)$$
$$= D_2 L_d(q_k, q_{k+1}) \xi_Q(q_{k+1}) - D_2 L_d(q_k, q_{k+1}) \xi_Q(q_{k+1}) = 0.$$

Moreover, when regarding the discrete mechanical model as an approximation to a continuous system, one can verify that the constant value of the discrete momentum map approaches the value of its continuous counterpart, as the time step decreases.

Consequently, variational integrators are symplectic-momentum integrators.

7.3 Discrete Lagrange-d'Alembert principle

IN this section, we introduce a discrete version of the Lagrange-d'Alembert principle and derive from it the nonholonomic integrators. We follow the idea that, by respecting the geometric structure of nonholonomic systems, one can create integrators capturing their essential features.

In Nonholonomic Mechanics, the symplectic form constructed from the Lagrangian is no longer preserved as in the unconstrained case. Moreover, in the case of a nonholonomic system with symmetry, the momentum map is not conserved in general, due to the presence of the constraint forces. On the other hand, at least in the case of homogeneous constraints, the energy is still a conservation law for the system. Consequently, two of the three cornerstones on which the construction of mechanical integrators for unconstrained systems relies (i.e. preservation of symplectic structure and momentum) are lacking in the nonholonomic case. In Section 7.5, we will show that in spite of this, the integrators that we propose in the following enjoy several interesting geometric properties that account for their good long time behavior in examples.

Consider as before a discrete Lagrangian $L_d : Q \times Q \longrightarrow \mathbb{R}$ and the associated action sum

$$S = \sum_{k=0}^{N-1} L_d(q_k, q_{k+1}), \tag{7.8}$$

where $q_k \in Q$ and $k \in \{0, 1, \ldots, N\}$ is the discrete time. In the unconstrained discrete mechanics case (cf. Section 7.2), we have seen that one extremizes the action sum with respect to all possible sequences of $N - 1$ points, given fixed end points q_0, q_N. This means that at each point $q \in Q$, the allowed variations are given by the whole tangent space $T_q Q$. However, in the nonholonomic case, we must restrict the allowed variations. These are exactly given by the distribution $\mathcal{D}$. In addition, we will consider a discrete constraint space $\mathcal{D}_d \subset Q \times Q$ with the same dimension as $\mathcal{D}$ and such that $(q, q) \in \mathcal{D}_d$ for all $q \in Q$. This discrete constraint space imposes constraints on the solution sequence $\{q_k\}$, namely, $(q_k, q_{k+1}) \in \mathcal{D}_d$. Later, when regarding the discrete principle as an approximation of the continuous one, we shall impose more conditions on the selection of $\mathcal{D}_d$ in order to obtain a consistent discretization of the continuous equations of motion.

Consequently, to develop the discrete nonholonomic mechanics, one needs three ingredients: a discrete Lagrangian L_d, a constraint distribution $\mathcal{D}$ on Q and a discrete constraint space $\mathcal{D}_d$. Notice that the discrete mechanics can also be seen within this framework, where $\mathcal{D} = TQ$ and $\mathcal{D}_d = Q \times Q$.

Then, we define the discrete Lagrange-d'Alembert principle to be the extremization of (7.8) among the sequence of points (q_k) with given fixed end points q_0 and q_N, where the variations must satisfy $\delta q_k \in \mathcal{D}_{q_k}$ and $(q_k, q_{k+1}) \in \mathcal{D}_d$, for all $k \in \{0, \ldots, N-1\}$. This leads to the set of equations

$$(D_1 L_d(q_k, q_{k+1}) + D_2 L_d(q_{k-1}, q_k))_i \, \delta q_k^i = 0, \ 1 \le k \le N-1,$$

where $\delta q_k \in \mathcal{D}_{q_k}$, along with $(q_k, q_{k+1}) \in \mathcal{D}_d$. If $\omega_d^a : Q \times Q \to \mathbb{R}$, $a \in \{1, \ldots, m\}$, are functions whose annihilation defines $\mathcal{D}_d$, what we have got is the following *discrete Lagrange-d'Alembert* (DLA) algorithm

$$\begin{cases} D_1 L_d(q_k, q_{k+1}) + D_2 L_d(q_{k-1}, q_k) = \lambda_a \omega^a(q_k) \\ \omega_d^a(q_k, q_{k+1}) = 0. \end{cases} \tag{7.9}$$

Notice that the discrete Lagrange-d'Alembert principle is not truly variational, as it also happens with the continuous principle. Alternatively, we will refer to the DLA algorithm (7.9) as a *nonholonomic integrator*, by analogy with the unconstrained case. Note also that, under appropriate regularity assumptions, the implicit function theorem ensures us that we have obtained a well-defined algorithm $\Phi : Q \times Q \longrightarrow Q \times Q$, $\Phi(q_{k-1}, q_k) = (q_k, q_{k+1})$. In fact, this is guaranteed if the matrix

$$\begin{pmatrix} D_1 D_2 L_d(q_k, q_{k+1}) & \omega^a(q_k) \\ D_2 \omega_d^a(q_k, q_{k+1}) & 0 \end{pmatrix} \tag{7.10}$$

is invertible for each (q_k, q_{k+1}) in a neighborhood of the diagonal of $Q \times Q$.

Remark 7.3.1. Assume we are given a continuous nonholonomic problem with data $L : TQ \longrightarrow \mathbb{R}$ and $\mathcal{D} \subset TQ$. In the following section, we shall discuss some types of discretizations of this problem. To guarantee that the DLA algorithm approximates the continuous flow within a desired order of accuracy, one should select the discrete Lagrangian $L_d : Q \times Q \longrightarrow \mathbb{R}$ and the discrete constraint space $\mathcal{D}_d$ in a consistent way. This essentially means that, if $\omega^1, \ldots, \omega^m$ are 1-forms on Q whose annihilation locally define the constraint distribution $\mathcal{D}$, one performs the same type of discretization of both the Lagrangian $L : TQ \longrightarrow \mathbb{R}$ and the 1-forms (interpreted as functions linear in the velocities, $\omega^a : TQ \longrightarrow \mathbb{R}$). For instance, if L_d is constructed by means of a discretization mapping $\Psi : Q \times Q \longrightarrow TQ$ defined on a neighborhood of the diagonal of $Q \times Q$, that is, $L_d = L \circ \Psi$, then $\mathcal{D}_d$ must locally be defined by the annihilation of the functions $\omega_d^a = \omega^a \circ \Psi$.

Remark 7.3.2. Consider the continuous nonholonomic problem given by L and $\mathcal{D}$, and let L_d and $\mathcal{D}_d$ be appropriate discrete versions of them. Then, if the matrix

$$\begin{pmatrix} D_1 D_2 L_d(q_k, q_k) & \omega^a(q_k) \\ D_2 \omega_d^a(q_k, q_k) & 0 \end{pmatrix}$$

is invertible for each $q_k \in Q$, a sufficiently small stepsize h guarantees that the matrix (7.10) is also nonsingular and hence the DLA algorithm is solvable for q_{k+1}.

Remark 7.3.3 (The holonomic case).

Let us examine the nonholonomic integrator when the constraints are holonomic, that is, the case when the distribution $\mathcal{D}$ is integrable. Assume that there exists a function

$$g : Q \longrightarrow \mathbb{R}^l,$$

whose level surfaces are precisely the integral manifolds of $\mathcal{D}$, i.e. for each $r \in \mathbb{R}^l$, $N_r = g^{-1}(r)$ is a submanifold of Q such that $T_q N_r = \mathcal{D}_q$ for all $q \in N_r$. Then, we can consider as discrete constraint space the following subspace of $Q \times Q$,

$$\mathcal{D}_d = \cup_{r \in \mathbb{R}^l} N_r \times N_r.$$

Observe that if we take $q_0 \in N_0$, then $(q_0, q_1) \in \mathcal{D}_d$ is equivalent to $q_1 \in N_0$. We then find that the nonholonomic integrator for an initial pair $q_0, q_1 \in N_0$ becomes

$$\begin{cases} D_1 L_d(q_k, q_{k+1}) + D_2 L_d(q_{k-1}, q_k) = \lambda_a D g^a(q_k) \\ g(q_{k+1}) = 0, \end{cases} \tag{7.11}$$

where $g^a : Q \longrightarrow \mathbb{R}$ denotes the a component of g. Notice that (7.11) is just a variational integrator [256]. It is in this sense that we say that nonholonomic integrators are an extention of variational integrators, since for integrable constraints we do recover the latter ones. It is known, for instance, that an appropriate choice of the discrete Lagrangian in (7.11) gives rise to the Shake algorithm used in molecular dynamics simulation [134, 209], written in terms of position variables.

7.4 Construction of integrators

$\mathbb{A}$SSUME that we have a continuous nonholonomic problem given by $L : TQ \longrightarrow \mathbb{R}$ and $\mathcal{D} \subset TQ$. In the unconstrained case [256], there are mainly two ways of constructing mechanical integrators, depending on whether Q is seen as a manifold in its own right (the "intrinsic" point of view) or as being embedded in a larger space (the "extrinsic" point of view).

When adopting the intrinsic point of view, one makes use of coordinate charts on Q to construct the discrete Lagrangian. Let $\varphi : U \subset Q \longrightarrow \mathbb{R}^n$ be a local chart whose coordinate domain U contains q_k and assume that $q_{k+1} \in U$ (a condition guaranteed by a sufficiently small time step h). A choice of discrete Lagrangian is the following

$$L_d^\alpha(q_k, q_{k+1}) = \tag{7.12}$$
$$L\left(\varphi^{-1}((1-\alpha)\varphi(q_k) + \alpha\varphi(q_{k+1})), (\varphi^{-1})_* \left(\frac{\varphi(q_{k+1}) - \varphi(q_k)}{h}\right)\right),$$

where $0 \leq \alpha \leq 1$ is an interpolation parameter and the differential $(\varphi^{-1})_*$ is taken at the point $x = (1 - \alpha)\varphi(q_k) + \alpha\varphi(q_{k+1})$. Of course there are other possible choices of discretizations, as for instance,

$$L_d^{sym,\alpha}(q_k, q_{k+1}) =$$
$$\frac{1}{2} L\left(\varphi^{-1}((1-\alpha)\varphi(q_k) + \alpha\varphi(q_{k+1})), (\varphi^{-1})_* \left(\frac{\varphi(q_{k+1}) - \varphi(q_k)}{h}\right)\right)$$
$$+ \frac{1}{2} L\left(\varphi^{-1}(\alpha\varphi(q_k) + (1-\alpha)\varphi(q_{k+1})), (\varphi^{-1})_* \left(\frac{\varphi(q_{k+1}) - \varphi(q_k)}{h}\right)\right).$$
$$\tag{7.13}$$

In the unconstrained case, the choice (7.13) always yields second order accurate numerical methods, whereas in general this is only guaranteed for the discretization (7.12) if $\alpha = \frac{1}{2}$ (although for natural Lagrangians of the form $L = \frac{1}{2}\dot{q}M\dot{q} - V(q)$, the discrete Lagrangian (7.12) also gives second order numerical methods [110]).

This approach is called the Generalized Coordinate Formulation.

Remark 7.4.1. In general, this viewpoint is necessarily local, since the discretizations are only valid in the coordinate domain U of φ. If we choose an atlas of charts covering the whole manifold Q, we cannot guarantee that the construction of the discrete Lagrangian L_d will coincide on the chart overlaps. There are certain cases, however, in which this is indeed possible. For example, if we can find an atlas $\{(U_s, \varphi_s)\}$ such that for any two overlapping charts, φ_{s_1} and φ_{s_2}, the local diffeomorphism $\varphi_{s_1 s_2} = \varphi_{s_1} \circ \varphi_{s_2}^{-1}$ verifies $\varphi_{s_1 s_2}((1-\alpha)x + \alpha y) = (1-\alpha)\varphi_{s_1 s_2}(x) + \alpha\varphi_{s_1 s_2}(y)$, for any $x, y \in \varphi_{s_2}(U_{s_2})$

and $(\varphi_{s_1 s_2})_* = id$, then it is easy to see that one can "paste" the local constructions (7.12) (respectively (7.13)) to have a well-defined discrete Lagrangian on a neighborhood of the diagonal of $Q \times Q$. [A simple example of this situation is given by the manifold $\mathbb{S}^1$, with the local charts $\varphi_1(z_1, z_2) = \arcsin(z_2/z_1) \in (0, 2\pi)$ and $\varphi_2(z_1, z_2) = \arcsin(z_2/z_1) \in (-\pi, \pi)$].

Remark 7.4.2. Another way of constructing a well-defined discrete Lagrangian on a neighborhood of the diagonal of $Q \times Q$ is the following. Assume that there exist a $q_0 \in Q$ and a differentiable mapping $\Upsilon : Q \longrightarrow \mathrm{Diff}(Q)$ such that $\Upsilon(q)(q) = q_0$, for all $q \in Q$. In this case, we can define L_d for each (q, q') according to either (7.12) or (7.13) by means of $\varphi = \varphi_0 \circ \Upsilon(q)$, where φ_0 is a local chart whose coordinate domain contains q_0. It is important to note that in this construction the mapping $\varphi_0 \circ \Upsilon(q)$ *varies* with the pair (q, q'). This is the case for instance of finite dimensional Lie groups $Q = G$, where one can take $q_0 = e$, the identity element, $\Upsilon(g) = L_{g^{-1}}$ for each $g \in G$ and $\varphi_0 = \exp_e^{-1}$ (see [161]). We shall make use of this construction in Section 7.5.3.

The extrinsic point of view assumes that Q is embedded in some linear space V and that we have a Lagrangian $L : TV \longrightarrow \mathbb{R}$ such that $L_{|TQ} = L$. In addition, it is assumed that there exists a vector valued constraint function $g : V \longrightarrow \mathbb{R}^l$, such that $g^{-1}(0) = Q \subset V$, with 0 a regular value of g. According to (7.12) and (7.13), we can consider the following discrete Lagrangians on $V \times V$

$$L_d^\alpha(v_k, v_{k+1}) = \mathbb{L}\left((1-\alpha)v_k + \alpha v_{k+1}, \frac{v_{k+1} - v_k}{h} \right), \qquad (7.14)$$

and

$$L_d^{sym,\alpha}(v_k, v_{k+1}) = \frac{1}{2}\mathbb{L}\left((1-\alpha)v_k + \alpha v_{k+1}, \frac{v_{k+1} - v_k}{h} \right)$$
$$+ \frac{1}{2}\mathbb{L}\left(\alpha v_k + (1-\alpha)v_{k+1}, \frac{v_{k+1} - v_k}{h} \right). \quad (7.15)$$

In this way, we can sum and subtract points in Q because we are regarding them as vectors in V by means of the natural inclusion $j : Q \hookrightarrow V$. Of course, we must ensure that the points obtained by the algorithm all belong to Q. Then, the solution sequence (v_k) will extremize the action sum $S = \sum_{k=0}^{N-1} L_d(v_k, v_{k+1})$ subject to the holonomic constraints imposed by g. This leads to the discrete equations

$$\begin{cases} D_1 L_d(v_k, v_{k+1}) + D_2 L_d(v_{k-1}, v_k) = \lambda_l Dg^l(v_k) \\ g(v_{k+1}) = 0. \end{cases}$$

This approach is called the Constrained Coordinate Formulation.

Both formulations are shown to be equivalent in the domain of definition of the local chart φ selected in the Generalized Coordinate Formulation [256], whereby the following identification is understood: $L_d(q_k, q_{k+1}) =$
$\mathbb{L}_d(j(q_k), j(q_{k+1}))$, which is valid for choices of the chart (U, φ) in the definition of L_d such that the map $J = j \circ \varphi^{-1} : \varphi(U) \subset \mathbb{R}^n \longrightarrow V$ is linear. Notice
that this assumption is not at all restrictive, since j is an injective immersion
and such a chart (U, φ) can always be chosen.

In the nonholonomic case, we can construct an appropriate adaptation of
both formulations. In the Generalized Coordinate Formulation, we introduce
$\mathcal{D}_d$ as follows. Take a local basis of 1-forms of the annihilator of the constraint
distribution $\mathcal{D}$, $\{\omega^1, \ldots, \omega^m\} \in \mathcal{D}^o$. These 1-forms can be interpreted as
functions linear in the velocities, locally defined on TQ. Then, we discretize
them according to the previous discretizations of the Lagrangian, that is, we
take either

$$\omega_d^a(q_k, q_{k+1}) = \tag{7.16}$$
$$\omega^a \left(\varphi^{-1}((1-\alpha)\varphi(q_k) + \alpha\varphi(q_{k+1})), (\varphi^{-1})_* \left(\frac{\varphi(q_{k+1}) - \varphi(q_k)}{h} \right) \right),$$

or

$$\omega_d^a(q_k, q_{k+1}) =$$
$$\frac{1}{2}\omega^a \left(\varphi^{-1}((1-\alpha)\varphi(q_k) + \alpha\varphi(q_{k+1})), (\varphi^{-1})_* \left(\frac{\varphi(q_{k+1}) - \varphi(q_k)}{h} \right) \right)$$
$$+ \frac{1}{2}\omega^a \left(\varphi^{-1}(\alpha\varphi(q_k) + (1-\alpha)\varphi(q_{k+1})), (\varphi^{-1})_* \left(\frac{\varphi(q_{k+1}) - \varphi(q_k)}{h} \right) \right).$$
$$\tag{7.17}$$

In this way we obtain the functions $\omega_d^a : Q \times Q \longrightarrow \mathbb{R}$ whose annihilation
defines $\mathcal{D}_d \subset Q \times Q$. As in the unconstrained case, it is not hard to prove
that the discretization (7.13) together with (7.17) yields second order accurate
approximations to the continuous flow, whereas this is only guaranteed for
the discretization (7.12), (7.16) if $\alpha = \frac{1}{2}$.

In the Constrained Coordinate Formulation, we assume that there exist
local 1-forms on V defining $\mathcal{D}^o$, $\{\tilde{\omega}^1, \ldots, \tilde{\omega}^m\}$ such that $\tilde{\omega}^a(q)_{|T_qQ} = \omega^a(q)$
for $q \in Q$. Then, we discretize them according to

$$\tilde{\omega}_d^a(v_k, v_{k+1}) = \tilde{\omega}^a \left((1-\alpha)v_k + \alpha v_{k+1}, \frac{v_{k+1} - v_k}{h} \right), \tag{7.18}$$

and

$$\tilde{\omega}_d^a(v_k, v_{k+1}) = \frac{1}{2}\tilde{\omega}^a \left((1-\alpha)v_k + \alpha v_{k+1}, \frac{v_{k+1} - v_k}{h} \right)$$
$$+ \frac{1}{2}\tilde{\omega}^a \left(\alpha v_k + (1-\alpha)v_{k+1}, \frac{v_{k+1} - v_k}{h} \right). \tag{7.19}$$

Observe that we can identify $\omega_d^a(q_k, q_{k+1}) = \tilde{\omega}_d^a(j(q_k), j(q_{k+1}))$ in the same way as we have done for the discrete Lagrangians. Then, the discrete Lagrange-d'Alembert principle with the holonomic constraints g and the non-holonomic constraints $\tilde{\omega}^1, \ldots, \tilde{\omega}^m$ leads us to the equations

$$\begin{cases} D_1 \mathbb{L}_d(v_k, v_{k+1}) + D_2 \mathbb{L}_d(v_{k-1}, v_k) = \lambda_l D g^l(v_k) + \mu_a \tilde{\omega}^a(v_k) \\ g(v_{k+1}) = 0 \\ \tilde{\omega}_d^a(v_k, v_{k+1}) = 0 \,. \end{cases} \tag{7.20}$$

The following theorem, analogous to the one presented in [256], ensures that both formulations (7.9) and (7.20) are indeed equivalent in the same sense as before, as one might expect. We prove it for the discretizations (7.12), (7.16). The proof for the symmetric discretizations (7.13), (7.17) is analogous.

Theorem 7.4.3. *Let $\varphi : U \subset Q \longrightarrow \mathbb{R}^n$ be a local chart of Q such that $J = j \circ \varphi^{-1}$ is linear. Identify U with $\varphi(U)$ and $j_{|U}$ with $J_{|U}$ through φ. Let q_{k-1}, q_k be two initial points in the coordinate chart and let $v_{k-1} = J(q_{k-1})$, $v_k = J(q_k)$. Then, the Generalized Coordinate Formulation (7.9) has a solution $(q_{k+1}, \mu_a^{(k)})$ if and only if the Constrained Coordinate Formulation (7.20) has a solution $(v_{k+1}, \lambda_l^{(k)}, \bar{\mu}_a^{(k)})$. Indeed, $v_{k+1} = J(q_{k+1})$ and $\bar{\mu}_a^{(k)} = \mu_a^{(k)}$.*

Proof. To establish the equivalence, we first expand equations (7.9) and (7.20) in terms of $\mathbb{L}$ and the 1-forms $\{\tilde{\omega}^a\}_{a=1}^m$. Let $(v, \dot{v})$ denote the canonical coordinates of TV. Equations (7.9) become, when written in matrix form,

$$D^T J(q_k) \left\{ \frac{1}{h} \left[\frac{\partial \mathbb{L}}{\partial \dot{v}}(a_k, b_k) - \frac{\partial \mathbb{L}}{\partial \dot{v}}(a_{k+1}, b_{k+1}) \right] + (1 - \alpha) \frac{\partial \mathbb{L}}{\partial v}(a_{k+1}, b_{k+1}) \right.$$
$$\left. + \alpha \frac{\partial \mathbb{L}}{\partial v}(a_k, b_k) + \mu_a^{(k)} \tilde{\omega}^a(J(q_k)) \right\} = 0 \,, \tag{7.21}$$

$$\tilde{\omega}_d^a(J(q_k), J(q_{k+1})) = \tilde{\omega}^a(a_{k+1}, b_{k+1}) = 0 \,,$$

where $a_k = \alpha J(q_k) + (1 - \alpha) J(q_{k-1})$ and $b_k = (J(q_k) - J(q_{k-1}))/h$. Note that we are using the identifications $\omega_d^a = (\tilde{\omega}_d^a)_{|Q \times Q}$, $(\mathbb{L}_d^\alpha)_{|Q \times Q} = L_d^\alpha$ and $\tilde{\omega}_{|TQ}^a = \omega^a$. If $\tilde{\omega}^a = \tilde{\omega}_l^a dv^l$ and $\omega^a = \omega_i^a dq^i$, we have that

$$\omega_i^a(q_k) dq^i = J^*(\tilde{\omega}_l^a dv^l)(q_k) = \frac{\partial J^l}{\partial q^i}(q_k) \tilde{\omega}_l^a(J(q_k)) dq^i \,,$$

which can be written in a more compact way as $\omega^a(q_k) = D^T J(q_k) \tilde{\omega}^a(J(q_k))$. Here and in the following the superscript T refers to the transpose of a matrix.

On the other hand, equation (7.20) can be written as

$$\frac{1}{h}\left[\frac{\partial \mathbb{L}}{\partial \dot{v}}\left(v_{k-1+\alpha}, \frac{v_k - v_{k-1}}{h}\right) - \frac{\partial \mathbb{L}}{\partial \dot{v}}\left(v_{k+\alpha}, \frac{v_{k+1} - v_k}{h}\right)\right]$$

$$+ (1-\alpha)\frac{\partial \mathbb{L}}{\partial v}\left(v_{k+\alpha}, \frac{v_{k+1} - v_k}{h}\right) + \alpha\frac{\partial \mathbb{L}}{\partial v}\left(v_{k-1+\alpha}, \frac{v_k - v_{k-1}}{h}\right)$$

$$+ \bar{\mu}_a^{(k)}\tilde{\omega}^a(v_k) = \lambda_l^{(k)}Dg^l(v_k),$$

$$(7.22)$$

$$g(v_{k+1}) = 0,$$

$$\tilde{\omega}_d^a(v_k, v_{k+1}) = \tilde{\omega}_l^a(v_{k+\alpha})\left(\frac{v_{k+1} - v_k}{h}\right)^l = 0,$$

where the shorthand notation $v_{k+\alpha} = (1-\alpha)v_k + \alpha v_{k+1}$ is used. Now, assume that $(v_{k+1}, \lambda_l^{(k)}, \bar{\mu}_a^{(k)})$ is a solution of (7.22) with $v_k = J(q_k)$ and $v_{k-1} = J(q_{k-1})$. The fact that $g(v_{k+1}) = 0$ implies that v_{k+1} belongs to the image of J. Let $q_{k+1} = J^{-1}(v_{k+1})$. Multiplying the first equation of (7.22) by $D^T J(q_k)$ and making the corresponding substitutions, one obtains for the pair $(q_{k+1}, \bar{\mu}_a^{(k)})$ just the first equation of (7.21), since the term $D^T J(q_k)D^T g(v_k)$ cancels due to $g \circ J = 0$.

Conversely, if $(q_{k+1}, \mu_a^{(k)})$ is a solution of (7.21), then one can find Lagrange multipliers $\lambda_l^{(k)}$, such that $(v_{k+1} = J(q_{k+1}), \lambda_l^{(k)}, \mu_a^{(k)})$ is a solution of (7.22) as follows. The second and the third equation of (7.22) are automatically satisfied because of $v_{k+1} \in Q$ and the second equation of (7.21). Moreover, as $DJ(q_k)$ and $Dg(q_k)$ are assumed to have full rank, we have that $T_{v_k}V = \mathcal{R}(DJ(q_k)) \oplus \mathcal{N}(D^T J(q_k))$, where $\mathcal{R}(DJ(q_k))$ and $\mathcal{N}(D^T J(q_k))$ refer to the range and the kernel, respectively, of the operator under consideration. Since $\mathcal{R}(D^T g(q_k)) \subset \mathcal{N}(D^T J(q_k))$ and $\dim \mathcal{R}(D^T g(q_k)) = \dim \mathcal{N}(D^T J(q_k))$, we have $T_{v_k}V = \mathcal{R}(DJ(q_k)) \oplus \mathcal{R}(D^T g(q_k))$. Now, the left-hand side of the first equation of (7.22) can be decomposed into a part belonging to $\mathcal{R}(DJ(q_k))$ and a part belonging to $\mathcal{R}(D^T g(q_k))$. But the part in $\mathcal{R}(DJ(q_k))$ is zero, because of the first equation of (7.21). Consequently, the entire expression belongs to $\mathcal{R}(D^T g(q_k))$, and thus there exist some $\lambda_l^{(k)}$ such that

$$\frac{1}{h}\left[\frac{\partial \mathbb{L}}{\partial \dot{v}}\left(v_{k-1+\alpha}, \frac{v_k - v_{k-1}}{h}\right) - \frac{\partial \mathbb{L}}{\partial \dot{v}}\left(v_{k+\alpha}, \frac{v_{k+1} - v_k}{h}\right)\right]$$

$$+ (1-\alpha)\frac{\partial \mathbb{L}}{\partial v}\left(v_{k+\alpha}, \frac{v_{k+1} - v_k}{h}\right) + \alpha\frac{\partial \mathbb{L}}{\partial v}\left(v_{k-1+\alpha}, \frac{v_k - v_{k-1}}{h}\right)$$

$$+ \bar{\mu}_a^{(k)}\tilde{\omega}^a(v_k) = \lambda_l^{(k)}Dg^l(v_k),$$

which is precisely the first equation of (7.22).   □

The relevance of Theorem 7.4.3 becomes apparent when handling concrete examples. Generally, it is easier to treat the nonholonomic integrator following the Constrained Coordinate Formulation, since points in Q can be treated as points in some $\mathbb{R}^s$, and this is a definite advantage for the numerical implementation. On the other hand, the geometric study of the properties of discrete nonholonomic mechanics is carried out from the "intrinsic" point of view.

7.5 Geometric invariance properties

IN the unconstrained case, one can study discrete mechanics by itself, starting from a given discrete Lagrangian L_d and investigating the geometric properties that the discrete flow enjoys, such as the preservation of the symplectic form or of the momentum in the presence of symmetry. Furthermore, when one regards a discrete mechanical system as an approximation of a continuous one, it turns out that the symplectic-momentum nature of the variational integrators makes the difference in capturing the essential features of Lagrangian systems.

In the following, we provide some geometric explanations for the good performance of the DLA algorithm when compared with other standard higher order numerical methods, such as the 4th order Runge-Kutta, as will be shown in Section 7.6. Of course, a more thorough error analysis would be of interest, but here we focus our attention on the invariance properties that the discrete nonholonomic mechanics possesses, as a sign of its appropriateness for approximating the continuous counterpart.

As we have shown in Section 3.4.1, in Nonholonomic Mechanics the symplectic form is not preserved by the flow of the system, so one can not expect the discrete version to preserve it. However, we will show in Section 7.5.1 that the discrete flow preserves the structure of the evolution of the symplectic form along the trajectories of the system. This property generalizes the symplectic character of variational integrators and, in fact, one precisely recovers the preservation of the symplectic form in the absence of constraints.

Moreover, under the action of a Lie group G on the configuration manifold Q, leaving invariant the Lagrangian $L : TQ \longrightarrow \mathbb{R}$ and the constraints $\mathcal{D} \subset TQ$, the associated momentum $J : TQ \longrightarrow \mathfrak{g}^*$ in general will not be conserved either, as we discussed in Chapter 4. However, we can "measure" the evolution of the nonholonomic momentum mapping along the integral curves of the Lagrange-d'Alembert equations by means of the nonholonomic momentum equation (cf. equations (4.19) and (4.20) in Section 4.4). We shall show in Section 7.5.2 that the DLA algorithm satisfies a discrete version of the nonholonomic momentum equation. In addition, in the presence of horizontal symmetries, we shall prove that the associated momenta are actually conservation laws.

In the vertical or purely kinematic case, there are no symmetry directions lying in the constraint distribution and one does not have any nonholonomic momentum (see Section 4.2). Nonholonomic systems of Chaplygin type form the most representative class of systems falling into this category. In Section 7.5.3, we will discuss how, for Chaplygin systems, the DLA algorithm passes to the reduced space Q/G and yields a variational integrator in the sense of [110]. In some cases (in agreement with the continuous counterpart), this reduced formulation exactly yields a standard variational integrator.

7.5.1 The symplectic form

In this section, we investigate the behavior of the DLA algorithm with respect to the discrete symplectic form Ω_{L_d} defined in Section 7.2. The reader is referred to Section 3.4.1 for the geometric picture in the continuous case. We just recall here that the evolution of the symplectic form along the trajectories of the system is given by equation (3.14)

$$\mathcal{L}_X \omega_L = d\beta \,, \tag{7.23}$$

with $\beta \in F^o = S^*((T\mathcal{D})^o)$.

The DLA algorithm preserves this structure for the evolution of the discrete symplectic form Ω_{L_d}. Indeed we have that

$$\Phi^* \Omega_{L_d} = -\Phi^* \mathcal{F} L_d^* d\Theta_Q = -d(\mathcal{F} L_d \circ \Phi)^* \Theta_Q$$
$$= -d(\tilde{\mathcal{F}} L_d \circ \Phi)^* \Theta_Q \overset{DLA}{=} -d(\mathcal{F} L_d \Theta_Q - \beta_d) \,,$$

where $\beta_d \in \mathcal{D}^o$ and in the last equality we have used the definition of the discrete principle (7.9). Finally, we get

$$\Phi^* \Omega_{L_d} = \Omega_{L_d} + d\beta_d \,, \tag{7.24}$$

which is the discrete version of (7.23). Note that in the absence of constraints, we precisely recover the conservation of the discrete symplectic form (cf. equation (7.4)).

7.5.2 The momentum

What we develop in the following is a discrete version of the nonholonomic momentum map and show that the nonholonomic integrator (7.9) fulfills a discrete version of the momentum equation.

Firstly, given the discrete Lagrangian $L_d : Q \times Q \longrightarrow \mathbb{R}$, define the *discrete nonholonomic momentum map* by

$$J_d^{nh} : \quad Q \times Q \quad \longrightarrow \quad \mathfrak{g}^{\mathcal{D}*}$$
$$(q_{k-1}, q_k) \quad \longmapsto \quad J_d^{nh}(q_{k-1}, q_k) : \mathfrak{g}_q \to \mathbb{R}$$
$$\xi \mapsto \langle D_2 L_d(q_{k-1}, q_k), \xi_Q(q_k) \rangle .$$

Take, as in the continuous case (cf. Section 4.4), a smooth section $\tilde{\xi}$ of the bundle $\mathfrak{g}^{\mathcal{D}} \longrightarrow Q$ and consider the function on $Q \times Q$, $(J_d^{nh})_{\tilde{\xi}} : Q \times Q \longrightarrow \mathbb{R}$, given by $(J_d^{nh})_{\tilde{\xi}} = \langle J_d^{nh}, \tilde{\xi} \rangle$. Then, one finds that the nonholonomic integrator fullfils the following discrete version of the nonholonomic momentum equation (4.20).

Theorem 7.5.1. *The flow of the discrete Lagrange-d'Alembert equations,* $(q_{k-1}, q_k) \longmapsto (q_k, q_{k+1})$, *verifies*

$$(J_d^{nh})_{\tilde{\xi}}(q_k, q_{k+1}) - (J_d^{nh})_{\tilde{\xi}}(q_{k-1}, q_k)$$
$$= D_2 L_d(q_k, q_{k+1}) \left(\tilde{\xi}(q_{k+1}) - \tilde{\xi}(q_k) \right)_Q (q_{k+1}) . \quad (7.25)$$

Proof. The invariance of the discrete Lagrangian L_d implies that

$$L(\exp(s\tilde{\xi}(q_k))q_k, \exp(s\tilde{\xi}(q_k))q_{k+1}) = L(q_k, q_{k+1}) .$$

Differentiating with respect to s and setting $s = 0$ yields

$$D_1 L_d(q_k, q_{k+1}) \left(\tilde{\xi}(q_k) \right)_Q (q_k) + D_2 L_d(q_k, q_{k+1}) \left(\tilde{\xi}(q_k) \right)_Q (q_{k+1}) = 0 .$$
$$(7.26)$$

On the other hand, the discretization of the Lagrange-d'Alembert principle (7.9) implies that

$$D_1 L_d(q_k, q_{k+1}) \left(\tilde{\xi}(q_k) \right)_Q (q_k) + D_2 L_d(q_{k-1}, q_k) \left(\tilde{\xi}(q_k) \right)_Q (q_k) = 0 . \quad (7.27)$$

Subtracting equation (7.26) from equation (7.27), we find that

$$D_2 L_d(q_k, q_{k+1}) \left(\tilde{\xi}(q_k) \right)_Q (q_{k+1}) = D_2 L_d(q_{k-1}, q_k) \left(\tilde{\xi}(q_k) \right)_Q (q_k) . \quad (7.28)$$

Finally, the result follows from (7.28) since

$$(J_d^{nh})_{\tilde{\xi}}(q_k, q_{k+1}) - (J_d^{nh})_{\tilde{\xi}}(q_{k-1}, q_k)$$
$$= D_2 L_d(q_k, q_{k+1}) \left(\tilde{\xi}(q_{k+1}) \right)_Q (q_{k+1}) - D_2 L_d(q_{k-1}, q_k) \left(\tilde{\xi}(q_k) \right)_Q (q_k)$$
$$= D_2 L_d(q_k, q_{k+1}) \left(\tilde{\xi}(q_{k+1}) \right)_Q (q_{k+1}) - D_2 L_d(q_k, q_{k+1}) \left(\tilde{\xi}(q_k) \right)_Q (q_{k+1}) .$$

$\square$

In the presence of horizontal symmetries we find that the algorithm (7.9) exactly preserves the associated components of the momentum.

Corollary 7.5.2. *If ξ is a horizontal symmetry, then $(J_d^{nh})_\xi$ is conserved by the nonholonomic integrator.*

7.5.3 Chaplygin systems

The study of the previous section on the discrete nonholonomic momentum mapping does not cover all the possible cases that can occur when dealing with nonholonomic systems with symmetry, as we already know from the discussion in Chapter 4. Consider the situation in which the action of the Lie group G has no symmetry direction lying in $\mathcal{D}$, i.e. $\mathcal{V} \cap \mathcal{D} = 0$, where $\mathcal{V}$ denotes the vertical bundle of the projection $\pi : Q \longrightarrow Q/G$. In this case there is no nonholonomic momentum mapping. Under the common assumption that $\mathcal{D}_q + \mathcal{V}_q = T_q Q$ for all $q \in Q$ (*dimension assumption*), one has indeed a splitting of the tangent bundle at each point $q \in Q$, $T_q Q = \mathcal{D}_q \oplus \mathcal{V}_q$. The distribution $\mathcal{D}$ being G-invariant, this situation corresponds precisely to the notion of a principal connection γ on the principal fiber bundle $\pi : Q \longrightarrow Q/G$. Hence, we are dealing with a generalized Chaplygin system [49, 120].

From the discussion in Section 4.2, we know that one of the peculiarities of nonholonomic Chaplygin systems is that, after reduction by the Lie group G, they take on the form of an unconstrained system, subject to an "external" force of a special type. In this section, we investigate what happens with the nonholonomic integrator in this situation.

Reduction of the discrete principle Next, we examine the possibility of passing the discrete nonholonomic principle to the reduced space Q/G. Consider a discrete Lagrangian $L_d : Q \times Q \longrightarrow \mathbb{R}$ and a discrete space $\mathcal{D}_d$, described by the annihilation of some constraint functions $\omega_d^a : Q \times Q \longrightarrow \mathbb{R}$, $a \in \{1, \ldots, m\}$. Assume that both the Lagrangian and the constraints are G-invariant under the diagonal action of the Lie group on the manifold $Q \times Q$. The DLA algorithm then becomes

$$
\begin{cases}
D_1 L_d(r_k, g_k, r_{k+1}, g_{k+1}) + D_2 L_d(r_{k-1}, g_{k-1}, r_k, g_k) = \lambda_a \omega^a(r_k, g_k) \\
\omega_d^a(r_k, g_k, r_{k+1}, g_{k+1}) = 0,
\end{cases}
$$
(7.29)

where it should be recalled that D_1 denotes the derivative with respect to $q_k = (r_k, g_k)$. These equations can be rewritten, using the expression (4.12) for the constraint 1-forms, in the following way

$$
\begin{cases}
\dfrac{\partial L_d}{\partial r_k^\beta}(r_k, g_k, r_{k+1}, g_{k+1}) + \dfrac{\partial L_d}{\partial r_k^\beta}(r_{k-1}, g_{k-1}, r_k, g_k) = \\
\qquad = \left(\dfrac{\partial L_d}{\partial g_k}(r_k, g_k, r_{k+1}, g_{k+1}) + \dfrac{\partial L_d}{\partial g_k}(r_{k-1}, g_{k-1}, r_k, g_k) \right) \mathcal{A}_\beta^b(r_k) L_{g_k *} e_b \\
\omega_d^a(r_k, g_k, r_{k+1}, g_{k+1}) = 0,
\end{cases}
$$
(7.30)

where $\beta \in \{1, \ldots, n - m\}$, $b \in \{1, \ldots, m\}$ and L_g denotes the left multiplication by g in G. It must be noted that in the right-hand side of the first equation a shorthand notation is used to denote the natural pairing between tangent vectors and covectors on G.

Observe that $\mathcal{D}_d$ can be locally identified with $Q/G \times Q/G \times G$ via the assignment

$$(r_k, g_k, r_{k+1}, g_{k+1}) \in \mathcal{D}_d \longmapsto (r_k, g_k, r_{k+1}),$$

since g_{k+1} is uniquely determined by the equations $\omega_d^a(r_k, g_k, r_{k+1}, g_{k+1}) = 0$, $a \in \{1, \ldots, m\}$, that is, $g_{k+1} = g_{k+1}(r_k, g_k, r_{k+1})$. In addition, the G-invariance of the constraint functions implies that $g_{k+1}(r_k, g_k, r_{k+1}) = g_k \cdot g_{k+1}(r_k, e, r_{k+1})$.

Let us consider the restriction of $L_d : Q \times Q \longrightarrow \mathbb{R}$ to $\mathcal{D}_d$, $L_d^c : \mathcal{D}_d \longrightarrow \mathbb{R}$. The G-invariance of L_d and $\mathcal{D}_d$ implies the G-invariance of L_d^c. Define a discrete Lagrangian L_d^* on the reduced manifold as

$$L_d^* : Q/G \times Q/G \longrightarrow \mathbb{R}$$
$$(r_k, r_{k+1}) \longmapsto L_d^c(r_k, e, r_{k+1}).$$

Now, we shall write the DLA algorithm (7.30) in terms of the constrained discrete Lagrangian L_d^c and then examine the possibility of passing the equations to Q/G, in terms of the reduced discrete Lagrangian L_d^*. First, we have that

$$\frac{\partial L_d^c}{\partial r_k^\beta} = \frac{\partial L_d}{\partial r_k^\beta} + \frac{\partial L_d}{\partial g_{k+1}} \frac{\partial g_{k+1}}{\partial r_k^\beta},$$

$$\frac{\partial L_d^c}{\partial r_{k+1}^\beta} = \frac{\partial L_d}{\partial r_{k+1}^\beta} + \frac{\partial L_d}{\partial g_{k+1}} \frac{\partial g_{k+1}}{\partial r_{k+1}^\beta}.$$

Secondly, we also have

$$0 = \frac{\partial L_d^c}{\partial g_k} = \frac{\partial L_d}{\partial g_k} + \frac{\partial L_d}{\partial g_{k+1}} \frac{\partial g_{k+1}}{\partial g_k} = \frac{\partial L_d}{\partial g_k} + R_{g_{k+1}}^* \frac{\partial L_d}{\partial g_{k+1}},$$

where R_g denotes the right multiplication in the Lie group by the element $g \in G$.

In view of this, we see that the nonholonomic integrator can be expressed in the following way

$$D_1 L_d^c(r_k, g_k, r_{k+1}) + D_2 L_d^c(r_{k-1}, g_{k-1}, r_k) = F^-(q_k, q_{k+1}) + F^+(q_{k-1}, q_k),$$

where

$$F^-(q_k, q_{k+1}) = \frac{\partial L_d}{\partial g_{k+1}} \frac{\partial g_{k+1}}{\partial r_k^\beta}(r_k, g_k, r_{k+1})$$

$$+ \frac{\partial L_d}{\partial g_k}(r_k, g_k, r_{k+1}, g_{k+1}) \mathcal{A}_\beta^b(r_k) L_{g_{k*}} e_b$$

$$= \left(\frac{\partial L_d}{\partial g_{k+1}} \frac{\partial g_{k+1}}{\partial r_k^\beta}(r_k, r_{k+1}) + \frac{\partial L_d}{\partial g_k} \mathcal{A}_\beta^b(r_k) \right) L_{g_{k*}} e_b,$$

$$F^+(q_{k-1}, q_k) = \frac{\partial L_d}{\partial g_k} \frac{\partial g_k}{\partial r_k^\beta}(r_{k-1}, g_{k-1}, r_k)$$

$$+ \frac{\partial L_d}{\partial g_k}(r_{k-1}, g_{k-1}, r_k, g_k) \mathcal{A}_\beta^b(r_k) L_{g_k *} e_b$$

$$= \frac{\partial L_d}{\partial g_k} \frac{\partial g_k}{\partial r_k^\beta}(r_{k-1}, r_k) L_{g_{k-1} *} e_b$$

$$- \frac{\partial L_d}{\partial g_{k-1}} \mathcal{A}_\beta^b(r_k) L_{g_{k-1} *} Ad_{g_k(r_{k-1}, r_k)} e_b .$$

Note that both discrete forces, F^- and F^+, are G-invariant. This can be seen as follows. As L_d is G-invariant, we have that

$$L_d(r_k, g_k, r_{k+1}, g_{k+1}) = L_d(r_k, e, r_{k+1}, g_k^{-1} g_{k+1}) = \ell_d(r_k, r_{k+1}, f_{k,k+1}),$$

where we use the shorthand notation $f_{k,k+1} = g_k^{-1} g_{k+1}$. From here, one can derive

$$\frac{\partial L_d}{\partial g_k}(r_k, g_k, r_{k+1}, g_{k+1}) = -L_{g_k^{-1}}^* R_{f_{k,k+1}}^* \frac{\partial \ell_d}{\partial f_{k,k+1}},$$

$$\frac{\partial L_d}{\partial g_{k+1}}(r_k, g_k, r_{k+1}, g_{k+1}) = L_{g_k^{-1}}^* \frac{\partial \ell_d}{\partial f_{k,k+1}}.$$

Moreover, if $(r_k, g_k, r_{k+1}, g_{k+1}) \in \mathcal{D}_d$, then $f_{k,k+1} = g_k^{-1} g_{k+1} = g_{k+1}(r_k, r_{k+1})$. Therefore, substituting in the expressions for the discrete forces, one verifies that

$$F^-(q_k, q_{k+1}) = \frac{\partial \ell_d}{\partial f_{k,k+1}}(r_k, r_{k+1}, f_{k,k+1}) \frac{\partial g_{k+1}}{\partial r_k^\beta}(r_k, r_{k+1}) e_b$$

$$- R_{f_{k,k+1}}^* \frac{\partial \ell_d}{\partial f_{k,k+1}}(r_k, r_{k+1}, f_{k,k+1}) \mathcal{A}_\beta^b(r_k) e_b ,$$

$$F^+(q_{k-1}, q_k) = \frac{\partial \ell_d}{\partial f_{k-1,k}}(r_{k-1}, r_k, f_{k-1,k}) \frac{\partial g_k}{\partial r_k^\beta}(r_{k-1}, r_k) e_b$$

$$+ L_{g_k(r_{k-1}, r_k)}^* \frac{\partial \ell_d}{\partial f_{k-1,k}}(r_{k-1}, r_k, f_{k-1,k}) \mathcal{A}_\beta^b(r_k) e_b .$$

Therefore, we can write a well-defined algorithm on Q/G of the form

$$D_1 L_d^*(r_k, r_{k+1}) + D_2 L_d^*(r_{k-1}, r_k) = F^-(r_k, r_{k+1}) + F^+(r_{k-1}, r_k). \quad (7.31)$$

Equation (7.31) belongs to the type of discretization generalizing variational integrators for systems with external forces developed in [110] (see also [194]),

$$\delta \sum L_d(q_k, q_{k+1}) + \sum (F_d^-(q_k, q_{k+1})\delta q_k + F_d^+(q_k, q_{k+1})\delta q_{k+1}) = 0, \quad (7.32)$$

where F_d^-, F_d^+ are the left and right discrete friction forces. Equation (7.32) defines an integrator $(q_{k-1}, q_k) \longmapsto (q_k, q_{k+1})$ given implicitly by the forced discrete Euler-Lagrange equations

$$D_1 L_d(q_k, q_{k+1}) + D_2 L_d(q_{k-1}, q_k) + F_d^-(q_k, q_{k+1}) + F_d^+(q_{k-1}, q_k) = 0. \quad (7.33)$$

Theorem 7.5.3. *Consider a discrete nonholonomic problem with data L_d : $Q \times Q \longrightarrow \mathbb{R}$, a distribution $\mathcal{D}$ on Q and a discrete constraint space $\mathcal{D}_d$. Let G be a Lie group acting freely and properly on Q, leaving $\mathcal{D}$ invariant and such that $T_q Q = \mathcal{D}_q \oplus V_q$, for all $q \in Q$, where V denotes the vertical bundle of the G-action. Assume further that both L_d and $\mathcal{D}_d$ are invariant under the diagonal action of the Lie group G on the manifold $Q \times Q$. Then, the DLA algorithm (7.30) passes to the reduced space Q/G, yielding a generalized variational integrator in the sense of [110]. We call the algorithm on Q/G the reduced discrete Lagrange-d'Alembert algorithm (RDLA).*

So far, we have obtained that the DLA algorithm respects the structure of the evolution of the symplectic form along the flow of the system (cf. equation (7.24)) and that, in the presence of symmetries, it satisfies a discrete version of the nonholonomic momentum equation. In addition, we have been able to establish in the two extreme cases (horizontal and vertical) Corollary 7.5.2 and Theorem 7.5.3, respectively. These results are important, both from a geometrical and from a numerical perspective. On the one hand, they show interesting interactions between the discrete unconstrained and nonholonomic mechanics, similar to those occuring in the continuous case. On the other hand, when regarding the discrete version of mechanics as an approximation of the continuous one, they provide good arguments to consider the proposed DLA algorithm (7.9) as an appropriate (in a symplectic-momentum sense) discretization of the continuous flow.

It is worth noting, though, that when regarding the discrete nonholonomic mechanics as an approximation of the continuous one, we cannot expect the diagram (R here stands for 'reduced')

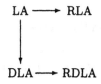

to be commutative in general, because the two horizontal arrows symbolize processes that are of a different mathematical nature (discrete and continuous, respectively). For instance, as we know (cf. Sections 4.2 and 5.3), there exist some special situations in which the reduced Chaplygin system admits a Hamiltonian description, that is, the gyroscopic force F vanishes [49]. But in general, the RDLA will not be a standard variational integrator. In the following section we show that, under strong assumptions on the linearity of the geometric operations involved, the midpoint RDLA algorithm [which corresponds to taking $\alpha = 1/2$ in equation (7.12) (or in equation (7.13), since in this case both discretizations coincide)] yields indeed a variational integrator, i.e. the diagram is commutative. The hypothesis on the linearity are justified by the fact that the diagram involves both discrete and continuous systems. This result provides an additional reason (since we already know that this type of discretization always guarantees a second order accurate numerical approximation to the continuous flow) to consider the midpoint rule as a reliable integrator.

The midpoint RDLA algorithm Consider a nonholonomic Chaplygin system, with the following data: a principal G-bundle $\pi : Q \longrightarrow Q/G$, associated with a free and proper action Ψ of G on Q, a Lagrangian $L : TQ \longrightarrow \mathbb{R}$ which is G-invariant with respect to the lifted action on TQ, and linear nonholonomic constraints determined by the horizontal distribution $\mathcal{D}$ of a principal connection γ on π. In this section, we will focus our attention on the midpoint RDLA algorithm.

For each $q = (r_k, g_k) \in Q$, take a product chart $\varphi = \varphi_1 \times \varphi_2$, given by a chart φ_1 in Q/G and a chart φ_2 in G. For the latter, we take (see [161]): $\varphi_2 = \exp^{-1} \circ L_{g_k^{-1}}$, which is defined in a neighborhood of g_k, where $\exp : \mathfrak{g} \longrightarrow G$ is the exponential mapping. Denote by

$$\eta = \frac{1}{2}\varphi_2(g_k) + \frac{1}{2}\varphi_2(g_{k+1}) = \frac{1}{2}\log(g_k^{-1}g_{k+1}),$$

$$\zeta = \frac{\varphi_2(g_{k+1}) - \varphi_2(g_k)}{h} = \frac{\log(g_k^{-1}g_{k+1})}{h}.$$

We assume that Q/G is itself a linear space, so that we can always take the identity chart $\varphi_1 = \mathrm{Id}_{Q/G}$. With this type of charts, we can construct the discrete Lagrangian and the discrete constraint distribution as explained in Remark 7.4.2.

The discrete Lagrangian then reads

$$L_d^{\frac{1}{2}}(r_k, g_k, r_{k+1}, g_{k+1}) = L\left(r_{k+\frac{1}{2}}, \varphi_2^{-1}(\eta), \frac{r_{k+1} - r_k}{h}, (\varphi_2^{-1})_*(\zeta)\right),$$

and the discrete nonholonomic constraints

$$\zeta + A_\beta^b(r_{k+\frac{1}{2}})\left(\frac{r_{k+1} - r_k}{h}\right)^\beta e_b = 0.$$

As before, the shorthand notation $r_{k+\frac{1}{2}} = \frac{1}{2}r_k + \frac{1}{2}r_{k+1}$ is understood. The above discretizations of the Lagrangian and of the constraints are G-invariant under the diagonal action of the Lie group on the manifold $Q \times Q$.

Here, we will make a different identification between $\mathcal{D}_d$ and $Q/G \times Q/G \times G$, taking into account the specific structure of the constraint functions. More precisely, we identify $\mathcal{D}_d$ with $Q/G \times Q/G \times G$ via the assignment

$$(r_k, g_k, r_{k+1}, g_{k+1}) \in \mathcal{D}_d \longmapsto (r_k, r_{k+1}, \hat{g}),$$

where

$$\hat{g} = \varphi_2^{-1}(\eta) = L_{g_k} \exp(\frac{1}{2}h\zeta) = g_k \exp(-\frac{1}{2}\mathcal{A}(r_{k+\frac{1}{2}})(r_{k+1} - r_k)).$$

The inverse mapping $(r_k, r_{k+1}, \hat{g}) \longmapsto (r_k, g_k, r_{k+1}, g_{k+1}) \in \mathcal{D}_d$ is given by

$$g_{k+1} = \hat{g} \exp(\frac{1}{2}h\zeta), \quad g_k = \hat{g} \exp(-\frac{1}{2}h\zeta). \tag{7.34}$$

Consider the restriction of $L_d^{\frac{1}{2}} : Q \times Q \longrightarrow \mathbb{R}$ to $\mathcal{D}_d$, $L_d^c : \mathcal{D}_d \longrightarrow \mathbb{R}$. Define, as before, the discrete Lagrangian L_d^* on the reduced manifold as

$$L_d^* : Q/G \times Q/G \longrightarrow \mathbb{R}$$
$$(r_k, r_{k+1}) \longmapsto L_d^c(r_k, r_{k+1}, e).$$

Then, we have that

$$\frac{\partial L_d^c}{\partial r_k^\beta} = \frac{\partial L_d^{\frac{1}{2}}}{\partial r_k^\beta} + \frac{\partial L_d^{\frac{1}{2}}}{\partial g_k}\frac{\partial g_k}{\partial r_k^\beta} + \frac{\partial L_d^{\frac{1}{2}}}{\partial g_{k+1}}\frac{\partial g_{k+1}}{\partial r_k^\beta},$$

where, from (7.34),

$$\frac{\partial g_k}{\partial r_k^\beta} = -\frac{1}{2}\left(A_\beta^b(r_{k+\frac{1}{2}}) - \frac{1}{2}\frac{\partial A_\gamma^b}{\partial r^\beta}(r_{k+\frac{1}{2}})(r_{k+1}^\gamma - r_k^\gamma)\right)L_{g_k *}e_b,$$

$$\frac{\partial g_{k+1}}{\partial r_k^\beta} = \frac{1}{2}\left(A_\beta^b(r_{k+\frac{1}{2}}) - \frac{1}{2}\frac{\partial A_\gamma^b}{\partial r^\beta}(r_{k+\frac{1}{2}})(r_{k+1}^\gamma - r_k^\gamma)\right)L_{g_{k+1} *}e_b.$$

Analogously, we see that

$$\frac{\partial L_d^c}{\partial r_{k+1}^\beta} = \frac{\partial L_d^{\frac{1}{2}}}{\partial r_{k+1}^\beta} + \frac{\partial L_d^{\frac{1}{2}}}{\partial g_k}\frac{\partial g_k}{\partial r_{k+1}^\beta} + \frac{\partial L_d^{\frac{1}{2}}}{\partial g_{k+1}}\frac{\partial g_{k+1}}{\partial r_{k+1}^\beta},$$

where

$$\frac{\partial g_k}{\partial r_{k+1}^{\beta}} = -\frac{1}{2}\left(-\mathcal{A}_{\beta}^b(r_{k+\frac{1}{2}}) - \frac{1}{2}\frac{\partial \mathcal{A}_{\gamma}^b}{\partial r^{\beta}}(r_{k+\frac{1}{2}})(r_{k+1}^{\gamma} - r_k^{\gamma})\right) L_{g_{k*}} e_b,$$

$$\frac{\partial g_{k+1}}{\partial r_{k+1}^{\beta}} = \frac{1}{2}\left(-\mathcal{A}_{\beta}^b(r_{k+\frac{1}{2}}) - \frac{1}{2}\frac{\partial \mathcal{A}_{\gamma}^b}{\partial r^{\beta}}(r_{k+\frac{1}{2}})(r_{k+1}^{\gamma} - r_k^{\gamma})\right) L_{g_{k+1*}} e_b.$$

Secondly, we also have

$$0 = \frac{\partial L_d^c}{\partial \hat{g}} = \frac{\partial L_d^{\frac{1}{2}}}{\partial g_k}\frac{\partial g_k}{\partial \hat{g}} + \frac{\partial L_d^{\frac{1}{2}}}{\partial g_{k+1}}\frac{\partial g_{k+1}}{\partial \hat{g}} = R^*_{\exp(-\frac{1}{2}h\zeta)}\frac{\partial L_d^{\frac{1}{2}}}{\partial g_k} + R^*_{\exp(\frac{1}{2}h\zeta)}\frac{\partial L_d^{\frac{1}{2}}}{\partial g_{k+1}}.$$
$$(7.35)$$

Now, we expand the term $\dfrac{\partial L_d^{\frac{1}{2}}}{\partial g_k}(r_k, g_k, r_{k+1}, g_{k+1})$ on the right-hand side of the first equation in the DLA algorithm (7.30) as

$$\frac{\partial L_d^{\frac{1}{2}}}{\partial g_k}(r_k, g_k, r_{k+1}, g_{k+1}) = \frac{1}{2}\frac{\partial L_d^{\frac{1}{2}}}{\partial g_k}(r_k, g_k, r_{k+1}, g_{k+1})$$
$$+ \frac{1}{2}\frac{\partial L_d^{\frac{1}{2}}}{\partial g_k}(r_k, g_k, r_{k+1}, g_{k+1}),$$

and then make use of (7.35) to get the expression

$$\frac{\partial L_d^{\frac{1}{2}}}{\partial g_k}(r_k, g_k, r_{k+1}, g_{k+1}) = \frac{1}{2}\frac{\partial L_d^{\frac{1}{2}}}{\partial g_k}(r_k, g_k, r_{k+1}, g_{k+1})$$
$$- \frac{1}{2}R^*_{\exp(h\zeta)}\frac{\partial L_d^{\frac{1}{2}}}{\partial g_{k+1}}(r_k, g_k, r_{k+1}, g_{k+1}).$$

Analogously, we find for the other term

$$\frac{\partial L_d^{\frac{1}{2}}}{\partial g_k}(r_{k-1}, g_{k-1}, r_k, g_k) = \frac{1}{2}\frac{\partial L_d^{\frac{1}{2}}}{\partial g_k}(r_{k-1}, g_{k-1}, r_k, g_k)$$
$$- \frac{1}{2}R^*_{\exp(-h\zeta)}\frac{\partial L_d^{\frac{1}{2}}}{\partial g_{k-1}}(r_{k-1}, g_{k-1}, r_k, g_k).$$

Then, the discrete forces in the RDLA algorithm take the form

$$F^-(q_k, q_{k+1}) =$$

$$= \frac{\partial L_d^{\frac{1}{2}}}{\partial g_k} \frac{\partial g_k}{\partial r_k^\beta} + \frac{\partial L_d^{\frac{1}{2}}}{\partial g_{k+1}} \frac{\partial g_{k+1}}{\partial r_k^\beta} + \frac{\partial L_d^{\frac{1}{2}}}{\partial g_k}(r_k, g_k, r_{k+1}, g_{k+1})\mathcal{A}_\beta^b(r_k)L_{g_{k\,*}}e_b$$

$$= \frac{1}{2}\frac{\partial L_d^{\frac{1}{2}}}{\partial g_k}\left(-\mathcal{A}_\beta^b(r_{k+\frac{1}{2}}) + \frac{1}{2}\frac{\partial \mathcal{A}_\gamma^b}{\partial r^\beta}(r_{k+\frac{1}{2}})(r_{k+1}^\gamma - r_k^\gamma) + \mathcal{A}_\beta^b(r_k)\right)L_{g_{k\,*}}e_b$$

$$+ \frac{1}{2}\frac{\partial L_d^{\frac{1}{2}}}{\partial g_{k+1}}\left(\left(\mathcal{A}_\beta^b(r_{k+\frac{1}{2}}) - \frac{1}{2}\frac{\partial \mathcal{A}_\gamma^b}{\partial r^\beta}(r_{k+\frac{1}{2}})(r_{k+1}^\gamma - r_k^\gamma)\right)L_{g_{k+1\,*}}e_b$$

$$- \mathcal{A}_\beta^b(r_k)L_{g_{k+1}}Ad_{\exp(-h\zeta)}e_b\right),$$

$$F^+(q_{k-1}, q_k) =$$

$$= \frac{\partial L_d^{\frac{1}{2}}}{\partial g_{k-1}} \frac{\partial g_{k-1}}{\partial r_k^\beta} + \frac{\partial L_d^{\frac{1}{2}}}{\partial g_k} \frac{\partial g_k}{\partial r_k^\beta} + \frac{\partial L_d^{\frac{1}{2}}}{\partial g_k}(r_{k-1}, g_{k-1}, r_k, g_k)\mathcal{A}_\beta^b(r_k)L_{g_{k\,*}}e_b$$

$$= \frac{1}{2}\frac{\partial L_d^{\frac{1}{2}}}{\partial g_{k-1}}\left(\left(\mathcal{A}_\beta^b(r_{k-\frac{1}{2}}) + \frac{1}{2}\frac{\partial \mathcal{A}_\gamma^b}{\partial r^\beta}(r_{k-\frac{1}{2}})(r_k^\gamma - r_{k-1}^\gamma)\right)L_{g_{k-1\,*}}e_b\right.$$

$$- \mathcal{A}_\beta^b(r_k)L_{g_{k-1}}Ad_{\exp(h\zeta)}e_b\right)$$

$$+ \frac{1}{2}\frac{\partial L_d^{\frac{1}{2}}}{\partial g_k}\left(-\left(\mathcal{A}_\beta^b(r_{k-\frac{1}{2}}) + \frac{1}{2}\frac{\partial \mathcal{A}_\gamma^b}{\partial r^\beta}(r_{k-\frac{1}{2}})(r_k^\gamma - r_{k-1}^\gamma)\right) + \mathcal{A}_\beta^b(r_k)\right)L_{g_{k\,*}}e_b.$$

By the linear dependence of $\mathcal{A}(r)$ on r, we have that

$$\mathcal{A}_\beta^b(r_k) = \mathcal{A}_\beta^b(r_{k+\frac{1}{2}}) - \frac{1}{2}\frac{\partial \mathcal{A}_\beta^b}{\partial r^\gamma}(r_{k+\frac{1}{2}})(r_{k+1}^\gamma - r_k^\gamma),$$

$$\mathcal{A}_\beta^b(r_k) = \mathcal{A}_\beta^b(r_{k-\frac{1}{2}}) + \frac{1}{2}\frac{\partial \mathcal{A}_\beta^b}{\partial r^\gamma}(r_{k-\frac{1}{2}})(r_k^\gamma - r_{k-1}^\gamma).$$

Substituting into the expressions for the discrete forces, we get

$$F^-(q_k, q_{k+1}) =$$

$$= \frac{1}{2}\frac{\partial L_d}{\partial g_k}\left(\frac{1}{2}\frac{\partial \mathcal{A}^b_\gamma}{\partial r^\beta}(r_{k+\frac{1}{2}})(r^\gamma_{k+1} - r^\gamma_k) - \frac{1}{2}\frac{\partial \mathcal{A}^b_\beta}{\partial r^\gamma}(r_{k+\frac{1}{2}})(r^\gamma_{k+1} - r^\gamma_k)\right) L_{g_{k*}}e_b$$

$$+ \frac{1}{2}\frac{\partial L_d}{\partial g_{k+1}}\left(\left(\mathcal{A}^b_\beta(r_{k+\frac{1}{2}}) - \frac{1}{2}\frac{\partial \mathcal{A}^b_\gamma}{\partial r^\beta}(r_{k+\frac{1}{2}})(r^\gamma_{k+1} - r^\gamma_k)\right) L_{g_{k+1*}}e_b\right.$$

$$\left. - \mathcal{A}^b_\beta(r_k)L_{g_{k+1*}}(e_b - [h\zeta, e_b])\right)$$

$$= \frac{1}{4}\frac{\partial L_d}{\partial g_k}\left(-\frac{\partial \mathcal{A}^b_\beta}{\partial r^\gamma}(r_{k+\frac{1}{2}})(r^\gamma_{k+1} - r^\gamma_k) + \frac{\partial \mathcal{A}^b_\gamma}{\partial r^\beta}(r_{k+\frac{1}{2}})(r^\gamma_{k+1} - r^\gamma_k)\right) L_{g_{k*}}e_b$$

$$+ \frac{1}{4}\frac{\partial L_d}{\partial g_{k+1}}\left(\frac{\partial \mathcal{A}^b_\beta}{\partial r^\gamma}(r_{k+\frac{1}{2}})(r^\gamma_{k+1} - r^\gamma_k) - \frac{\partial \mathcal{A}^b_\gamma}{\partial r^\beta}(r_{k+\frac{1}{2}})(r^\gamma_{k+1} - r^\gamma_k)\right.$$

$$\left. - 2\mathcal{A}^a_\beta(r_k)\mathcal{A}^c_\gamma(r_{k+\frac{1}{2}})c^b_{ca}(r^\gamma_{k+1} - r^\gamma_k)\right) L_{g_{k+1*}}e_b,$$

$$F^+(q_{k-1}, q_k) =$$

$$= \frac{1}{2}\frac{\partial L_d}{\partial g_{k-1}}\left(\left(\mathcal{A}^b_\beta(r_{k-\frac{1}{2}}) + \frac{1}{2}\frac{\partial \mathcal{A}^b_\gamma}{\partial r^\beta}(r_{k-\frac{1}{2}})(r^\gamma_k - r^\gamma_{k-1})\right) L_{g_{k-1*}}e_b\right.$$

$$\left. - \mathcal{A}^b_\beta(r_k)L_{g_{k-1*}}(e_b + [h\zeta, e_b])\right)$$

$$+ \frac{1}{2}\frac{\partial L_d}{\partial g_k}\left(\frac{1}{2}\frac{\partial \mathcal{A}^b_\beta}{\partial r^\gamma}(r_{k-\frac{1}{2}})(r^\gamma_k - r^\gamma_{k-1}) - \frac{1}{2}\frac{\partial \mathcal{A}^b_\gamma}{\partial r^\beta}(r_{k-\frac{1}{2}})(r^\gamma_k - r^\gamma_{k-1})\right) L_{g_{k*}}e_b$$

$$= \frac{1}{4}\frac{\partial L_d}{\partial g_{k-1}}\left(-\frac{\partial \mathcal{A}^b_\beta}{\partial r^\gamma}(r_{k-\frac{1}{2}})(r^\gamma_k - r^\gamma_{k-1}) + \frac{\partial \mathcal{A}^b_\gamma}{\partial r^\beta}(r_{k-\frac{1}{2}})(r^\gamma_k - r^\gamma_{k-1})\right.$$

$$\left. + 2\mathcal{A}^a_\beta(r_k)\mathcal{A}^c_\gamma(r_{k-\frac{1}{2}})c^b_{ca}(r^\gamma_k - r^\gamma_{k-1})\right) L_{g_{k-1*}}e_b$$

$$+ \frac{1}{4}\frac{\partial L_d}{\partial g_k}\left(-\frac{\partial \mathcal{A}^b_\gamma}{\partial r^\beta}(r_{k-\frac{1}{2}})(r^\gamma_k - r^\gamma_{k-1}) + \frac{\partial \mathcal{A}^b_\beta}{\partial r^\gamma}(r_{k-\frac{1}{2}})(r^\gamma_k - r^\gamma_{k-1})\right) L_{g_{k*}}e_b,$$

where the c^b_{ca} denote the structure constants of the Lie algebra $\mathfrak{g}$, defined by $[e_c, e_a] = c^b_{ca}e_b$. From the G-invariance of the continuous Lagrangian, one can derive that

$$\frac{\partial L_d}{\partial g_k} = -\frac{1}{h}L^*_{g_k^{-1}}\frac{\partial \ell}{\partial \xi}, \qquad \frac{\partial L_d}{\partial g_{k+1}} = \frac{1}{h}L^*_{g_{k+1}^{-1}}\frac{\partial \ell}{\partial \xi}.$$

Therefore, we find that the discrete forces can be rewritten as

$F^-(q_k, q_{k+1}) =$

$$= -\frac{1}{4}\frac{\partial \ell}{\partial \xi}\left(\frac{r_{k+1}^\gamma - r_k^\gamma}{h}\right)\left(-\frac{\partial A_\beta^b}{\partial r^\gamma}(r_{k+\frac{1}{2}}) + \frac{\partial A_\gamma^b}{\partial r^\beta}(r_{k+\frac{1}{2}})\right)e_b$$

$$+ \frac{1}{4}\frac{\partial \ell}{\partial \xi}\left(\frac{r_{k+1}^\gamma - r_k^\gamma}{h}\right)\left(\frac{\partial A_\beta^b}{\partial r^\gamma}(r_{k+\frac{1}{2}}) - \frac{\partial A_\gamma^b}{\partial r^\beta}(r_{k+\frac{1}{2}}) - 2A_\beta^a(r_k)A_\gamma^c(r_{k+\frac{1}{2}})c_{ca}^b\right)e_b$$

$$= \frac{1}{2}\frac{\partial \ell}{\partial \xi}\left(\frac{r_{k+1}^\gamma - r_k^\gamma}{h}\right)\left(\frac{\partial A_\beta^b}{\partial r^\gamma}(r_{k+\frac{1}{2}}) - \frac{\partial A_\gamma^b}{\partial r^\beta}(r_{k+\frac{1}{2}}) - A_\beta^a(r_k)A_\gamma^c(r_{k+\frac{1}{2}})c_{ca}^b\right)e_b,$$

$F^+(q_{k-1}, q_k) =$

$$= \frac{1}{4}\frac{\partial \ell}{\partial \xi}\left(\frac{r_k^\gamma - r_{k-1}^\gamma}{h}\right)\left(-\frac{\partial A_\gamma^b}{\partial r^\beta}(r_{k-\frac{1}{2}}) + \frac{\partial A_\beta^b}{\partial r^\gamma}(r_{k-\frac{1}{2}})\right)e_b$$

$$- \frac{1}{4}\frac{\partial \ell}{\partial \xi}\left(\frac{r_k^\gamma - r_{k-1}^\gamma}{h}\right)\left(\frac{\partial A_\gamma^b}{\partial r^\beta}(r_{k-\frac{1}{2}}) - \frac{\partial A_\beta^b}{\partial r^\gamma}(r_{k-\frac{1}{2}}) + 2A_\beta^a(r_k)A_\gamma^c(r_{k-\frac{1}{2}})c_{ca}^b\right)e_b$$

$$= \frac{1}{2}\frac{\partial \ell}{\partial \xi}\left(\frac{r_k^\gamma - r_{k-1}^\gamma}{h}\right)\left(\frac{\partial A_\beta^b}{\partial r^\gamma}(r_{k-\frac{1}{2}}) - \frac{\partial A_\gamma^b}{\partial r^\beta}(r_{k-\frac{1}{2}}) - A_\beta^a(r_k)A_\gamma^c(r_{k-\frac{1}{2}})c_{ca}^b\right)e_b,$$

Note that the sum of both forces $F^-(q_k, q_{k+1}) + F^+(q_{k-1}, q_k)$ is a discretization around the point q_k of the continuous force

$$F = -\frac{\partial \ell}{\partial \xi^a}\dot{r}^\gamma B_{\beta\gamma}^a = -\frac{\partial \ell}{\partial \xi^a}\dot{r}^\gamma\left(\frac{\partial A_\gamma^a}{\partial r^\beta} - \frac{\partial A_\beta^a}{\partial r^\gamma} - c_{bc}^a A_\beta^b A_\gamma^c\right),$$

which is precisely the gyroscopic force $\overline{\alpha_X}$ obtained in Section 4.2.1 (see the local expression in equation (4.14)). Now, we are in a position to prove the following

Theorem 7.5.4. *Consider a nonholonomic Chaplygin system with data: a free and proper action $\Psi : G \times Q \longrightarrow Q$, a G-invariant Lagrangian $L : TQ \longrightarrow \mathbb{R}$ and a G-invariant distribution D on Q. Assume that Q/G is a linear space, the Lie group G is Abelian and the constraints have a linear dependence on the base point. Then, if the reduced continuous Chaplygin system is Hamiltonian, the midpoint RDLA algorithm is a variational integrator.*

Proof. If the Lie group is Abelian, then the structural constants vanish, $c_{bc}^a = 0$. Therefore, we can see the discrete forces as

$$F^-(q_{k-1}, q_k) = F(\frac{r_{k-1} + r_k}{2}, \frac{r_k - r_{k-1}}{h})$$

$$F^+(q_k, q_{k+1}) = F(\frac{r_k + r_{k+1}}{2}, \frac{r_{k+1} - r_k}{h}).$$

As a consequence, the vanishing of F implies the vanishing of the discrete forces and hence the RDLA algorithm takes the form

$$D_1 L_d^*(r_k, r_{k+1}) + D_2 L_d^*(r_{k-1}, r_k) = 0,$$

which is the variational integrator derived from the discrete Lagrangian L_d^* : $Q/G \times Q/G \longrightarrow \mathbb{R}$.   □

7.6 Numerical examples

TO illustrate the performance of the algorithms obtained from the non-holonomic integrator, we simulate in this section two examples using Matlab: a nonholonomic particle with a quadratic potential and a mobile robot with fixed orientation.

When dealing with symplectic integrators, energy is commonly used as a fairly reliable indicator [61, 210]. This is further justified for variational integrators by Theorem 7.1.1: since variational integrators are already symplectic-momentum integrators, the "lack" of preservation of energy is measuring, in a sense, how far they are from the exact flow they approximate. Since the energy E_L is conserved for nonholonomic systems with homogeneous constraints and nonholonomic integrators are an extention of variational integrators, we will also take E_L as an indicator of their performance.

7.6.1 Nonholonomic particle

This example is a variation of the nonholonomic free particle treated in previous chapters (cf. Sections 4.2.1 and 4.4.1). Let a particle of unit mass be moving in space, $Q = \mathbb{R}^3$, with Lagrangian $L : TQ \longrightarrow \mathbb{R}$,

$$L = K - V = \frac{1}{2} \left(\dot{x}^2 + \dot{y}^2 + \dot{z}^2 \right) - \left(x^2 + y^2 \right),$$

and subject to the constraint

$$\Phi = \dot{z} - y\dot{x} = 0.$$

As shown in Section 4.2.1, this system is a Chaplygin system (the presence of the potential in this case does not affect this fact): indeed, considering the Lie group $G = \mathbb{R}$ and its trivial action by translation on Q,

$$\Psi : \quad \begin{array}{ccc} G \times Q & \longrightarrow & Q \\ (a, (x, y, z)) & \longmapsto & (x, y, z + a), \end{array}$$

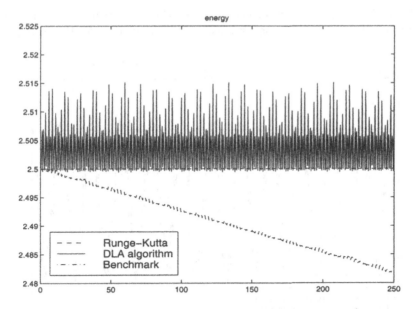

Fig. 7.1. Energy behavior of integrators for the nonholonomic particle with a quadratic potential. Note the long-time stable behavior of the nonholonomic integrator, as opposed to classical methods such as Runge Kutta.

we see that the constraint distribution $\mathcal{D}$ is the horizontal subspace of the principal connection $\gamma = (dz - ydx)e$, where e denotes the generator of the Lie algebra.

As we already know, since the constraint is linear, the energy is a conserved quantity. We are interested here in the extent to which the different integration schemes actually preserve this quantity, as well as the constraint.

The tested algorithms are the following:

- Nonholonomic integrator: L_d^α and $\omega_{d,\alpha}$ with $\alpha = 1/2$;
- Runge-Kutta: 4th order, time step fixed;
- Benchmark: Matlab 5.1 ODE 113 (Predictor-Corrector).

The 4th order Runge-Kutta method is a classical integrator which does not make use of the mechanical nature of the system. To implement it, we have first eliminated the Lagrange multiplier from the nonholonomic equations, so that one gets second order equations in (x, y, z), amenable to integration by RK4. On the other hand, the nonholonomic integrator has been designed taking into account the special structure of the problem.

The two algorithms are run with the same stepsize, $h = 0.2$ to provide a reasonable comparison between them. The Benchmark algorithm is a high order, multi-step, predictor-corrector method which has been carried out with a very small stepsize. It can be regarded as the true solution for this example.

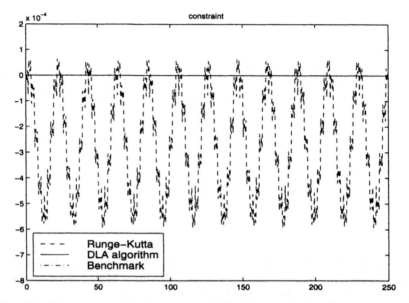

Fig. 7.2. Illustration of the extent to which the tested algorithms respect the constraint. The Runge Kutta technique does not take into account the special nature of nonholonomic systems which explains its bad behavior in this regard.

The results are shown in the figures. In Figure 7.1, we have plotted the energy behavior of the integrators for a relatively short time, but the same pattern is observed if we carry out the simulation for arbitrarily long periods of time. It is immediately apparent that the nonholonomic integrator and the Runge-Kutta method have qualitatively different behaviors. We take as a good indication the fluctuating energy behavior of the nonholonomic integrator, since this property is also observed in symplectic methods.

The extent to which the three algorithms respect the constraints is plotted in Figure 7.2. Notice that the results from the Benchmark algorithm and the nonholonomic integrator are indistinguishable, whereas the behavior of the Runge-Kutta technique is much less satisfactory.

7.6.2 Mobile robot with fixed orientation with a potential

Consider the planar mobile robot presented in Section 5.2.1. Recall that the configuration space for this system is $Q = \mathbb{R}^2 \times \mathbb{S}^1 \times \mathbb{S}^1$, where $(x, y) \in \mathbb{R}^2$

denote the position of the center of the body, $\theta \in \mathbb{S}^1$ the orientation angle of the wheels and $\psi \in \mathbb{S}^1$ the rotation angle of the wheels. The kinetic energy of the robot is given by

$$K = \frac{1}{2}m(\dot{x}^2 + \dot{y}^2) + \frac{1}{2}I\dot{\theta}^2 + \frac{3}{2}I_\omega\dot{\psi}^2,$$

where m is the mass of the robot, I its moment of inertia and I_ω the axial moment of inertia of each wheel, respectively. In addition, we introduce an artificial potential $V = 10\sin\psi$ in order to "force" the behavior of the numerical methods. The Lagrangian of the system is then

$$L = K - V.$$

The constraints, induced by the conditions of no lateral sliding and rolling without sliding of the wheels, are

$$\dot{x} - R\cos\theta\dot{\psi} = 0, \quad \dot{y} - R\sin\theta\dot{\psi} = 0,$$

where R is the radius of the wheels.

Again, this example is a Chaplygin system with Lie group $G = (\mathbb{R}^2, +)$,

$$\Psi: \quad \begin{matrix} G \times Q & \longrightarrow Q \\ ((a,b),(x,y,\theta,\psi)) & \longmapsto (x+a, y+b, \theta, \psi), \end{matrix}$$

and principal connection $\gamma = (dx - R\cos\theta d\psi)e_1 + (dy - R\sin\theta d\psi)e_2$. Since the constraints are linear, the energy is a conserved quantity for the continuous system. It can be immediately checked (cf. Section 5.2.1) that the reduced system on $Q/G = \mathbb{S}^1 \times \mathbb{S}^1$ is the free system determined by the reduced Lagrangian

$$L^* = \frac{1}{2}I\dot{\theta}^2 + \frac{1}{2}(3I_\omega + mR^2)\dot{\psi}^2 - 10\sin\psi.$$

From the discussion in Section 7.5.3, we know that the midpoint DLA scheme can be passed to Q/G and that the RDLA algorithm is indeed variational in the sense defined in [110]. However, the hypotheses of Theorem 7.5.4 are not fulfilled, and hence we cannot assure that the RDLA is a variational integrator. Nevertheless, the comparison of the DLA algorithm with the 4th order Runge-Kutta method in the approximation of the energy and the constraints turns out to be very satisfactory (see Figures 7.3 and 7.4). Again, the 4th order Runge-Kutta method is implemented by elimination of the Lagrange multipliers, while the DLA scheme is implemented using the Newton-Raphson technique. The stepsize employed is $h = 0.2$.

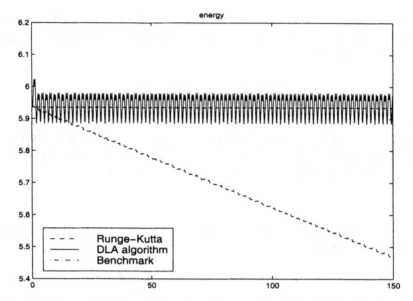

Fig. 7.3. Energy behavior of integrators for a mobile robot with fixed orientation.

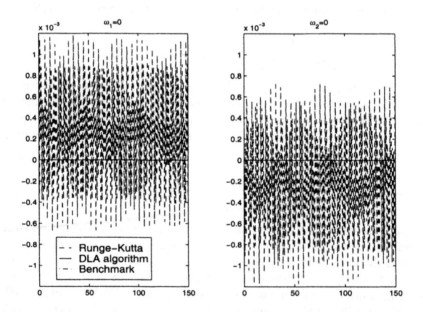

Fig. 7.4. Illustration of the extent to which the tested algorithms respect the constraints $\omega_1 = 0$ and $\omega_2 = 0$. The behavior of the nonholonomic integrator and the Benchmark algorithm are indistinguishable.

8 Control of mechanical systems

IN this chapter, we introduce an important novelty with respect to the previous ones. So far, the emphasis has been on the *analysis* of the dynamics of the systems under consideration. Here, however, the focus is on *design*. Accordingly, we explicitly handle external forces acting on the system, which we call *controls* or *inputs*. Typically, there are less controls than degrees of freedom, i.e. the system is *underactuated*. Throughout the chapter, the Lagrangian is assumed to be of natural type.

Along the exposition, we exploit the affine connection control formalism, which in the last years has revealed to be very appropriate for the modeling and control of mechanical systems. On the one hand, it perfectly captures the dynamic character of the equations of motion both for unconstrained and constrained systems. On the other hand, new interesting geometric structures emerge in a natural way within this framework. Among them, we have to make special mention to the *symmetric product* defined in Chapter 3, which plays a key role in a variety of problems ranging from the controllability analysis and series expansions to motion planning and trajectory tracking algorithms.

The chapter is organized as follows. In Section 8.1, we review some basic facts about simple mechanical control systems, paying special attention to homogeneity and the controllability notions we shall consider. Section 8.2 reviews the existing results concerning configuration accessibility and controllability [146, 147] and series expansion [41]. In Section 8.3 we briefly recall the characterization of controllability in the single-input case found in [142]. Section 8.4 contains the characterization for mechanical systems underactuated by one control. In Section 8.5 we treat two examples to illustrate the results. Finally, in Section 8.6 we extend the previous results to systems subject to isotropic dissipation.

8.1 Simple mechanical control systems

THROUGHOUT this chapter, the manifold Q and the mathematical objects defined on it will be assumed analytic.

A *simple mechanical control system* is defined by a quadruple $(Q, g, V, \mathcal{F})$, where Q is a n-dimensional manifold defining the configuration space of the system, g is a Riemannian metric on Q, V is a smooth function on Q (the potential energy) and $\mathcal{F} = \{F^1, \ldots, F^m\}$ is a set of m linearly independent 1-forms on Q, which physically correspond to forces or torques. We shall simplify the treatment by assuming that the system has no potential energy, that is, $V \equiv 0$, although we remark that the controllability analysis of Section 8.2.1 can be adapted to account for the presence of potential [145].

Instead of the input forces $F^1, \ldots, F^m$, we shall make use of the vector fields $Y_1, \ldots, Y_m$, defined as $Y_i = b_g^{-1}(F^i)$. Roughly speaking, this corresponds to consider "accelerations" rather than forces.

Resorting to the discussion in Chapter 3.1, we have that, if $Y_i = Y_i^A(q)\frac{\partial}{\partial q^A}$, then the control equations for the simple mechanical control system read in coordinates

$$\dot{q}^A = v^A$$
$$\dot{v}^A = -\Gamma_{BC}^A \dot{q}^B \dot{q}^C + \sum_{i=1}^m u_i(t)Y_i^A(q), \quad 1 \le A \le n.$$

These equations can be written in a coordinate-free way as (cf. equation (3.15))

$$\nabla_{\dot{c}(t)}^g \dot{c}(t) = \sum_{i=1}^m u_i(t)Y_i(c(t)). \tag{8.1}$$

The inputs we will consider come from the set $\mathcal{U} = \{u : [0, T] \to \mathbb{R}^m | T > 0, u \text{ is measurable and } \|u\| \le 1\}$, where

$$\|u\| = \sup_{t \in [0,T]} \|u(t)\|_\infty = \sup_{t \in [0,T]} \max_{l=1,\ldots,m} |u_l(t)|.$$

It is clear that we can use a general affine connection in (8.1) instead of the Levi-Civita connection without changing the structure of the equation. This is particularly interesting, since nonholonomic mechanical control systems give also rise to equations of the form (8.1) by means of the nonholonomic affine connection (cf. Section 3.4.2). Indeed, inspecting equation (3.17), we see that if we substitute the affine connection ∇^g by $\overline{\nabla}$ and the vector fields Y_i by $\mathcal{P}(Y_i)$ in (8.1), we exactly get the equations for the nonholonomic control problem. Therefore, the discussion throughout the chapter is carried out for a general affine connection ∇.

We can turn (8.1) into a general affine control system with drift

$$\dot{x}(t) = f(x(t)) + \sum u_i(t)g_i(x(t)), \tag{8.2}$$

which is the typical class of systems considered by the Nonlinear Control community [105, 189, 224]. Recall the definition of vertical lift of a vector field given in Section 2.9. The second-order equation (8.1) on Q can be written as the first-order system on TQ

$$\dot{v} = Z(v) + \sum_{i=1}^{m} u_i(t) Y_i^{\text{lift}}(v), \qquad (8.3)$$

where Z is the geodesic spray associated with the affine connection ∇.

8.1.1 Homogeneity and Lie algebraic structure

One fundamental feature of the control systems in equation (8.1) (or equivalently, in equation (8.3)) is the polynomial dependence of the geodesic spray Z and the vertical lifts Y^{lift} on the velocity variables v^A. As shown in [43], this structure leads to remarkable simplifications in the iterated Lie brackets between the vector fields $\{Z, Y_1^{\text{lift}}, \ldots, Y_m^{\text{lift}}\}$. This fact is indeed the enabling property for the proof of the controllability and the series expansion results that we will review in Section 8.2.

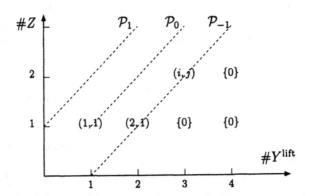

Fig. 8.1. Table of Lie brackets between the drift vector field Z and the input vector field Y^{lift}. The (i, j)th position contains Lie brackets with i copies of Y^{lift} and j copies of Z. The corresponding homogeneous degree is $j - i$. All Lie brackets to the right of $\mathcal{P}_{-1}$ exactly vanish. All Lie brackets to the left of $\mathcal{P}_{-1}$ vanish when evaluated at $v_q = 0_q$. Figure courtesy of Francesco Bullo.

Here, we start by introducing the notion of *geometric homogeneity* as described in [113]: given two vector fields X and X_E, the vector field X is *homogeneous with degree $m \in \mathbb{Z}$ with respect to X_E* if

$$[X_E, X] = mX.$$

Lemma 8.1.1. *Let ∇ be an affine connection on Q with geodesic spray Z, and let Y be a vector field on Q. Denote by Δ the Liouville vector field on TQ. Then*

1. $[\Delta, Z] = (+1)Z$,
2. $[\Delta, Y^{lift}] = (-1)Y^{lift}$.

Hence, the vector field Z is homogeneous of degree $+1$, and the vector field Y^{lift} is homogeneous of degree -1 with respect to the Liouville vector field. In the following, a vector field X on TQ is simply *homogeneous of degree* $m \in \mathbb{Z}$ if it is homogeneous of degree m with respect to Δ.

Let $\mathcal{P}_j$ be the set of vector fields on TQ of homogeneous degree j, so that $Z \in \mathcal{P}_1$ and $Y^{lift} \in \mathcal{P}_{-1}$ (see Figure 8.1). One can see that $[\Delta, X] = 0$, for all $X \in \mathcal{P}_0$, and that $[\mathcal{P}_i, \mathcal{P}_j] \subset \mathcal{P}_{i+j}$. By convention, $\mathcal{P}_j = \{0\}$, for $j \leq -2$.

8.1.2 Controllability notions

The control equations for the mechanical system (8.3) are nonlinear. The standard techniques in control theory [189], as for example the linearization around an equilibrium point or linearization by feedback, do not yield satisfactory results in the analysis of its controllability properties, in the sense that they do not provide necessary and sufficient conditions characterizing them.

The point in the approach of Lewis and Murray [146, 147] to simple mechanical control systems is precisely to focus on what is happening to configurations, rather than to states, since in many of these systems configurations may be controlled, but not configurations and velocities at the same time. The basic question they pose is "what is the set of configurations which are attainable from a given configuration starting from rest?" Moreover, since we are dealing with objects defined on the configuration manifold Q, we expect to find answers on Q, although the control system (8.3) lives in TQ.

Definition 8.1.2. *A solution of (8.1) is a pair (c, u), where $c : [0, T] \longrightarrow Q$ is a piecewise smooth curve and $u \in \mathcal{U}$ such that $(\dot{c}, u)$ satisfies the first order control system (8.3).*

Consider $q_0 \in Q$, $(q_0, 0_{q_0}) \in T_{q_0}Q$ and let $U \subset Q$, $\bar{U} \subset TQ$ be neighborhoods of q_0 and $(q_0, 0_{q_0})$, respectively. Define

$$\mathcal{R}_Q^U(q_0, T) = \{q \in Q \mid \text{there exists a solution } (c, u) \text{ of } (8.1) \text{ such that}$$
$$\dot{c}(0) = 0_{q_0}, c(t) \in U \text{ for } t \in [0, T] \text{ and } \dot{c}(T) \in T_qQ\}$$

$$\mathcal{R}_{TQ}^{\bar{U}}(q_0, T) = \{(q, v) \in TQ \mid \text{there exists a solution } (c, u) \text{ of } (8.1) \text{ such that}$$
$$\dot{c}(0) = 0_{q_0}, (c(t), \dot{c}(t)) \in \bar{U} \text{ for } t \in [0, T] \text{ and } \dot{c}(T) = v \in T_qQ\}$$

and denote

$$\mathcal{R}_Q^U(q_0, \leq T) = \cup_{0 \leq t \leq T} \mathcal{R}_Q^U(q_0, t) \,, \quad \mathcal{R}_{TQ}^{\bar{U}}(q_0, \leq T) = \cup_{0 \leq t \leq T} \mathcal{R}_{TQ}^{\bar{U}}(q_0, t) \,.$$

Now, we recall the notions of accessibility considered in [146].

Definition 8.1.3. *The system* (8.1) *is* locally configuration accessible (LCA) *at* $q_0 \in Q$ *if there exists* $T > 0$ *such that* $\mathcal{R}_Q^U(q_0, \leq t)$ *contains a non-empty open set of* Q, *for all neighborhoods* U *of* q_0 *and all* $0 \leq t \leq T$. *If this holds for any* $q_0 \in Q$, *then the system is called* locally configuration accessible.

Definition 8.1.4. *The system* (8.1) *is* locally accessible (LA) *at* $q_0 \in Q$ *and zero velocity if there exists* $T > 0$ *such that* $\mathcal{R}_{TQ}^{\bar{U}}(q_0, \leq t)$ *contains a non-empty open set of* TQ, *for all neighborhoods* $\bar{U}$ *of* $(q_0, 0_{q_0})$ *and all* $0 \leq t \leq T$. *If this holds for any* $q_0 \in Q$, *then the system is called* locally accessible at zero velocity.

We shall focus our attention on the following concepts of controllability [146].

Definition 8.1.5. *The system* (8.1) *is* small-time locally configuration controllable (STLCC) *at* $q_0 \in Q$ *if there exists* $T > 0$ *such that* $\mathcal{R}_Q^U(q_0, \leq t)$ *contains a non-empty open set of* Q *to which* q_0 *belongs, for all neighborhoods* U *of* q_0 *and all* $0 \leq t \leq T$. *If this holds for any* $q_0 \in Q$, *then the system is called* small-time locally configuration controllable.

Definition 8.1.6. *The system* (8.1) *is* small-time locally controllable (STLC) *at* $q_0 \in Q$ *and zero velocity if there exists* $T > 0$ *such that* $\mathcal{R}_{TQ}^{\bar{U}}(q_0, \leq t)$ *contains a non-empty open set of* TQ *to which* $(q_0, 0_{q_0})$ *belongs, for all neighborhoods* $\bar{U}$ *of* $(q_0, 0_{q_0})$ *and all* $0 \leq t \leq T$. *If this holds for any* $q_0 \in Q$, *then the system is called* small-time locally controllable at zero velocity.

Therefore, the notions of configuration accessibility and controllability concern the reachable set restricted to the configuration space Q, and are weaker than full-state accessibility and controllability, respectively.

8.2 Existing results

HERE we briefly review some accessibility and controllability results obtained in [146, 147] and expose the work by Bullo [41] in describing the evolution of mechanical control systems via a series expansion.

8.2.1 On controllability

Given an affine connection ∇ on Q, recall the definition of symmetric product of two vector fields (cf. Section 3.4.2).

Given the input vector fields $\mathcal{Y} = \{Y_1, \ldots, Y_m\}$, let us denote by $\overline{Sym}(\mathcal{Y})$ the distribution obtained by closing the set $\mathcal{Y}$ under the symmetric product or *symmetric closure* and by $\overline{Lie}(\mathcal{Y})$ the *involutive closure* of $\mathcal{Y}$. With these ingredients, one can prove

Theorem 8.2.1 ([146]). *The control system* (8.1) *is LCA at q (resp. LA at q and zero velocity) if* $\overline{Lie}(\overline{Sym}(\mathcal{Y}))_q = T_qQ$ *(resp.* $\overline{Sym}(\mathcal{Y})_q = T_qQ$*).*

If P is a symmetric product of vector fields in $\mathcal{Y}$, we let $\gamma_i(P)$ denote the number of occurrences of Y_i in P. The *degree* of P will be $\gamma_1(P) + \cdots + \gamma_m(P)$. We shall say that P is *bad* if $\gamma_i(P)$ is even for each $1 \leq i \leq m$. We say that P is *good* if it is not bad. Of course, to make precise sense of these notions (degree, bad, good) one must use the notion of free symmetric algebra, but it should be clear what we mean here. See [146] for details.

The following theorem gives sufficient conditions for STLCC.

Theorem 8.2.2. *Suppose that the system is LCA at q (respectively, LA at q and zero velocity) and that $\mathcal{Y}$ is such that every bad symmetric product P at q in $\mathcal{Y}$ can be written as a linear combination of good symmetric products at q of lower degree than P. Then* (8.1) *is STLCC at q (resp. STLC at q and zero velocity).*

This result was proved in [146], adapting previous work by Sussmann [232] on general control systems of the form (8.2). Throughout the chapter, we will often refer to the conditions of every bad symmetric product at q being a linear combination of good symmetric products at q of lower degree as the *sufficient conditions for STLCC at q*.

8.2.2 Series expansions

Within the realm of geometric control theory, series expansions play a key role in the study of nonlinear controllability [4, 112, 231, 232], trajectory generation and motion planning problems [42, 131, 140, 169, 170], etc. In [155], Magnus describes the evolution of systems on a Lie group. In [64, 85, 114, 233] a general framework is developed to describe the evolution of a nonlinear system via the so-called Chen-Fliess series and its factorization.

In the context of mechanical control systems, the work by Bullo [41] describes the evolution of the trajectories with zero initial velocity via a series

expansion on the configuration manifold Q. In this section we describe this series expansions, which will be key in the subsequent discussion. Before doing
so, however, we need to introduce some notation on analyticity over complex
neighborhoods.

Let $q_0 \in Q$. By selecting a coordinate chart around q_0, we locally identify
$Q \equiv \mathbb{R}^n$. In this way, we write $q_0 \in \mathbb{R}^n$. Let σ be a positive scalar, and define
the complex σ-neighborhood of q_0 in $\mathbb{C}^n$ as $B_\sigma(q_0) = \{z \in \mathbb{C}^n \mid \|z - q_0\| < \sigma\}$.
Let f be a real analytic function on $\mathbb{R}^n$ that admits a bounded analytic
continuation over $B_\sigma(q_0)$. The norm of f is defined as

$$\|f\|_\sigma \triangleq \max_{z \in B_\sigma(q_0)} |f(z)|,$$

where f denotes both the function over $\mathbb{R}^n$ and its analytic continuation.
Given a time-varying vector field $(q, t) \mapsto \mathcal{Z}(q, t) = \mathcal{Z}_t(q)$, let $\mathcal{Z}_t^A$ be its
Ath component with respect to the usual basis on $\mathbb{R}^n$. Assuming $t \in [0, T]$,
and assuming that every component function $\mathcal{Z}_t^A$ is analytic over $B_\sigma(q_0)$, we
define the norm of $\mathcal{Z}$ as

$$\|\mathcal{Z}\|_{\sigma,T} \triangleq \max_{t \in [0,T]} \max_{A \in \{1,\ldots,n\}} \|\mathcal{Z}_t^A\|_\sigma .$$

In what follows, we will often simplify notation by neglecting the subscript
T in the norm of a time-varying vector field. Finally, given an affine connec-
tion ∇ with Christoffel symbols $\{\Gamma_{BC}^A \mid A, B, C \in \{1, \ldots, n\}\}$, introduce the
notation,

$$\|\Gamma\|_\sigma \triangleq \max_{ABC} \|\Gamma_{BC}^A\|_\sigma .$$

In the sequel, we let

$$\mathcal{Z}(q, t) = \sum_{i=1}^m u_i(t) Y_i(q) .$$

Theorem 8.2.3 ([41]). *Let $c(t)$ be the solution of equation (8.1) with input
given by $\mathcal{Z}(q, t)$ and with initial conditions $c(0) = q_0$, $\dot{c}(0) = 0$. Let the
Christoffel symbols $\Gamma_{BC}^A(q)$ and the vector field $\mathcal{Z}(q, t)$ be uniformly integrable
and analytic in Q. Define recursively the time varying vector fields*

$$V_1(q, t) = \int_0^t \mathcal{Z}(q, s) ds ,$$

$$V_k(q, t) = -\frac{1}{2} \sum_{j=1}^{k-1} \int_0^t \langle V_j(q, s) : V_{k-j}(q, s) \rangle \, ds , \quad k \geq 2 ,$$

*where q is maintained fixed at each integral. Select a coordinate chart around
the point $q_0 \in Q$, let $\sigma > \sigma'$ be two positive constants, and assume that*

$$\|\mathcal{Z}\|_\sigma T^2 < L \triangleq \min\left\{ \frac{\sigma - \sigma'}{2^4 n^2 (n+1)}, \frac{1}{2^4 n(n+1)\|\Gamma\|_\sigma}, \frac{\eta^2(\sigma' n^2 \|\Gamma\|_{\sigma'})}{n^2 \|\Gamma\|_{\sigma'}} \right\}.$$

(8.4)

Then the series $(q,t) \longmapsto \sum_{k=1}^\infty V_k(q,t)$ converges absolutely and uniformly in t and q, for all $t \in [0,T]$ and for all $q \in B_{\sigma'}(q_0)$, with the V_k satisfying the bound

$$\|V_k\|_{\sigma'} \leq L^{1-k} \|\mathcal{Z}\|_\sigma^k \, t^{2k-1},$$

(8.5)

Over the same interval, the solution $c(t)$ satisfies

$$\dot{c}(t) = \sum_{k=1}^\infty V_k(c(t), t).$$

(8.6)

This theorem generalizes various previous results obtained in [42] under the assumption of small amplitude forcing. The first few terms of the series (8.6) can be computed to obtain

$$\dot{c}(t) = \overline{\mathcal{Z}}(c(t), t) - \frac{1}{2}\overline{\langle \overline{\mathcal{Z}} : \overline{\mathcal{Z}} \rangle}(c(t), t)$$

$$+ \frac{1}{2}\overline{\langle \overline{\langle \overline{\mathcal{Z}} : \overline{\mathcal{Z}} \rangle} : \overline{\mathcal{Z}} \rangle}(c(t), t) - \frac{1}{2}\overline{\langle \overline{\langle \overline{\langle \overline{\mathcal{Z}} : \overline{\mathcal{Z}} \rangle} : \overline{\mathcal{Z}} \rangle} : \overline{\mathcal{Z}} \rangle}(c(t), t) \quad (8.7)$$

$$- \frac{1}{8}\overline{\langle \overline{\langle \overline{\mathcal{Z}} : \overline{\mathcal{Z}} \rangle} : \overline{\langle \overline{\mathcal{Z}} : \overline{\mathcal{Z}} \rangle} \rangle}(c(t), t) + O(\|\mathcal{Z}\|_\sigma^5 t^9),$$

where $\overline{\mathcal{Z}}(q,t) \equiv \int_0^t \mathcal{Z}(q,s)ds$ and so on.

8.3 The one-input case

THEOREM 8.2.2 gives us sufficient conditions for small-time local configuration controllability. A natural concern both from the theoretical and the practical point of view is to try to sharpen this controllability test. Lewis [142] investigated the single-input case and proved the next result.

Theorem 8.3.1. *Let (Q,g) be an analytic manifold with an affine connection ∇. Let Y be an analytic vector field on Q and $q_0 \in Q$. Then the system*

$$\nabla_{\dot{c}(t)}\dot{c}(t) = u(t)Y(c(t))$$

is STLCC at $q_0 \in Q$ if and only if $\dim Q = 1$.

The fact of being able to completely characterize STLCC in the single-input case (something which has not been accomplished yet for general control systems of the form (8.2)) suggests that understanding local configuration controllability for mechanical systems may be possible. More precisely, examining the single-input case, one can deduce that if (8.1) is STLCC at q_0 then $\dim Q = 1$, which implies $\langle Y : Y \rangle (q_0) \in \mathrm{span}\{Y(q_0)\}$, i.e. sufficient conditions for STLCC are also necessary. Can this be extrapolated to the multi-input case? The following conjecture was posed by Lewis,

> Let a mechanical control system of the form (8.1) be LCA at $q_0 \in Q$. Then it is STLCC at q_0 if and only if there exists a basis of input vector fields which satisfies the sufficient conditions for STLCC at q_0.

Theorem 8.3.1 implies that this claim is true for $m = 1$. In the following section we prove that it is also valid for $m = n - 1$.

8.4 Mechanical systems underactuated by one control

HERE we focus our attention on mechanical control systems of the form (8.1) which has n degrees of freedom and $m = n - 1$ control input vector fields. The following lemma, taken from [231], will be helpful in the proof of Theorem 8.4.2.

Lemma 8.4.1. *Let Q be a n-dimensional analytic manifold. Given $q_0 \in Q$ and $X_1, \ldots, X_p \in \mathfrak{X}(Q)$, $p \leq n$, linearly independent vector fields, there exists a function $\phi : Q \longrightarrow \mathbb{R}$ satisfying the properties*

1. ϕ is analytic,

2. $\phi(q_0) = 0$,

3. $X_1(\phi) = \cdots = X_{p-1}(\phi) = 0$ on a neighborhood V of q_0,

4. $X_p(\phi)(q_0) = -1$,

5. Within any neighborhood of q_0 there exists points q where $\phi(q) < 0$ and $\phi(q) > 0$.

Proof. Let $Z_1, \ldots, Z_n$ be vector fields defined in a neighborhood of q_0 such that $\{Z_1(q_0), \ldots, Z_n(q_0)\}$ form a basis for $T_{q_0}Q$ and $Z_i = X_i$, $1 \leq i \leq p - 1$, $Z_p = -X_p$. Let $t_i \longmapsto \Psi_i(t)$ be the flow of Z_i, $1 \leq i \leq n$. In a sufficiently small neighborhood V of q_0, any point q may be expressed as $q = \Psi_1(t_1) \circ \cdots \circ \Psi_n(t_n)(q_0)$ for some unique n-tuple $(t_1, \ldots, t_n) \in \mathbb{R}^n$. Define $\phi(q) = t_p$. It is a simple exercise to verify that ϕ satisfies the required properties. $\square$

Next, we state and prove the following important result.

Theorem 8.4.2. *Let Q be a n-dimensional analytic manifold and let $Y_1, \ldots, Y_{n-1}$ be analytic vector fields on Q. Consider the control system*

$$\nabla_{\dot{c}(t)}\dot{c}(t) = \sum_{i=1}^{n-1} u_i(t)Y_i(c(t)), \qquad (8.8)$$

and assume that it is LCA at $q_0 \in Q$. Then the system is STLCC at q_0 if and only if there exists a basis of input vector fields satisfying the sufficient conditions for STLCC at q_0.

Proof. We only need to prove one implication (the other one is Theorem 8.2.2). Let us suppose that the system is locally configuration controllable at q_0. Let $\mathcal{I}$ denote the distribution generated by the input vector fields. Either one of the following is true,

1. For all $Y_1, Y_2 \in \mathcal{I}$, $\langle Y_1 : Y_2 \rangle (q_0) \in \mathcal{I}_{q_0}$.

2. There exist $Y_1, Y_2 \in \mathcal{I}$ such that $\langle Y_1 : Y_2 \rangle (q_0) \notin \mathcal{I}_{q_0}$.

In the first case, there is nothing to prove since any basis of input vector fields will satisfy the sufficient conditions for STLCC at q_0. In the second one, it is clear that one can always choose $Y_1, Y_2 \in \mathcal{I}$, linearly independent at q_0 and such that $\langle Y_1 : Y_2 \rangle (q_0) \notin \mathcal{I}_{q_0}$ (if Y_1, Y_2 in (ii) are linearly dependent, then $\langle Y_1 : Y_1 \rangle (q_0) \notin \mathcal{I}_{q_0}$. Take any Y_2 linearly independent with Y_1. If $\langle Y_1 : Y_2 \rangle (q_0) \in \mathcal{I}_{q_0}$, define a new Y_2' by $Y_1 + Y_2$ and we are done).

Then, we can complete the set $\{Y_1(q_0), Y_2(q_0)\}$ to a basis of $\mathcal{I}_{q_0}$,

$$\{Y_1(q_0), Y_2(q_0), \ldots, Y_m(q_0)\}$$

and we have that

$$\mathrm{span}\{Y_1(q_0), Y_2(q_0), \ldots, Y_m(q_0), \langle Y_1 : Y_2 \rangle (q_0)\} = T_{q_0}Q.$$

In this basis, the symmetric products of degree two of the vector fields $\{Y_1, \ldots, Y_m\}$ at q_0 are expressed as

$$\langle Y_1 : Y_1 \rangle (q_0) = lc(Y_1(q_0), \ldots, Y_m(q_0)) + a_{11} \langle Y_1 : Y_2 \rangle (q_0)$$

$$\vdots$$

$$\langle Y_m : Y_m \rangle (q_0) = lc(Y_1(q_0), \ldots, Y_m(q_0)) + a_{mm} \langle Y_1 : Y_2 \rangle (q_0)$$
$$\langle Y_1 : Y_2 \rangle (q_0) = \qquad\qquad\qquad\qquad\qquad a_{12} \langle Y_1 : Y_2 \rangle (q_0)$$
$$\langle Y_1 : Y_3 \rangle (q_0) = lc(Y_1(q_0), \ldots, Y_m(q_0)) + a_{13} \langle Y_1 : Y_2 \rangle (q_0)$$

$$\vdots$$

$$\langle Y_{m-1} : Y_m \rangle (q_0) = lc(Y_1(q_0), \ldots, Y_m(q_0)) + a_{m-1m} \langle Y_1 : Y_2 \rangle (q_0),$$

where $lc(Y_1(q_0), \ldots, Y_m(q_0))$ means a linear combination of $Y_1(q_0), \ldots, Y_m(q_0)$. The coefficients a_{ij} define a symmetric matrix $A = (a_{ij}) \in \mathbb{R}^{m \times m}$. Observe that if $a_{11} = \cdots = a_{mm} = 0$, then the bad symmetric products $\langle Y_i : Y_i \rangle(q_0)$ are in $\mathcal{I}_{q_0}$ and we have finished. Suppose then that the opposite situation is true, that is, there exists $s = s_1$ such that $a_{s_1 s_1} \neq 0$.

What we are going to prove now is that, under the hypothesis of STLCC at q_0, there exists a change of basis $B = (b_{jk})$, $\det B \neq 0$, providing new vector fields in $\mathcal{I}$,

$$Y'_j = \sum_{k=1}^{m} b_{jk} Y_k, \quad 1 \leq j \leq m,$$

which satisfy the sufficient conditions for STLCC at q_0. Since

$$\langle Y'_j : Y'_j \rangle (q_0) = \sum_{k,l=1}^{m} b_{jk} b_{jl} \langle Y_k : Y_l \rangle (q_0)$$

$$= \sum_{k=1}^{m} b_{jk}^2 \langle Y_k : Y_k \rangle (q_0) + 2 \sum_{1 \leq k < l \leq m} b_{jk} b_{jl} \langle Y_k : Y_l \rangle (q_0) \qquad (8.9)$$

$$= lc(Y'_1(q_0), \ldots, Y'_m(q_0))$$

$$+ \left(\sum_{k=1}^{m} b_{jk}^2 a_{kk} + 2 \sum_{1 \leq k < l \leq m} b_{jk} b_{jl} a_{kl} \right) \langle Y_1 : Y_2 \rangle (q_0),$$

the matrix B we are looking for must fulfill

$$\sum_{k=1}^{m} b_{jk}^2 a_{kk} + 2 \sum_{1 \leq k < l \leq m} b_{jk} b_{jl} a_{kl} = 0, \quad 1 \leq j \leq m, \qquad (8.10)$$

or, equivalently,

$$(BAB^T)_{jj} = 0, \quad 1 \leq j \leq m.$$

Note that, since $a_{s_1 s_1} \neq 0$, this is equivalent to

$$b_{js_1} = \frac{-\sum_{k \neq s_1} b_{jk} a_{ks_1}}{a_{s_1 s_1}}$$

$$\pm \frac{\sqrt{(\sum_{k \neq s_1} b_{jk} a_{ks_1})^2 - a_{s_1 s_1}(\sum_{k \neq s_1} b_{jk}^2 a_{kk} + 2 \sum_{k < l, k, l \neq s_1} b_{jk} b_{jl} a_{kl})}}{a_{s_1 s_1}},$$

for each $1 \leq j \leq m$. After some computations, the radicand of this expression becomes

$$\sum_{k \neq s_1} b_{jk}^2 (a_{ks_1}^2 - a_{s_1 s_1} a_{kk}) + 2 \sum_{k < l, k, l \neq s_1} b_{jk} b_{jl} (a_{ks_1} a_{ls_1} - a_{s_1 s_1} a_{kl}).$$

If this radicand is zero, it would imply that the matrix B should be singular in order to satisfy (8.10). We must ensure then that it is possible to select B such that the radicand is different from zero. We do this in the following, studying several cases that can occur. Denoting by

$$a_{kl}^{(2)} = a_{ks_1}a_{ls_1} - a_{s_1s_1}a_{kl}, \quad k, l \in \{1, \ldots, m\}/\{s_1\},$$

we have that the radicand would vanish if

$$\sum_{k \neq s_1} b_{jk}^2 a_{kk}^{(2)} + 2 \sum_{k < l, k, l \neq s_1} b_{jk}b_{jl}a_{kl}^{(2)} = 0. \tag{8.11}$$

Note the similarity between (8.10) and (8.11). Define recursively

$$a_{kl}^{(1)} = a_{kl}, \tag{8.12}$$

$$a_{kl}^{(i)} = a_{ks_{i-1}}^{(i-1)} a_{ls_{i-1}}^{(i-1)} - a_{s_{i-1}s_{i-1}}^{(i-1)} a_{kl}^{(i-1)}, \quad i \geq 2, \; k, l \in \{1, \ldots, m\}/\{s_1, \ldots, s_{i-1}\}.$$

Case A: Here we treat the case when for each i there exists s_i such that $a_{s_is_i}^{(i)} \neq 0$. Several subcases are discussed.

Reasoning as before, (8.11) would imply that for $1 \leq j \leq m$

$$b_{js_2} = lc(b_{j1}, \ldots, \hat{b}_{js_1}, \ldots, \hat{b}_{js_2}, \ldots, b_{jm})$$

$$\pm \frac{1}{a_{s_2s_2}^{(2)}} \sqrt{\sum_{k \neq s_1, s_2} b_{jk}^2 a_{kk}^{(3)} + 2 \sum_{k < l, k, l \neq s_1, s_2} b_{jk}b_{jl}a_{kl}^{(3)}},$$

where the symbol $\hat{b}$ means that the term b has been removed. Iterating this procedure, we finally obtain the following equations for the $b_{js_{m-1}}$,

$$b_{js_{m-1}} = b_{js_m} \frac{-a_{s_{m-1}s_m}^{(m-1)} \pm \sqrt{(a_{s_{m-1}s_m}^{(m-1)})^2 - a_{s_{m-1}s_{m-1}}^{(m-1)} a_{s_ms_m}^{(m-1)}}}{a_{s_{m-1}s_{m-1}}^{(m-1)}}, \quad 1 \leq j \leq m.$$

Let $(b_{js_m})_{1 \leq j \leq m}$ be a non-zero vector in $\mathbb{R}^m$. Now, we distinguish three possibilities.

Case A1: We show that if the radicand $(a_{s_{m-1}s_m}^{(m-1)})^2 - a_{s_{m-1}s_{m-1}}^{(m-1)} a_{s_ms_m}^{(m-1)}$ is positive, then it is possible to obtain the desired change of basis.

If $(a_{s_{m-1}s_m}^{(m-1)})^2 - a_{s_{m-1}s_{m-1}}^{(m-1)} a_{s_ms_m}^{(m-1)} > 0$, then the quadratic polynomial in $b_{js_{m-1}}$

$$a_{s_{m-1}s_{m-1}}^{(m-1)} b_{js_{m-1}}^2 + 2a_{s_{m-1}s_m}^{(m-1)} b_{js_{m-1}}b_{js_m} + a_{s_ms_m}^{(m-1)}b_{js_m}^2, \tag{8.13}$$

has two real roots and we can choose $(b_{js_{m-1}})_{1 \leq j \leq m} \in \mathbb{R}^m$ linearly independent with $(b_{js_m})_{1 \leq j \leq m}$ and such that (8.13) be positive for all $1 \leq j \leq m$. As this polynomial is the radicand of the preceding one,

$$\sum_{k \neq s_1, \dots, s_{m-3}} b_{jk}^2 a_{kk}^{(m-2)} + 2 \sum_{k<l, k, l \neq s_1, \dots, s_{m-3}} b_{jk} b_{jl} a_{kl}^{(m-2)} , \tag{8.14}$$

our choice of $(b_{js_{m-1}})_{1 \leq j \leq m}$ ensures that we can again take $(b_{js_{m-2}})_{1 \leq j \leq m} \in \mathbb{R}^m$, linearly independent with $(b_{js_{m-1}})_{1 \leq j \leq m}$ and $(b_{js_m})_{1 \leq j \leq m}$ such that the expression in (8.14) is positive for all $1 \leq j \leq m$. This is inherited step by step through the iteration process and we are able to choose a non-singular matrix (b_{jk}) satisfying (8.10).

Case A2: We show that when the radicand $(a_{s_{m-1}s_m}^{(m-1)})^2 - a_{s_{m-1}s_{m-1}}^{(m-1)} a_{s_m s_m}^{(m-1)}$ *is negative, then either it is possible to find the change of basis or the system is not STLCC at* q_0.

If $(a_{s_{m-1}s_m}^{(m-1)})^2 - a_{s_{m-1}s_{m-1}}^{(m-1)} a_{s_m s_m}^{(m-1)} < 0$, then (8.13) does not change its sign for all $b_{js_{m-1}}$, b_{js_m}. If this sign is positive, the same argument as in case $A1$ ensures us the choice of the desired matrix. If negative, it implies that (8.14) does not change its sign for all $b_{js_{m-2}}$, $b_{js_{m-1}}$, b_{js_m}. Then, the unique problem we must face is when, through the iteration process, all the radicands are negative. In the following, we discard this latter case by contradiction with the hypothesis of controllability. Apply Lemma 8.4.1 to the vector fields $\{Y_1, \dots, Y_m, \langle Y_1 : Y_2 \rangle\}$ to find a function ϕ satisfying the properties (i)-(v). By (8.7), we have that

$$\dot{c}(t) = \sum_{i=1}^m \bar{u}_i Y_i - \frac{1}{2} \overline{\langle \sum_{j=1}^m \bar{u}_j Y_j : \sum_{k=1}^m \bar{u}_k Y_k \rangle} + O(\|\mathcal{Z}\|_\sigma^3 t^5)$$

$$= \sum_{i=1}^m \bar{u}_i Y_i - \frac{1}{2} \overline{\sum_{j=1}^m \bar{u}_j^2 \langle Y_j : Y_j \rangle + 2 \sum_{j<k} \bar{u}_j \bar{u}_k \langle Y_j : Y_k \rangle} + O(\|\mathcal{Z}\|_\sigma^3 t^5) ,$$

where $\mathcal{Z} = \sum_{i=1}^m u_i Y_i$. Now, observe that $\frac{d}{dt}(\phi(c(t))) = \dot{c}(t)(\phi)$. Then, using properties *(iii)* and *(iv)* of ϕ, we get

$$\frac{d}{dt}(\phi(c(t))) = \frac{1}{2} \overline{\sum_{j=1}^m a_{jj} \bar{u}_j^2 + 2 \sum_{j<k} a_{jk} \bar{u}_j \bar{u}_k} + O(\|\mathcal{Z}\|_\sigma^3 t^5) .$$

The expression $\sum_{j=1}^m a_{jj} \bar{u}_j^2 + 2 \sum_{j<k} a_{jk} \bar{u}_j \bar{u}_k$ does not change its sign, whatever the functions $u_1(t), \dots, u_m(t)$ might be, because as a quadratic polynomial in $\bar{u}_{s_1}$ its radicand is always negative. Therefore,

$$\frac{d}{dt}(\phi(c(t)))$$

has constant sign for sufficiently small t, since $\overline{\sum_{j=1}^m a_{jj} \bar{u}_j^2 + 2 \sum_{j<k} a_{jk} \bar{u}_j \bar{u}_k} = O(\|u\|^2 t^3)$ and dominates $O(\|\mathcal{Z}\|_\sigma^3 t^5) = O(\|u\|^3 t^5)$ when $t \to 0$. Finally,

$$\phi(c(t)) = \phi(q_0) + \int_0^t \frac{d}{ds}(\phi(c(s))) = \int_0^t \frac{d}{ds}(\phi(c(s)))$$

will have constant sign for t small enough. As a consequence, all the points in a neighborhood of q_0 where ϕ has the opposite sign (property (v)) are unreachable in small time, which contradicts the hypothesis of controllability.

Case A3: We show that if the radicand $(a_{s_{m-1}s_m}^{(m-1)})^2 - a_{s_{m-1}s_{m-1}}^{(m-1)} a_{s_m s_m}^{(m-1)}$ vanishes, then an intermediate change of basis reduces the problem to considering $m-1$ input vector fields. The preceding discussion can be then reproduced.

The situation now is similar to that of case $A2$. However, the argument employed above to discard the possibility of all the radicands being negative does not apply, since in this case there *do* exist controls such that $\sum_{j=1}^m a_{jj}\bar{u}_j^2 + 2\sum_{j<k} a_{jk}\bar{u}_j\bar{u}_k$ is zero and hence we should really investigate the sign of $O(\|Z\|_\sigma^3 t^5)$ to reach a contradiction. Instead, what we are going to do is to get a new basis $\{Y_j'\}$ such that $\langle Y_1' : Y_j'\rangle(q_0) \in \mathcal{I}_{q_0}, 1 \le j \le m$, and thus remove one vector field (Y_1') from the discussion. By repeating this procedure, we finally come to consider a limit case, which we will discard by contradiction with the controllability hypothesis.

For $j = 1$, we choose $b_{1s_m} \ne 0$ and

$$b_{1s_{m-1}} = -b_{1s_m}\frac{a_{s_{m-1}s_m}^{(m-1)}}{a_{s_{m-1}s_{m-1}}^{(m-1)}} = C_{s_{m-1}}b_{1s_m}$$

$$b_{1s_{m-2}} = -\frac{a_{s_{m-2}s_{m-1}}^{(m-2)}b_{1s_{m-1}} + a_{s_{m-2}s_m}^{(m-2)}b_{1s_m}}{a_{s_{m-2}s_{m-2}}^{(m-2)}} = C_{s_{m-2}}b_{1s_m}$$

$$\vdots \tag{8.15}$$

$$b_{1s_1} = -\frac{\sum_{k \ne s_1} b_{1k}a_{ks_1}}{a_{s_1 s_1}} = C_{s_1}b_{1s_m}.$$

We denote $C_{s_m} = 1$. For $j > 1$, we select the $(b_{jk})_{1 \le k \le m}$ such that the matrix B be non-singular. Consequently, we change our original basis $\{Y_1, \ldots, Y_m\}$ to a new one $\{Y_1', \ldots, Y_m'\}$. In this basis, following (8.9), one has

$$\langle Y_1' : Y_1'\rangle(q_0) = lc(Y_1'(q_0), \ldots, Y_m'(q_0))$$
$$\langle Y_j' : Y_j'\rangle(q_0) = lc(Y_1'(q_0), \ldots, Y_m'(q_0)) + a_{jj}' \langle Y_1 : Y_2\rangle(q_0), \quad 2 \le j \le m.$$

In addition, one can check that for each $2 \le j \le m$,

$$\langle Y_1' : Y_j'\rangle(q_0) = lc(Y_1'(q_0), \ldots, Y_m'(q_0)) + \left(\sum_{k,l} a_{kl}b_{1k}b_{jl}\right)\langle Y_1 : Y_2\rangle(q_0)$$

$$= lc(Y_1'(q_0), \ldots, Y_m'(q_0)) + b_{1s_m}\left(\sum_l b_{jl}\left(\sum_k a_{kl}C_k\right)\right)\langle Y_1 : Y_2\rangle(q_0).$$

Now, when the C_k are given by (8.15), we have

$$\sum_k a_{kl} C_k = 0, \quad 1 \le l \le m,$$

(see Lemma 8.4.4 below) and this guarantees that

$$\langle Y_1' : Y_j' \rangle(q_0) = lc(Y_1'(q_0), \ldots, Y_m'(q_0)), \quad 2 \le j \le m.$$

If the $a_{jj}' = 0$, $2 \le j \le m$, we are done. Assume then that $a_{33}' \ne 0$, reordering the input vector fields if necessary. Assume further that $\langle Y_2' : Y_3' \rangle(q_0)$ is not a linear combination of $\{Y_1', \ldots, Y_m'\}$ (otherwise, redefine a new Y_2'' as $Y_2' + Y_3'$). Then we have,

$$\langle Y_2' : Y_2' \rangle (q_0) = lc(Y_1'(q_0), \ldots, Y_m'(q_0)) + a_{22}' \langle Y_2' : Y_3' \rangle (q_0)$$

$$\vdots$$

$$\langle Y_m' : Y_m' \rangle (q_0) = lc(Y_1'(q_0), \ldots, Y_m'(q_0)) + a_{mm}' \langle Y_2' : Y_3' \rangle (q_0)$$
$$\langle Y_2' : Y_3' \rangle (q_0) = \qquad\qquad\qquad\qquad\qquad a_{23}' \langle Y_2' : Y_3' \rangle (q_0)$$
$$\langle Y_2' : Y_4' \rangle (q_0) = lc(Y_1'(q_0), \ldots, Y_m'(q_0)) + a_{24}' \langle Y_2' : Y_3' \rangle (q_0)$$

$$\vdots$$

$$\langle Y_{m-1}' : Y_m' \rangle (q_0) = lc(Y_1'(q_0), \ldots, Y_m'(q_0)) + a_{m-1m}' \langle Y_2' : Y_3' \rangle (q_0),$$

where we have denoted with a slight abuse of notation by a_{jk}' the new coefficients corresponding to $\langle Y_2' : Y_3' \rangle$. Consequently, we can now reproduce the preceding discussion, but with the $m-1$ vector fields $\{Y_2', \ldots, Y_m'\}$. That is, we look for one change of basis B' in the vector fields $\{Y_2', \ldots, Y_m'\}$ such that the new ones $\{Y_2'', \ldots, Y_m''\}$ together with Y_1' verify the sufficient conditions for STLCC. Accordingly, we must consider the vanishing of the new polynomials

$$\sum_{k=2}^{m} b_{jk}^2 {}' a_{kk}' + 2 \sum_{2 \le k < l \le m} b_{jk}' b_{jl}' a_{kl}' = 0, \quad 2 \le j \le m.$$

The cases in which the last radicand $(a_{s_{m-1}s_m}^{(m-1)}{}')^2 - a_{s_{m-1}s_{m-1}}^{(m-1)}{}' a_{s_m s_m}^{(m-1)}{}'$ does not vanish are treated as before (cases $A1$ and $A2$). When it vanishes, we obtain a new basis $\{Y_1'' = Y_1', Y_2'', \ldots, Y_m''\}$ such that

$$\langle Y_1'' : Y_1'' \rangle(q_0), \langle Y_2'' : Y_2'' \rangle(q_0) \in \mathcal{I}_{q_0}$$
$$\langle Y_j'' : Y_j'' \rangle(q_0) = lc(Y_1''(q_0), \ldots, Y_m''(q_0)) + c_{jj}' \langle Y_2' : Y_3' \rangle (q_0), \quad 3 \le j \le m$$
$$\langle Y_1'' : Y_j'' \rangle, \langle Y_2'' : Y_{j+1}'' \rangle \in \mathcal{I}_{q_0}, \quad 2 \le j \le m,$$

where there could exit some $3 \le j \le m$ such that $c_{jj}' \ne 0$. By an induction procedure, we come to consider discarding the case of a certain basis $\{Z_1 = Y_1'', Z_2 = Y_2'', \ldots, Z_m\}$ of $\mathcal{I}$ satisfying $\langle Z_i : Z_j \rangle(q_0) \in$

span$\{Z_1(q_0), \ldots, Z_m(q_0)\}$, $1 \le i < j \le m$, and the sufficient conditions for STLCC at q_0 for $Z_1, \ldots, Z_{m-1}$, but such that $\langle Z_m : Z_m \rangle (q_0) \notin$ span$\{Z_1(q_0), \ldots, Z_m(q_0)\}$. Similarly as we have done above, the application of Lemma 8.4.1 with the vector fields $\{Z_1, \ldots, Z_m, \langle Z_m : Z_m \rangle\}$ implies that the system is not controllable at q_0, yielding a contradiction.

Case B: Finally, we prove that if there exists an $i \ge 2$ such that $a_{kk}^{(i)} = 0$, for all $k \in \{1, \ldots, m\}/\{s_1, \ldots, s_{i-1}\}$, then either the desired change of basis is straightforward or an intermediate step can be done that reduces the problem to considering $i - 1$ input vector fields.

In this case, the polynomial

$$\sum_{k \ne s_1, \ldots, s_{i-1}} b_{jk}^2 a_{kk}^{(i)} + 2 \sum_{k < l, k, l \ne s_1, \ldots, s_{i-1}} b_{jk} b_{jl} a_{kl}^{(i)}$$

takes the form

$$2 \sum_{k < l, k, l \ne s_1, \ldots, s_{i-1}} b_{jk} b_{jl} a_{kl}^{(i)}. \qquad (8.16)$$

If any of the $a_{kl}^{(i)}$ is different from zero, then it is clear that we can choose the b_{jk}, $k \notin \{s_1, \ldots, s_{i-1}\}$, such that (8.16) be positive. Then, reasoning as before, we find a regular matrix B yielding the desired change of basis. If this is not the case, i.e. $a_{kl}^{(i)} = 0$, for all $k < l, k, l \notin \{s_1, \ldots, s_{i-1}\}$, we can do the following. Choose $\{(b_{jk})_{1 \le j \le m}\}$, with $k \notin \{s_1, \ldots, s_{i-1}\}$, $m - i + 1$ linearly independent vectors in $\mathbb{R}^m$ such that the minor $\{b_{jk}\}_{1 \le j \le m-i+1}^{k \ne s_1, \ldots, s_{i-1}}$ is regular. Now, let j in equation (8.15) vary between 1 and $m - i + 1$; that is, take

$$b_{js_{i-1}} = -\frac{\sum_{k \ne s_1, \ldots, s_{i-1}}^m b_{jk} a_{s_{i-1}k}^{(i-1)}}{a_{s_{i-1}s_{i-1}}^{(i-1)}}$$

$$b_{js_{i-2}} = -\frac{\sum_{k \ne s_1, \ldots, s_{i-2}}^m b_{jk} a_{s_{i-2}k}^{(i-2)}}{a_{s_{i-2}s_{i-2}}^{(i-2)}}$$

$$\vdots \qquad (8.17)$$

$$b_{js_1} = -\frac{\sum_{k \ne s_1} b_{jk} a_{ks_1}}{a_{s_1 s_1}},$$

for $1 \le j \le m - i + 1$. Finally, for $j > m - i + 1$, we select the b_{jk} such that the matrix B is non-singular. In this manner, in a unique step, we would change to a new basis $\{Y_1', \ldots, Y_m'\}$ verifying

$\langle Y_1' : Y_1' \rangle (q_0), \ldots, \langle Y_{m-i+1}' : Y_{m-i+1}' \rangle (q_0) \in \mathcal{I}_{q_0}$

$\langle Y_j' : Y_j' \rangle (q_0) = lc(Y_1'(q_0), \ldots, Y_m'(q_0)) + a_{jj}' \langle Y_1 : Y_2 \rangle (q_0)$, $m - i + 1 \le j \le m$

$\langle Y_k' : Y_l' \rangle (q_0) \in \mathcal{I}_{q_0}$, $k < l, 1 \le k \le m - i + 1$, with possibly some of the $(a'_{jj})_{m-i+1\leq j\leq m}$ being different from zero. Now, the above discussion can be redone in this context to assert the validity of the theorem.   □

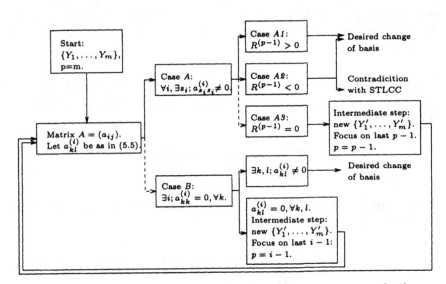

Fig. 8.2. Illustration of the proof of Theorem 8.4.2. $R^{(p-1)}$ denotes $(a^{(p-1)}_{s_{p-1}s_p})^2 - a^{(p-1)}_{s_{p-1}s_{p-1}}a^{(p-1)}_{s_p s_p}$. The dashed lines mean that one cannot fall repeatedly in cases $A3$ or B without contradicting STLCC.

To recap, the steps of the proof can be summarized as follows (see Figure 8.2): first, we have considered the case when there exists for all i a s_i such that $a^{(i)}_{s_i s_i} \neq 0$. We have seen that this case can be subdivided into three: one (case $A1$) ensuring the desired change of basis, another one (case $A2$) in which either one obtains the basis or one contradicts the hypothesis of small-time local configuration controllability, and a third one (case $A3$) where an intermediate change of basis is performed that allows us to focus on the search of a change of basis for $m-1$ of the new vector fields. Then, under the same assumption on the new coefficients, a'_{jk} (i.e. for all i, there exists a s_i such that $a^{(i)'}_{s_i s_i} \neq 0$), we can reproduce the former discussion. We cannot repeatedly fall into case $A3$, since we would contradict the controllability assumption. Finally, we have treated the case when this type of "circular" process is broken (case B): that is, when there exists an i such that $a^{(i)}_{kk} = 0$, for all $k \neq s_1, \ldots, s_{i-1}$. What we have shown then is that this leads to either a new basis of input vector fields satisfying the sufficient conditions for STLCC or a reduced situation where we can "get rid" at the same time of the problems associated with $m-i+1$ vector fields.

Remark 8.4.3. Notice that the proof of this result can be reproduced for the corresponding notions of accessibility and controllability at zero velocity. Indeed, a mechanical control system of the form (8.1) with $m = n - 1$, which is STLC at q_0 and zero velocity is in particular STLCC at q_0. Then, Theorem 8.4.2 implies that there exists a basis of input vector fields $\mathcal{Y}$ satisfying the sufficient conditions of Theorem 8.2.2, so the same result is valid for local controllability at zero velocity.

Lemma 8.4.4. *With the notation of Theorem 8.4.2, assume that we have* $(a_{s_{m-1}s_m}^{(m-1)})^2 - a_{s_{m-1}s_{m-1}}^{(m-1)} a_{s_m s_m}^{(m-1)} = 0$. *Then the coefficients C_k given by (8.15) verify*

$$\sum_{k=1}^{m} a_{kl} C_k = 0, \quad 1 \le l \le m.$$

Proof. From (8.15), one can obtain the following recurrence formula for the coefficients C_k,

$$C_{s_m} = 1, \quad C_{s_j} = -\frac{1}{a_{s_j s_j}^{(j)}} \left(\sum_{i=j+1}^{m} a_{s_i s_j}^{(j)} C_{s_i} \right), \quad 1 \le j \le m-1. \quad (8.18)$$

Let us denote

$$\Sigma(l) = \sum_{k=1}^{m} a_{kl} C_k.$$

It is easy to see that $\Sigma(s_1) = 0$. Indeed, using (8.18), we have that

$$\Sigma(s_1) = a_{s_1 s_1} C_{s_1} + \sum_{i=2}^{m} a_{s_i s_1} C_{s_i} = -\sum_{i=2}^{m} a_{s_i s_1} C_{s_i} + \sum_{i=2}^{m} a_{s_i s_1} C_{s_i} = 0.$$

To prove the result for the remaining indices we can do the following. First, note that

$$a_{s_1 s_j} C_{s_1} = -\frac{a_{s_1 s_j}}{a_{s_1 s_1}} \left(\sum_{i=2}^{m} a_{s_i s_j} C_{s_i} \right) = -\sum_{i=2}^{m} \left(\frac{a_{s_1 s_j} a_{s_i s_j}}{a_{s_1 s_1}} \right) C_{s_i}$$

Then, substituting in $\Sigma(s_j)$, we get

$$\Sigma(s_j) = -\sum_{i=2}^{m} \left(\frac{a_{s_1 s_j} a_{s_i s_j}}{a_{s_1 s_1}} \right) C_{s_i} + \sum_{i=2}^{m} a_{s_i s_j} C_{s_i}$$

$$= \sum_{i=2}^{m} \left(\frac{a_{s_i s_j} a_{s_1 s_1} - a_{s_1 s_j} a_{s_i s_j}}{a_{s_1 s_1}} \right) C_{s_i} = -\frac{1}{a_{s_1 s_1}} \left(\sum_{i=2}^{m} a_{s_i s_j}^{(2)} C_{s_i} \right),$$

where we have used the definition (8.12) for the coefficients $a_{kl}^{(j)}$. This procedure can be iterated to obtain the general expression

$$\Sigma(s_j) = \frac{(-1)^k}{a_{s_1 s_1}^{(2)} a_{s_2 s_2} \cdots a_{s_k s_k}^{(k)}} \left(\sum_{i=k+1}^{m} a_{s_i s_j}^{(k+1)} C_{s_i} \right), \qquad (8.19)$$

which is valid for any $1 \le k \le m-2$.

Now, consider the cases $2 \le j \le m-1$. Take $k = j-1$. Then, using (8.19),

$$\Sigma(s_j) = \frac{(-1)^{j-1}}{a_{s_1 s_1}^{(2)} a_{s_2 s_2} \cdots a_{s_{j-1} s_{j-1}}^{(j-1)}} \left(\sum_{i=j}^{m} a_{s_i s_j}^{(j)} C_{s_i} \right)$$

$$= \frac{(-1)^{j-1}}{a_{s_1 s_1}^{(2)} a_{s_2 s_2} \cdots a_{s_{j-1} s_{j-1}}^{(j-1)}} \left(a_{s_j s_j}^{(j)} C_{s_j} + \sum_{i=j+1}^{m} a_{s_i s_j}^{(j)} C_{s_i} \right) = 0,$$

where in the last equality we have used (8.18). Finally, if $j = m$, we have that

$$\Sigma(s_m) = \frac{(-1)^{m-2}}{a_{s_1 s_1}^{(2)} a_{s_2 s_2} \cdots a_{s_{m-2} s_{m-2}}^{(m-2)}} \left(a_{s_{m-1} s_m}^{(m-1)} C_{s_{m-1}} + a_{s_m s_m}^{(m-1)} C_{s_m} \right)$$

$$= \frac{(-1)^{m-2}}{a_{s_1 s_1}^{(2)} a_{s_2 s_2} \cdots a_{s_{m-2} s_{m-2}}^{(m-2)}} \left(-\frac{(a_{s_{m-1} s_m}^{(m-1)})^2}{a_{s_{m-1} s_{m-1}}^{(m-1)}} + a_{s_m s_m}^{(m-1)} \right).$$

From the hypothesis $(a_{s_{m-1} s_m}^{(m-1)})^2 - a_{s_{m-1} s_{m-1}}^{(m-1)} a_{s_m s_m}^{(m-1)} = 0$, we conclude that $\Sigma(s_m) = 0$, and this completes the proof.   □

Corollary 8.4.5. *Let Q be a 3-dimensional analytic manifold and let Y_1, Y_2 be analytic vector fields on Q. Consider the control system (8.8) and assume that it is locally configuration accessible at $q_0 \in Q$. Let A be the 2×2 symmetric matrix whose elements are given by*

$$\langle Y_1 : Y_1 \rangle (q_0) = lc(Y_1(q_0), Y_2(q_0)) + a_{11} \langle Y_1 : Y_2 \rangle (q_0)$$
$$\langle Y_2 : Y_2 \rangle (q_0) = lc(Y_1(q_0), Y_2(q_0)) + a_{22} \langle Y_1 : Y_2 \rangle (q_0)$$
$$\langle Y_1 : Y_2 \rangle (q_0) = a_{12} \langle Y_1 : Y_2 \rangle (q_0).$$

Then the system is STLCC at q_0 if and only if $\det A < 0$.

Proof. The results follows from the proof of Theorem 8.4.2 by noting that $\det A < 0$ corresponds to case A1, $\det A > 0$ to case A2 and $\det A = 0$ to case A3.   □

Remark 8.4.6. Note that Corollary 8.4.5 together with Theorem 8.3.1 completely characterize the configuration controllability properties of mechanical control systems with 3 degrees of freedom, since fully actuated systems are obviously STLCC.

8.5 Examples

8.5.1 The planar rigid body

Consider a planar rigid body [146]. Fix a point $P \in \mathbb{R}^2$ and let $\{e_1, e_2\}$ be the standard orthonormal frame at that point. Let $\{d_1, d_2\}$ be an orthonormal frame attached to the body at its center of mass. The configuration manifold is then $SE(2)$, with coordinates (x, y, θ), where (x, y) describe the position of the center of mass and θ the orientation of the frame $\{d_1, d_2\}$ with respect to $\{e_1, e_2\}$.

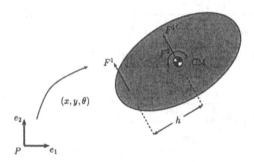

Fig. 8.3. The planar rigid body.

The inputs of the system consist of a force F^1 applied at a distance h from the center of mass CM and a torque, F^2, about CM (see Figure 8.3). In coordinates, the input forces are given by

$$F^1 = -\sin\theta dx + \cos\theta dy - hd\theta , \quad F^2 = d\theta .$$

The Riemannian metric is

$$g = mdx \otimes dx + mdy \otimes dy + Jd\theta \otimes d\theta ,$$

where m is the mass of the body and J its moment of inertia.

The input vector fields can be computed via b_g^{-1} as

$$Y_1 = -\frac{\sin\theta}{m}\frac{\partial}{\partial x} + \frac{\cos\theta}{m}\frac{\partial}{\partial y} - \frac{h}{J}\frac{\partial}{\partial\theta}d\theta , \quad Y_2 = \frac{1}{J}\frac{\partial}{\partial\theta} .$$

One can easily show that the planar body is locally configuration accessible [146]. However, the inputs Y_1, Y_2 fail to satisfy the sufficient conditions for STLCC. In fact,

$$\langle Y_1 : Y_1 \rangle = \frac{2h \cos \theta}{mJ} \frac{\partial}{\partial x} + \frac{2h \sin \theta}{mJ} \frac{\partial}{\partial y},$$

$$\langle Y_1 : Y_2 \rangle = -\frac{\cos \theta}{mJ} \frac{\partial}{\partial x} - \frac{\sin \theta}{mJ} \frac{\partial}{\partial y},$$

$$\langle Y_2 : Y_2 \rangle = 0.$$

Therefore, $\{Y_1, Y_2, \langle Y_1 : Y_2 \rangle\}$ are linearly independent and we have that $\langle Y_1 : Y_1 \rangle = -2h \langle Y_1 : Y_2 \rangle$. Theorem 8.4.2 ensures us STLCC if and only if there exist a basis of input vector fields satisfying the sufficient conditions. We have that

$$\det A = \det \begin{pmatrix} -2h & 1 \\ 1 & 0 \end{pmatrix} = -1 < 0,$$

and consequently, by Corollary 8.4.5, the system is locally configuration controllable. Indeed, this example falls into case $A1$ of the proof of Theorem 8.4.2. Accordingly, we obtain the change of basis: $Y_1' = Y_1 + hY_2$, $Y_2' = Y_2$. This yields

$$\langle Y_1' : Y_1' \rangle = \langle Y_2' : Y_2' \rangle = 0, \quad \langle Y_1' : Y_2' \rangle = \langle Y_1 : Y_2 \rangle,$$

which satisfies the sufficient conditions for STLCC. The new input vector field precisely corresponds to the force $F^{1'}$ in Figure 8.3.

8.5.2 A simple example

The following example does not necessarily correspond to a physical example, but illustrates the proof of Theorem 8.4.2. Consider a mechanical control system on $\mathbb{R}^3$, with coordinates (x, y, z). The Riemannian metric is given by

$$g = dx \otimes dx + dy \otimes dy + dz \otimes dz,$$

and the input vector fields

$$Y_1 = z\frac{\partial}{\partial x} + \frac{\partial}{\partial y} + \frac{1}{4}\frac{\partial}{\partial z}, \quad Y_2 = y\frac{\partial}{\partial x} + \frac{1}{4}\frac{\partial}{\partial y} - \frac{1}{2}\frac{\partial}{\partial z}.$$

In coordinates, we have the following control equations

$$\ddot{x} = u_1 z + u_2 y, \quad \ddot{y} = u_1 + \frac{u_2}{4}, \quad \ddot{z} = \frac{u_1}{4} - \frac{u_2}{2}. \tag{8.20}$$

Since

$$\langle Y_1 : Y_1 \rangle = \langle Y_1 : Y_2 \rangle = \langle Y_2 : Y_2 \rangle = \frac{1}{2}\frac{\partial}{\partial x},$$

we deduce that $\operatorname{span}\{Y_1(q), Y_2(q), \langle Y_1 : Y_2 \rangle(q)\} = T_q Q$ for all $q \in Q$ and the system (8.20) is locally configuration accessible. However, Corollary 8.4.5

implies that it is not STLCC, since $\det A = 0$. Going through the proof of Theorem 8.4.2, we see that this example falls into case $A3$. Choosing the change of basis

$$B = \begin{pmatrix} -1 & 1 \\ 1 & 1 \end{pmatrix},$$

we get the new input vector fields $Y_1' = -Y_1 + Y_2$ and $Y_2' = Y_1 + Y_2$. Now, we have

$$\langle Y_1' : Y_1' \rangle = 0, \quad \langle Y_1' : Y_2' \rangle = 0, \quad \langle Y_2' : Y_2' \rangle = 2\frac{\partial}{\partial x}.$$

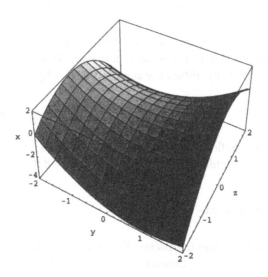

Fig. 8.4. The level surface $\phi(x, y, z) = 0$.

We can compute explicitely the function ϕ of Lemma 8.4.1 for this example. The flows of $Z_1 = Y_1'$, $Z_2 = Y_2'$, $Z_3 = -\langle Y_2' : Y_2' \rangle$ are given by

$$\Psi_1(t)(x, y, z) = (x + (y - z)t, y - 3t/4, z - 3t/4)$$
$$\Psi_2(t)(x, y, z) = (x + (y + z)t + t^2/2, y + 5t/4, z - t/4)$$
$$\Psi_3(t)(x, y, z) = (x - 2t, y, z)$$

Letting (x_0, y_0, z_0) be an arbitrary point, one verifies

$$\Psi_1(t_1) \circ \Psi_2(t_2) \circ \Psi_3(t_3)(x_0, y_0, z_0) =$$

$$\left(x_0 - 2t_3 + (y_0 + z_0 + \frac{1}{2}t_2)t_2 + t_1(y_0 - z_0 - \frac{3}{2}t_2), \right.$$

$$\left. y_0 - \frac{3}{4}t_1 + \frac{5}{4}t_2, z_0 - \frac{3}{4}t_1 - \frac{1}{4}t_2 \right).$$

We may solve for $\phi(x, y, z) = t_3$ as

$$\phi(x, y, z) =$$
$$\frac{1}{18}\left(-9(x - x_0) + 4(y^2 - yy_0 + yz - 5y_0z - 2z^2 + yz_0 + 3y_0z_0 + 5zz_0 - 3z_0^2)\right).$$

In Figure 8.4, we show the level set $\phi(x, y, z) = 0$ for $(x_0, y_0, z_0) = (0, 0, 0)$. The locally accessible configurations from $(0, 0, 0)$ are contained below the surface, where $\phi(x, y, z) \geq 0$.

8.6 Mechanical systems with isotropic damping

IN this section, we show that all of the previous results can be extended to the case of mechanical control systems subject to dissipative forces of linear isotropic nature.

The systems described by equations (8.1) are subject to no damping force. However, in a number of situations, friction and dissipation play a relevant role. Consider, for instance, a blimp experiencing the resistance of the air or an underwater vehicle moving in the sea. Introduce a linear isotropic term of dissipation into equations (8.1),

$$\nabla_{\dot{c}(t)}\dot{c}(t) = k_d\dot{c}(t) + u_i(t)Y_i(c(t)), \tag{8.21}$$

where $k_d \in \mathbb{R}$. In local coordinates,

$$\ddot{q}^A + \Gamma_{BC}^A(q)\dot{q}^B\dot{q}^C = k_d\dot{q}^A + u_i(t)Y_i^A(q). \tag{8.22}$$

This second-order system can be written as a first-order differential equation on the tangent bundle TQ, in the same way as we did in (8.2). Using the Liouville vector field Δ, the control system becomes

$$\dot{v}(t) = Z(v(t)) + k_d\Delta(v(t)) + u_i(t)Y_i^{\text{lift}}(v(t)), \tag{8.23}$$

where $t \mapsto v(t)$ is now a curve in TQ describing the evolution of a first-order control affine system.

In the following we extend the controllability analysis and the series expansion results to systems of the form (8.23). As before, the enabling property is the homogeneity property of the vector fields Z, Δ and Y_i^{lift}, which leads to several simplifications when computing the Lie brackets of the set $\{Z + k_d\Delta, Y_1^{\text{lift}}, \ldots, Y_m^{\text{lift}}\}$.

8.6.1 Local accessibility and controllability

Here we study conditions for accessibility and controllability of mechanical systems with dissipation. The next proposition show that the involutive closure of the system (8.1) at zero velocity is the same as the one of (8.21).

Proposition 8.6.1. *Consider the distributions*

$$\mathcal{D}_{(1)} = \text{span}\{Z, \mathcal{Y}^{lift}\}, \quad \mathcal{D}_{(1)}^\Delta = \text{span}\{Z + k_d \Delta, \mathcal{Y}^{lift}\}.$$

Define recursively

$$\mathcal{D}_{(k)} = \mathcal{D}_{(k-1)} + [\mathcal{D}_{(k-1)}, \mathcal{D}_{(k-1)}], \quad \mathcal{D}_{(k)}^\Delta = \mathcal{D}_{(k-1)}^\Delta + [\mathcal{D}_{(k-1)}^\Delta, \mathcal{D}_{(k-1)}^\Delta], \ k \geq 2.$$

Then, it holds that $\mathcal{D}_{(k)}(0_q) = \mathcal{D}_{(k)}^\Delta(0_q)$, for all k. Consequently, the accessibility distributions

$$\mathcal{D}_{(\infty)}(0_q) = \overline{\text{Lie}}(Z, \mathcal{Y}^{lift})_q \ and \ \mathcal{D}_{(\infty)}^\Delta(0_q) = \overline{\text{Lie}}(Z + k_d \Delta, \mathcal{Y}^{lift})_q$$

coincide.

Proof. Obviously $\mathcal{D}_{(1)}(0_q) = \mathcal{D}_{(1)}^\Delta(0_q)$. Moreover, we have $[\mathcal{D}_{(1)}, \mathcal{D}_{(1)}] \subset \mathcal{D}_{(2)}^\Delta$ and $[\mathcal{D}_{(1)}^\Delta, \mathcal{D}_{(1)}^\Delta] \subset \mathcal{D}_{(2)}$, since

$$[Z + k_d \Delta, Y^{lift}] = [Z, Y^{lift}] - k_d Y^{lift}.$$

Let us assume that

$$\mathcal{D}_{(k)}(0_q) = \mathcal{D}_{(k)}^\Delta(0_q), \tag{8.24}$$

$$[\mathcal{D}_{(k)}, \mathcal{D}_{(k)}] \subset \mathcal{D}_{(k+1)}^\Delta, \tag{8.25}$$

$$[\mathcal{D}_{(k)}^\Delta, \mathcal{D}_{(k)}^\Delta] \subset \mathcal{D}_{(k+1)}. \tag{8.26}$$

hold for k and let us show that (8.24-8.26) are valid for $k + 1$. We have

$$\mathcal{D}_{k+1} = \mathcal{D}_{(k)} + [\mathcal{D}_{(k)}, \mathcal{D}_{(k)}] \subset \mathcal{D}_{(k)} + \mathcal{D}_{(k+1)}^\Delta \implies$$
$$\mathcal{D}_{k+1}(0_q) \subset \mathcal{D}_{(k)}(0_q) + \mathcal{D}_{(k+1)}^\Delta(0_q) = \mathcal{D}_{(k)}^\Delta(0_q) + \mathcal{D}_{(k+1)}^\Delta(0_q) = \mathcal{D}_{(k+1)}^\Delta(0_q).$$

Similarly, we can prove $\mathcal{D}_{k+1}^\Delta(0_q) \subset \mathcal{D}_{(k+1)}(0_q)$, and thus $\mathcal{D}_{(k+1)}(0_q) = \mathcal{D}_{(k+1)}^\Delta(0_q)$. On the other hand,

$$[\mathcal{D}_{(k+1)}^\Delta, \mathcal{D}_{(k+1)}^\Delta] = [\mathcal{D}_{(k)}^\Delta + [\mathcal{D}_{(k)}^\Delta, \mathcal{D}_{(k)}^\Delta], \mathcal{D}_{(k)}^\Delta + [\mathcal{D}_{(k)}^\Delta, \mathcal{D}_{(k)}^\Delta]]$$
$$\subset [\mathcal{D}_{(k)}^\Delta + \mathcal{D}_{(k+1)}, \mathcal{D}_{(k)}^\Delta + \mathcal{D}_{(k+1)}]$$
$$\subset \mathcal{D}_{(k+1)} + [\mathcal{D}_{(k)}^\Delta, \mathcal{D}_{(k+1)}] + \mathcal{D}_{(k+2)} = [\mathcal{D}_{(k)}^\Delta, \mathcal{D}_{(k+1)}] + \mathcal{D}_{(k+2)}.$$

Thus, it remains to be checked that $[\mathcal{D}_{(k)}^{\Delta}, \mathcal{D}_{(k+1)}] \subset \mathcal{D}_{(k+2)}$. Observe that

$$[\mathcal{D}_{(k)}^{\Delta}, \mathcal{D}_{(k+1)}] = [\mathcal{D}_{(k-1)}^{\Delta} + [\mathcal{D}_{(k-1)}^{\Delta}, \mathcal{D}_{(k-1)}^{\Delta}], \mathcal{D}_{(k+1)}]$$
$$\subset [\mathcal{D}_{(k-1)}^{\Delta}, \mathcal{D}_{(k+1)}] + \mathcal{D}_{(k+2)},$$

where we have used the induction hypothesis on (8.26), i.e. $[\mathcal{D}_{(k-1)}^{\Delta}, \mathcal{D}_{(k-1)}^{\Delta}] \subset \mathcal{D}_{(k)}$. By a recursive argument, we find that what we must show is that $[\mathcal{D}_{(1)}^{\Delta}, \mathcal{D}_{(k+1)}] \subset \mathcal{D}_{(k+2)}$. Clearly, $[Y_i^{\mathrm{lift}}, \mathcal{D}_{(k+1)}] \subset \mathcal{D}_{(k+2)}$, $i \in \{1, \dots, m\}$. In addition,

$$[Z + k_d \Delta, \mathcal{D}_{(k+1)}] = [Z, \mathcal{D}_{(k+1)}] + [k_d \Delta, \mathcal{D}_{(k+1)}] \subset \mathcal{D}_{(k+2)},$$

since $[\Delta, X] \in \mathcal{D}_{(k+1)}$, for all $X \in \mathcal{D}_{(k+1)}$, by homogeneity. Finally, it can be similarly shown using (8.25) that $[\mathcal{D}_{(k+1)}, \mathcal{D}_{(k+1)}] \subset \mathcal{D}_{(k+2)}^{\Delta}$. Thus, (8.24-8.26) are satisfied for all k. $\qquad\square$

Corollary 8.6.2. *Consider a mechanical control system of the form* (8.21). *Then*

1. *the system is LA at q starting with zero velocity if* $\overline{\mathrm{Sym}}(\mathcal{Y})_q = T_q Q$,

2. *the system is LCA at $q \in Q$ if* $\overline{\mathrm{Lie}}(\overline{\mathrm{Sym}}(\mathcal{Y}))_q = T_q Q$.

Proof. The manifold Q can be identified with the set of zero vectors $Z(TQ)$ of TQ by the diffeomorphism $q \mapsto 0_q$. Consequently, the tangent space to $Z(TQ)$ at 0_q is isomorphic to $T_q Q$. On the other hand, the natural projection $\tau_Q(v_q) = q$ defines the set $\mathcal{V}$, which are those ones falling in the kernel of $T\tau_Q : TTQ \to TQ$. One has that $\mathcal{V}_{0_q}$ is isomorphic to $T_q Q$ for all $q \in Q$. Both parts give us the natural decomposition

$$T_{0_q} TQ = T_{0_q}(Z(TQ)) \oplus \mathcal{V}_{0_q} \simeq T_q Q \oplus T_q Q.$$

The first copy of $T_q Q$ corresponds to configurations, the second one to velocities. The result follows from the former proposition and Proposition 5.9 in [146] which asserts that

$$\mathcal{D}_{(\infty)}(0_q) \cap \mathcal{V}_{0_q} = \overline{\mathrm{Sym}}(\mathcal{Y})_q^{\mathrm{lift}}, \quad \mathcal{D}_{(\infty)}(0_q) \cap T_{0_q}(Z(TQ)) = \overline{\mathrm{Lie}}(\overline{\mathrm{Sym}}(\mathcal{Y}))_q.$$

$\qquad\square$

Next, we examine the small-time local controllability properties of the system in equation (8.21). We shall use the following conventions. Every Lie bracket B in $\{X_0, X_1, \dots, X_m\}$ has a unique decomposition as $B = [B_1, B_2]$. In turn, each of B_1 and B_2 may be uniquely expressed as $B_1 = [B_{11}, B_{12}]$ and $B_2 = [B_{21}, B_{22}]$. This process may be continued until we obtain elements which are not decomposable. All such elements $B_{i_1\ldots i_l}$, $i_b \in \{1,2\}$, shall be called *components* of B. The *length* of a component $B_{i_1\ldots i_l}$ is l. Recall that a Lie bracket B in $\{X_0, X_1, \ldots, X_m\}$ is *bad* if $\delta_0(B)$ is odd and $\delta_i(B)$ is even, $i \in \{1, \ldots, m\}$, where $\delta_a(B)$ denotes the number of times that X_a occurs in B. Otherwise, B is *good*. The *degree* of B is given by $\delta(B) = \delta_0(B) + \delta_1(B) + \cdots + \delta_m(B)$. As before for the symmetric product, to make precise sense of all these notions (component, good, bad, degree) one must use the notion of free Lie algebra, but we note again that it should be clear what we mean here. See [146] for details.

Let the system be LA at $q \in Q$ starting with zero velocity (resp. LCA at $q \in Q$). The system in equation (8.21) is STLC at q starting with zero velocity (resp. STLCC) if:

(Sussmann's criterium on the set $\{Z + k_d\Delta, \mathcal{Y}^{\text{lift}}\}$): Every bad bracket B in $\{Z + k_d\Delta, \mathcal{Y}^{\text{lift}}\}$ is a $\mathbb{R}$-linear combination of good brackets evaluated at 0_q of lower degree than B.

This fact is a consequence of the results in [146, 232] (see also Section 8.2.1).

We shall show that if the conditions for STLC and STLCC are satisfied for the set $\{Z, \mathcal{Y}^{\text{lift}}\}$, then they are also verified for the set $\{Z+k_d\Delta, \mathcal{Y}^{\text{lift}}\}$. We illustrate this fact by considering two low order settings. First, every bracket B of order 1 or 2, i.e., $\delta(B) \leq 2$, is good. In addition, $[Z + k_d\Delta, Y^{\text{lift}}] = [Z, Y^{\text{lift}}] - k_d Y^{\text{lift}}$, and therefore, every good bracket in $\{Z + k_d\Delta, \mathcal{Y}^{\text{lift}}\}$ of degree 2 is the sum of the corresponding good bracket in $\{Z, \mathcal{Y}^{\text{lift}}\}$ plus some good brackets of lower degree in $\{Z + k_d\Delta, \mathcal{Y}^{\text{lift}}\}$.

For $\delta(B) = 3$, the unique bad brackets are $[Y_i^{\text{lift}}, [Z+k_d\Delta, Y_i^{\text{lift}}]]$ such that

$$[Y_i^{\text{lift}}, [Z + k_d\Delta, Y_i^{\text{lift}}]] = [Y_i^{\text{lift}}, [Z, Y_i^{\text{lift}}]].$$

Consequently,

$$[Y_i^{\text{lift}}, [Z + k_d\Delta, Y_i^{\text{lift}}]](0_q) = [Y_i^{\text{lift}}, [Z, Y_i^{\text{lift}}]](0_q) = \sum \xi_l C_l(0_q),$$

where C_l are some good brackets in $\{Z, \mathcal{Y}^{\text{lift}}\}$ of degree ≤ 2. At the same time, these brackets are linear combinations of good brackets in $\{Z + k_d\Delta, \mathcal{Y}^{\text{lift}}\}$ of degree ≤ 3. In addition, observe that all the good brackets of degree 3 are either of the form

$$[Y_i^{\text{lift}}, [Z + k_d\Delta, Y_j^{\text{lift}}]] = [Y_i^{\text{lift}}, [Z, Y_j^{\text{lift}}]], \quad i \neq j,$$

or

$$[Z + k_d\Delta, [Z + k_d\Delta, Y_j^{\text{lift}}]] = [Z, [Z, Y_j^{\text{lift}}]] - [Z + k_d\Delta, Y_j^{\text{lift}}].$$

Thus, again, every good bracket in $\{Z + k_d\Delta, \mathcal{Y}^{\text{lift}}\}$ can be put as the sum of the corresponding good bracket in $\{Z, \mathcal{Y}^{\text{lift}}\}$ plus some good brackets in $\{Z + k_d\Delta, \mathcal{Y}^{\text{lift}}\}$ of lower degree.

Proposition 8.6.3. *Assume Sussmann's criterium on $\{Z, \mathcal{Y}^{\text{lift}}\}$. Then*

1. *every bad bracket B in $\{Z + k_d\Delta, \mathcal{Y}^{\text{lift}}\}$ of degree k, evaluated at 0_q, is a $\mathbb{R}$-linear combination of good brackets of lower degree,*

2. *every good bracket C in $\{Z + k_d\Delta, \mathcal{Y}^{\text{lift}}\}$ of degree k, evaluated at 0_q, is a $\mathbb{R}$-linear combination of the corresponding good bracket in $\{Z, \mathcal{Y}^{\text{lift}}\}$ and of some brackets in $\{Z + k_d\Delta, \mathcal{Y}^{\text{lift}}\}$ of lower degree, and*

3. *every good bracket in $\{Z, \mathcal{Y}^{\text{lift}}\}$ of degree k, evaluated at 0_q, is a $\mathbb{R}$-linear combination of good brackets in $\{Z + k_d\Delta, \mathcal{Y}^{\text{lift}}\}$ of degree $\leq k$.*

Proof. First, note that *(iii)* is an immediate consequence of *(i)* and *(ii)*. Next, we show *(i)* by induction. The result holds for $k \leq 3$. Suppose that it is valid for k and let us prove it for $k + 1$. Let B be a bad bracket in $\{Z + k_d\Delta, \mathcal{Y}^{\text{lift}}\}$ of degree $k + 1$. This means that $\delta_0(B)$ is odd and $\delta_i(B)$ is even, $i \in \{1, \ldots, m\}$. We first select a term of the form $Z + k_d\Delta$ which is in one of the longest components of B. We then write B as the sum of two Lie brackets, $B = B_1 + B_2$, by expanding the chosen term. By the homogeneity properties, we have that $\delta_0(B_2) = \delta_0(B) - 1$, $\delta_i(B_2) = \delta_i(B)$. Consequently, B_2 is a good bracket in $\{Z + k_d\Delta, \mathcal{Y}^{\text{lift}}\}$ of degree k. Expanding now all the possible terms $Z + k_d\Delta$ in B_1 as the sum of two Lie brackets, one with Z and the other with $k_d\Delta$ (going from the ones in the longest components of B_1 to those in the shortest ones), we finally obtain that B can be written as the sum of the corresponding bad bracket in $\{Z, \mathcal{Y}^{\text{lift}}\}$, plus good/bad brackets in $\{Z, \mathcal{Y}^{\text{lift}}\}$ of degree $\leq k$, plus B_2, which is a good bracket in $\{Z + k_d\Delta, \mathcal{Y}^{\text{lift}}\}$ of degree k. The induction hypothesis now implies *(i)*.

Let us prove *(ii)*. Let C be a good bracket in $\{Z + k_d\Delta, \mathcal{Y}^{\text{lift}}\}$ of degree $k + 1$. Expanding the terms $Z + k_d\Delta$ as before, we find that C can be written as the sum of the corresponding good bracket in $\{Z, \mathcal{Y}^{\text{lift}}\}$, plus brackets in $\{Z, \mathcal{Y}^{\text{lift}}\}$ of degree $\leq k$, plus brackets in $\{Z + k_d\Delta, \mathcal{Y}^{\text{lift}}\}$ of degree k. The induction hypothesis implies then *(ii)*. $\square$

Corollary 8.6.4. *Consider a mechanical control system as in (8.21). Then, we have*

1. *the system is STLC at $q \in Q$ starting with zero velocity if $\overline{\mathrm{Sym}}(\mathcal{Y})_q = T_qQ$ and every bad symmetric product B in $\overline{\mathrm{Sym}}(\mathcal{Y})_q$ is a linear combination of good symmetric products of lower degree, and*

2. *the system is STLCC at $q \in Q$ if $\overline{\text{Lie}}(\overline{\text{Sym}}(\mathcal{Y}))_q = T_qQ$ and every bad symmetric product B in $\overline{\text{Sym}}(\mathcal{Y})_q$ is a linear combination of good symmetric products of lower degree.*

Proof. It follows from the fact that there is a 1-1 correspondence between bad (resp. good) Lie brackets in $\{Z, \mathcal{Y}^{\text{lift}}\}$ and bad (resp. good) symmetric products in $\mathcal{Y}$; see [146]. □

8.6.2 Kinematic controllability

Kinematic controllability [45] has direct relevance to the trajectory planning problem for mechanical systems of the form (8.1). Here, we present a generalized notion of kinematic controllability for affine connection systems with isotropic dissipation. Consider a mechanical system as in (8.21), and let $\mathcal{I}$ be the distribution generated by the input vector fields $\{Y_1, \ldots, Y_m\}$. A controlled solution to equations (8.21) is a curve $t \mapsto q(t) \in Q$ satisfying

$$\nabla_{\dot{q}}\dot{q} - k_d\dot{q} \in \mathcal{I}_{q(t)} . \tag{8.27}$$

Let $s : [0, T] \to [0, 1]$ be a twice-differentiable function such that $s(0) = 0, s(T) = 1, \dot{s}(0) = \dot{s}(T) = 0$, and $\dot{s}(t) > 0$ for all $t \in (0, T)$. We call such a curve s a *time scaling*. A vector field V is a *decoupling vector field* for the mechanical system (8.21) if, for any time scaling s and for any initial condition q_0, the curve $t \mapsto q(t)$ on Q solving

$$\dot{q}(t) = \dot{s}(t)V(q(t)), \qquad q(0) = q_0, \tag{8.28}$$

satisfies the conditions in (8.27). Additionally, the integral curves of V defined on the time interval $[0, 1]$ are called *kinematic motions*.

Lemma 8.6.5. *The vector field V is decoupling for the mechanical system (8.21) if and only if $V \in \mathcal{I}$ and $\langle V : V \rangle \in \mathcal{I}$.*

Proof. Given a curve $\gamma : [0, T] \to Q$ satisfying equation (8.28), we compute

$$\nabla_{\dot{\gamma}}\dot{\gamma} = \ddot{s}V + \dot{s}\nabla_{\dot{\gamma}}V = \ddot{s}V + \dot{s}^2\nabla_V V.$$

Next, the curve γ is a kinematic motion if, for all time scalings s, the constraints (8.27) are satisfied. Thus

$$\nabla_{\dot{\gamma}}\dot{\gamma} - k_d\dot{\gamma} = (\ddot{s} - k_d\dot{s})V + \frac{\dot{s}^2}{2}\langle V : V \rangle \in \mathcal{I}.$$

Since s is an arbitrary time scaling and q_0 is an arbitrary point, V and $\langle V : V \rangle$ must separately belong to the input distribution $\mathcal{I}$. The other implication is trivial. □

We shall say that the system (8.21) is *locally kinematically controllable* if for any $q \in Q$ and any neighborhood U_q of q, the set of reachable configurations from q by kinematic motions remaining in U_q contains q in its interior.

The following sufficient test for local kinematic controllability was given in [45] for Levi-Civita affine connection systems without dissipation and remains valid for the class of systems under consideration.

Lemma 8.6.6. *The system (8.21) is locally kinematically controllable if there exist $p \in \{1, \ldots, m\}$ vector fields $\{V_1, \ldots, V_p\} \subset \mathcal{I}$ such that*

1. *$\langle V_c : V_c \rangle \in \mathcal{I}$, for all $c \in \{1, \ldots, p\}$, and*

2. *$\overline{\mathrm{Lie}}(V_1, \ldots, V_p)$ has rank n at all $q \in Q$.*

There are many interesting examples which are kinematically controllable. Among them, we mention the Snakeboard [44], three link planar robot manipulators with a passive joint [45] (see also [151]), and underwater vehicles [45].

8.6.3 Series expansion for the forced evolution starting from rest

The result in this section extends the treatment in [41]. Consider the system as in equation (8.21), with initial condition $\dot{q}(0) = 0$.

Proposition 8.6.7. *Given any integrable input vector field $(q, t) \mapsto Y(q, t)$, consider*

$$V_1(q, t) = \int_0^t e^{k_d(t-\tau)} Y(q, \tau) d\tau \,,$$

$$V_k(q, t) = -\frac{1}{2} \sum_{j=1}^{k-1} \int_0^t e^{k_d(t-\tau)} \langle V_j(q, \tau) : V_{k-j}(q, \tau) \rangle d\tau \,, \quad k \geq 2 \,.$$

There exists a $T > 0$ such that the series $(q, t) \mapsto \sum_{k=1}^{+\infty} V_k(q, t)$ converges absolutely and uniformly for $t \in [0, T]$ and for q in an appropriate neighborhood of q_0. Over the same interval, the solution $\gamma \colon [0, T] \to Q$ to the system (8.21) with $\dot{\gamma}(0) = 0$ satisfies

$$\dot{\gamma} = \sum_{k=1}^{+\infty} V(\gamma, t) \,. \tag{8.29}$$

The proof consists of four steps. First, we present the variation of constants formula. Then, we use it to manipulate the differential equation (8.23). Third, we write the flow of the resulting equation as the composition of more elementary flows. This procedure can be iterated to obtain the formal expansion (8.29). Finally, we discuss the convergence issue.

Proof. Step I. A time-varying vector field $(q,t) \mapsto X(q,t)$ gives rise to the initial value problem on Q

$$\dot{q}(t) = X(q,t), \qquad q(0) = q_0.$$

We denote its solution at time T via $q(T) = \Phi^X_{0,T}(q_0)$, and we refer to it as the flow of X. Consider the initial value problem

$$\dot{q}(t) = X(q,t) + Y(q,t), \qquad q(0) = q_0,$$

where X and Y are analytic (in q) time-varying vector fields. If we regard X as a perturbation to the vector field Y, we can describe the flow of $X + Y$ in terms of a nominal and perturbed flow. The following relationship is referred to as the *variation of constants formula* [3] and describes the perturbed flow,

$$\Phi^{X+Y}_{0,t} = \Phi^Y_{0,t} \circ \Phi^{(\Phi^Y_{0,t})^* X}_{0,t}, \qquad (8.30)$$

where, given any vector field X and any diffeomorphism ϕ, the $\phi^* X$ is the pull-back of X along ϕ. In particular, the pull-back along the flow of a vector field admits the following series expansion representation [3]

$$(\Phi^Y_{0,t})^* X(q,t) = X(q,t)$$
$$+ \sum_{k=1}^{+\infty} \int_0^t \cdots \int_0^{s_{k-1}} \left(\mathrm{ad}_{Y(q,s_k)} \cdots \mathrm{ad}_{Y(q,s_1)} X(q,t) \right) ds_k \ldots ds_1. \quad (8.31)$$

Step II. In equation (8.23), let the Liouville vector field play the role of the perturbation to the vector field $Z + Y^{\text{lift}}$. Then the application of (8.30) yields $\Phi^{Z+k_d\Delta+Y^{\text{lift}}} = \Phi^{k_d\Delta} \circ \Phi^{\Pi}$, where we compute

$$\Phi^{k_d\Delta}(q_0, v_0) = (q_0, e^{k_d t} v_0),$$

and where the homogeneity leads to

$$\Pi = \sum_{k=0}^{+\infty} \frac{t^k}{k!} \mathrm{ad}^k_{k_d\Delta}(Z + Y^{\text{lift}}) = \sum_{k=0}^{+\infty} \frac{(k_d t)^k}{k!} \mathrm{ad}^k_{\Delta}(Z + Y^{\text{lift}})$$

$$= \sum_{k=0}^{+\infty} \frac{(k_d t)^k}{k!}(Z + (-1)^k Y^{\text{lift}}) = \sum_{k=0}^{+\infty} \left(\frac{(k_d t)^k}{k!} Z + \frac{(-k_d t)^k}{k!} Y^{\text{lift}} \right)$$

$$= e^{k_d t} Z + e^{-k_d t} Y^{\text{lift}}.$$

Let $Z' = e^{k_d t} Z$, and accordingly $\langle X_1 : X_2 \rangle' = e^{k_d t} \langle X_1 : X_2 \rangle$. The initial value problem associated with Π is therefore

$$\dot{y} = Z'(y) + e^{-k_d t} Y(y, t)^{\text{lift}}, \tag{8.32}$$

where we let $y = (r, \dot{r})$.

Step III. Let $k \in \mathbb{N}$ and consider the differential equation

$$\dot{y}_k = (Z' + [X_k^{\text{lift}}, Z'] + Y_k^{\text{lift}}) (y_k, t). \tag{8.33}$$

We recover (8.32) by setting $k = 1$, $X_1 = 0$, $Y_1 = e^{-k_d t} Y(q, t)$, and accordingly $y(t) = y_1(t)$. We can now see the vector field $Z' + [X_k^{\text{lift}}, Z']$ as the perturbation to Y_k^{lift}. Using equations (8.30) and (8.31), we set

$$y_k(t) = \Phi_{0,t}^{Y_k^{\text{lift}}} (y_{k+1}(t)).$$

Some straightforward manipulations using the homogeneity properties of the vector fields lead to

$$\dot{y}_{k+1}(t) = \left(\left(\Phi_{0,t}^{Y_k^{\text{lift}}} \right)^* (Z' + [X_k^{\text{lift}}, Z']) \right) (y_{k+1}(t))$$

$$= Z' + [X_k^{\text{lift}} + \overline{Y}_k^{\text{lift}}, Z'] - e^{-k_d t} \langle \overline{Y}_k : X_k \rangle^{\text{lift}} - \frac{e^{k_d t}}{2} \langle \overline{Y}_k : \overline{Y}_k \rangle^{\text{lift}}.$$

Therefore, the differential equation for $y_{k+1}(t)$ is of the same form as (8.33), where

$$X_{k+1} = X_k + \overline{Y}_k, \quad Y_{k+1} = -e^{k_d t} \langle \overline{Y}_k : X_k + \frac{1}{2} \overline{Y}_k \rangle.$$

We easily compute $X_k = \sum_{m=1}^{k-1} \overline{Y}_m$ and set

$$Y_{k+1} = -e^{k_d t} \langle \overline{Y}_k : \sum_{m=1}^{k-1} \overline{Y}_m + \frac{1}{2} \overline{Y}_k \rangle.$$

One can iterate this procedure for an infinite number of times as in the case of no dissipation [41] to obtain the formal expansion

$$\dot{r} = \sum_{k=1}^{+\infty} V'(r, t), \quad V_1'(r, t) = \int_0^t e^{-k_d \tau} Y(r, \tau) d\tau,$$

$$V_k'(r, t) = -\frac{1}{2} \sum_{j=1}^{k-1} \int_0^t e^{k_d \tau} \langle V_j'(r, \tau) : V_{k-j}'(r, \tau) \rangle d\tau.$$

To obtain the flow of $Z + k_d \Delta + Y^{\text{lift}}$, we compose the flow of Π with that of $k_d \Delta$ to compute

$$\dot{q} = e^{k_d t}\dot{r} = \sum_{k=1}^{+\infty} V(q,t), \quad V_1(q,t) = \int_0^t e^{k_d(t-\tau)}Y(q,\tau)d\tau,$$

$$V_k(q,t) = -\frac{1}{2}\sum_{j=1}^{k-1}\int_0^t e^{k_d(\tau-2\tau+t)}\langle V_j(q,\tau):V_{k-j}(q,\tau)\rangle d\tau.$$

Step IV. Select a coordinate chart around q_0. In this way, we can locally identify Q with $\mathbb{R}^n$. Resorting step by step to the analysis in [41], it can be proven that there exists a $L > 0$ such that $\|V_k\|_{\sigma'} \leq L^{1-k}\|Y\|_{\sigma,t}\left(te^{k_d t}\right)^{2k-1}$, where $\sigma' < \sigma$. And immediate consequence is that for $\|Y\|_{\sigma,T} T^2 e^{2k_d T} < L$, the previous expansion converges absolutely and uniformly in $t \in [0,T]$ and $q \in B_{\sigma'}(q_0)$. □

8.6.4 Systems underactuated by one control

Given the discussion in Sections 8.6.1 and 8.6.3, we can conclude that Theorem 8.4.2 and Corollary 8.4.5 remain valid for mechanical control systems with isotropic dissipation underactuated by one control. We state them here for the sake of completeness.

Theorem 8.6.8. *Let Q be a n-dimensional analytic manifold and let $Y_1, \ldots, Y_{n-1}$ be analytic vector fields on Q. Consider the control system*

$$\nabla_{\dot{c}(t)}\dot{c}(t) = k_d\dot{c}(t) + \sum_{i=1}^{n-1} u_i(t)Y_i(c(t)), \tag{8.34}$$

and assume that it is LCA at $q_0 \in Q$. Then the system is STLCC at q_0 if and only if there exists a basis of input vector fields satisfying the sufficient conditions for STLCC at q_0.

Corollary 8.6.9. *Let Q be a 3-dimensional analytic manifold and let Y_1, Y_2 be analytic vector fields on Q. Consider the control system (8.34) and assume that it is LCA at $q_0 \in Q$. Let A be the 2×2 symmetric matrix whose elements are given by*

$$\langle Y_1:Y_1\rangle(q_0) = lc(Y_1(q_0),Y_2(q_0)) + a_{11}\langle Y_1:Y_2\rangle(q_0)$$
$$\langle Y_2:Y_2\rangle(q_0) = lc(Y_1(q_0),Y_2(q_0)) + a_{22}\langle Y_1:Y_2\rangle(q_0)$$
$$\langle Y_1:Y_2\rangle(q_0) = a_{12}\langle Y_1:Y_2\rangle(q_0).$$

Then the system is STLCC at q_0 if and only if $\det A < 0$.

References

1. R. Abraham, J.E. Marsden: *Foundations of Mechanics*. 2nd ed., Benjamin-Cummings, Reading, Ma, 1978.
2. R. Abraham, J.E. Marsden, T.S. Ratiu: *Manifolds, Tensor Analysis and Applications*. 2nd ed., Springer-Verlag, New-York-Heidelberg-Berlin, 1988.
3. A.A. Agračhev, R.V. Gamkrelidze: The exponential representation of flows and the chronological calculus. *Math. USSR Sbornik* **35** (6) (1978), 727-785.
4. A.A. Agračhev, R.V Gamkrelidze: Local controllability and semigroups of diffeomorphisms. *Acta Appl. Math.* **32** (1993), 1-57.
5. H. Anderson: Rattle: A velocity version of the shake algorithm for molecular dynamics calculations. *J. Comput. Phys.* **52** (1983), 24-34.
6. P. Appell: *Traité de Mécanique Rationnelle*. Tome II, 6th ed., Paris, Gauthier-Villars, 1953.
7. S. Arimoto: *Control Theory of Non-linear Mechanical Systems: A Passivity-Based and Circuit-Theoretic Approach*. OESS **49**, Oxford University Press, Oxford, 1996.
8. F. Armero, J.C. Simo: A priori stability and unconditionally stable product formula algorithms for nonlinear coupled thermoplasticity. *Int. J. Plasticity* **9** (6) (1993), 749-782.
9. V.I. Arnold: *Mathematical Methods of Classical Mechanics*. Springer-Verlag, New York-Heidelberg-Berlin, 1978.
10. V.I. Arnold: *Dynamical Systems*. Vol. III, Springer-Verlag, New York-Heidelberg-Berlin, 1988.
11. A. Astolfi: Discontinuous control of nonholonomic systems. *Systems Control Lett.* **27** (1996), 37-45.
12. J.C. Baez, J.W. Gilliam: An algebraic approach to discrete mechanics. *Lett. Math. Phys.* **31** (1994), 205-212.
13. J. Baillieul: Stable average motions of mechanical systems subject to periodic forcing. In *Dynamics and Control of Mechanical Systems: The Falling Cat and Related Problems*, ed. M.J. Enos, Fields Institute Communications **1**, AMS, 1993, pp. 1-23.
14. J. Baillieul, S. Weibel: Scale dependence in the oscillatory control of micromechanisms. *Proc. IEEE Conf. Decision &Control*, Tampa, Florida, 1998, 3058-3063.
15. L. Bates: Problems and progress in nonholonomic reduction. *Rep. Math. Phys.* **49** (2/3) (2002), 143-149.

16. L. Bates, R. Cushman: What is a completely integrable nonholonomic dynamical system? *Rep. Math. Phys.* **44** (1/2) (1999), 29-35.

17. L. Bates, H. Graumann, C. MacDonnell: Examples of gauge conservation laws in nonholonomic systems. *Rep. Math. Phys.* **37** (3) (1996), 295-308.

18. L. Bates, J. Śniatycki: Nonholonomic reduction. *Rep. Math. Phys.* **32** (1) (1992), 99-115.

19. S. Benenti: Geometrical aspects of the dynamics of nonholonomic systems. In *Journeés relativistes*, Univ. de Chambery, 1987, pp. 1-15.

20. K.H. Bhaskara, K. Viswanath: Poisson algebras and Poisson manifolds. Pitman Research Notes in Math. **174**, Longman, New York, 1988.

21. E. Binz, J. Śniatycki, H. Fisher: *The Geometry of Classical Fields.* North-Holland Math. Ser. **154**, Amsterdam, 1988.

22. G. Blankenstein: Matching and stabilization of constrained systems. Preprint, 2002.

23. G. Blankenstein, R. Ortega, A.J. van der Schaft: The matching conditions of controlled Lagrangians and interconnection and damping assignment passivity based control. To appear in *Int. J. Control*, 2002.

24. G. Blankenstein, A.J. van der Schaft: Symmetry and reduction in implicit generalized Hamiltonian systems. *Rep. Math. Phys.* **47** (1) (2001), 57-100.

25. A.M. Bloch: *Nonholonomic Mechanics and Control.* Applied Mathematical Sciences Series, Springer-Verlag, New York, in press.

26. A.M. Bloch, P.E. Crouch: Nonholonomic and vakonomic control systems on Riemannian manifolds. In *Dynamics and Control of Mechanical Systems, The Falling Cat and Related Problems*, ed. M.J. Enos, Fields Institute Communications 1, AMS, 1993, pp. 25-52.

27. A.M. Bloch, P.E. Crouch: Newton's law and integrability of nonholonomic systems. *SIAM J. Control Optim.* **36** (1998), 2020-2039.

28. A.M. Bloch, S. Drakunov: Stabilization and tracking in the nonholonomic integrator via sliding modes. *Systems Control Lett.* **29** (1996), 91-99.

29. A.M. Bloch, P.S. Krishnaprasad, J.E. Marsden, R.M. Murray: Nonholonomic mechanical systems with symmetry. *Arch. Rational Mech. Anal.* **136** (1996), 21-99.

30. A.M. Bloch, P.S. Krishnaprasad, J.E. Marsden, T.S. Ratiu: Dissipation induced instabilities. *Ann. Inst. H. Poincaré (analyse non lineaire)* **11** (1) (1994), 37-90.

31. A.M. Bloch, N.E. Leonard, J.E. Marsden: Controlled Lagrangians and the stabilization of mechanical systems. I. The first matching theorem. *IEEE Trans. Automat. Control* **45** (12) (2000), 2253-2270.

32. A.M. Bloch, D.E. Chang, N.E. Leonard, J.E. Marsden: Controlled Lagrangians and the stabilization of mechanical systems. II. Potential shaping. *IEEE Trans. Automat. Control* **46** (10) (2001), 1556-1571.

33. A.M. Bloch, M. Reyhanoglu, N.H. McClamroch: Control and stabilization of nonholonomic dynamic systems. *IEEE Trans. Automat. Control* **37** (11) (1992), 1746-1757.

34. A.I. Bobenko, Y.B. Suris: Discrete time Lagrangian mechanics on Lie groups, with an application to the Lagrange top. *Comm. Math. Phys.* **204** (1998), 147-188.

35. A.V. Bocharov, A.M. Vinogradov: Appendix II of A.M. Vinogradov, B.A. Kuperschmidt: The structures of Hamiltonian mechanics. *Russ. Math. Surv.* **32** (4) (1977), 177-243.

36. H. Bondi: The rigid-body dynamics of unidirectional spin. *Proc. Roy. Soc. Lond.* **405** (1986), 265-274.

37. R.W. Brockett: Asymptotic stability and feedback stabilization. In *Geometric Control Theory*, eds. R.W. Brockett, R.S. Millman, H.J. Sussmann, Birkhäuser, Boston, Massachusetts, 1983, pp. 181-191.

38. R.W. Brockett, L.Dai: Nonholonomic kinematics and the role of elliptic functions in constructive controllability. In *Nonholonomic Motion Planning*, eds Z. Li, J.F. Canny, Kluwer, 1993, pp. 1-22.

39. B. Brogliato: *Nonsmooth Impact Mechanics: Models, Dynamics and Control.* Lectures Notes in Control and Information Science **220**, Springer, New York, 1996.

40. F. Bullo: *Nonlinear Control of Mechanical Systems: A Riemannian Geometry Approach.* PhD. Thesis, California Institute of Technology, 1998.

41. F. Bullo: Series expansions for the evolution of mechanical control systems. *SIAM J. Control Optim.* **40** (1) (2001), 166-190.

42. F. Bullo, N.E. Leonard, A.D. Lewis: Controllability and motion algorithms for underactuated Lagrangian systems on Lie groups. *IEEE Trans. Automat. Control* **45** (8) (2000), 1437-1454.

43. F. Bullo, A.D. Lewis: On the homogeneity of the affine connection model for mechanical control systems. *Proc. IEEE Conf. Decision & Control*, Sydney, Australia, 2000, 1260-1265.

44. F. Bullo, A.D. Lewis: Kinematic controllability and motion planning for the snakeboard. Submitted to *IEEE Trans. Robot. Automat.*, 2002.

45. F. Bullo, K.M. Lynch: Kinematic controllability for decoupled trajectory planning in underactuated mechanical systems. *IEEE Trans. Robot. Automat.* **17** (4) (2001), 402-412.

46. F. Bullo, M. Zefran: Modeling and controllability for a class of hybrid mechanical systems. To appear in *IEEE Trans. Robot. Automat.*, 2002.

47. F. Bullo, M. Zefran: On mechanical control systems with nonholonomic constraints and symmetries. *Systems Control Lett.* **45** (2) (2002), 133-143.

48. F. Cantrijn, J. Cortés: Cosymplectic reduction of constrained systems with symmetry. *Rep. Math. Phys.* **49** (2/3) (2002), 167-182.

49. F. Cantrijn, J. Cortés, M. de León, D. Martín de Diego: On the geometry of generalized Chaplygin systems. *Math. Proc. Cambridge Philos.* **132** (2) (2002), 323-351.

50. F. Cantrijn, M. de León, J. C. Marrero, D. Martín de Diego: Reduction of nonholonomic mechanical systems with symmetries. *Rep. Math. Phys.* **42** (1/2) (1998), 25-45.

51. F. Cantrijn, M. de León, J.C. Marrero, D. Martín de Diego: Reduction of constrained systems with symmetries. *J. Math. Phys.* **40** (1999), 795-820.

52. F. Cantrijn, M. de León, J.C. Marrero, D. Martín de Diego: On almost-Poisson structures in nonholonomic mechanics II. The time-dependent framework. *Nonlinearity* **13** (2000), 1379-1409.

53. F. Cantrijn, M. de León, D. Martín de Diego: On almost-Poisson structures in nonholonomic mechanics. *Nonlinearity* **12** (1999), 721-737.
54. C. Canudas de Wit, H. Khennouf: Quasi-continuous stabilizing controllers for nonholonomic systems: design and robustness considerations. *Proc. European Control Conf.*, Rome, Italy, 1995.
55. F. Cardin, M. Favretti: On nonholonomic and vakonomic dynamics of mechanical systems with nonintegrable constraints. *J. Geom. Phys.* **18** (1996), 295-325.
56. J.F. Cariñena: Theory of singular Lagrangians. *Fortschr. Phys.* **38** (9) (1990), 641-679.
57. J.J. Cariñena, M.F. Rañada: Lagrangian systems with constraints: A geometric approach to the method of Lagrange multipliers. *J. Phys. A: Math. Gen.* **26** (1993), 1335-1351.
58. M.P. do Carmo: *Riemannian geometry*. Birkhäuser, Boston-Basel-Berlin, 1992.
59. H. Cendra, J.E. Marsden, T.S. Ratiu: Geometric mechanics, Lagrangian reduction and nonholonomic systems. In *Mathematics Unlimited-2001 and Beyond*, eds. B. Enguist, W. Schmid, Springer-Verlag, New York, 2001, pp. 221-273.
60. P.J. Channell, F.R. Neri: An introduction to symplectic integrators. In *Integration Algorithms and Classical Mechanics*, ed. J.E. Marsden, G.W. Patrick, W.F. Shadwick, Fields Institute Communications **10**, AMS, 1996, pp. 45-58.
61. P.J. Channell, C. Scovel: Symplectic integrations of Hamiltonian Systems. *Nonlinearity* **3** (1990), 231-259.
62. S.A. Chaplygin: On some feasible generalization of the theorem of area, with an application to the problem of rolling spheres (in Russian). *Mat. Sbornik* **20** (1897), 1-32.
63. S.A. Chaplygin: On a rolling sphere on a horizontal plane (in Russian). *Mat. Sbornik* **24** (1903), 139-168.
64. K.T. Chen: Integration of paths, geometric invariants and a generalized Baker-Hausdorff formula. *Ann. of Math.* **67** (1957), 164-178.
65. B. Chen, L.S. Wang, S.S. Chu, W.T. Chou: A new classification of nonholonomic constraints. *Proc. R. Soc. Lond. A* **453** (1997), 631-642.
66. G.S. Chirikjian, J.W. Burdick: The kinematics of hyper-redundant locomotion. *IEEE Trans. Robot. Automat.* **11** (6) (1995), 781-793.
67. J.M. Coron: Global asymptotic stabilization for controllable systems without drift. *Mathematics of Control, Signals and Systems* **5** (1992), 295-312.
68. J. Cortés, M. de León: Reduction and reconstruction of the dynamics of nonholonomic systems. *J. Phys. A: Math. Gen.* **32** (1999), 8615-8645.
69. J. Cortés, M. de León, D. Martín de Diego, S. Martínez: Mechanical systems subjected to generalized constraints. *R. Soc. Lond. Proc. Ser. A Math. Phys. Eng. Sci.* **457** 2007 (2001), 651-670.
70. J. Cortés, M. de León, D. Martín de Diego, S. Martínez: Geometric description of vakonomic and nonholonomic dynamics. Comparison of solutions. Submitted to *SIAM J. Control Optim.*, 2000.
71. J. Cortés, S. Martínez: Optimal control for nonholonomic systems with symmetry. *Proc. IEEE Conf. Decision & Control*, Sydney, Australia, 2000, 5216-5218.

72. J. Cortés, S. Martínez: Configuration controllability of mechanical systems underactuated by one control. Submitted to *SIAM J. Control Optim.*, 2000.

73. J. Cortés, S. Martínez: Nonholonomic integrators. *Nonlinearity* **14** (2001), 1365-1392.

74. J. Cortés, S. Martínez, F. Bullo: On nonlinear controllability and series expansions for Lagrangian systems with dissipative forces. To appear in *IEEE Trans. Automat. Control*, 2002.

75. J. Cortés, S. Martínez, J.P. Ostrowski, K.A. McIsaac: Optimal gaits for dynamic robotic locomotion. *Int. J. Robotics Research* **20** (9) (2001), 707-728.

76. J. Cortés, S. Martínez, J.P. Ostrowski, H. Zhang: Simple mechanical control systems with constraints and symmetry. To appear in *SIAM J. Control Optim.*, 2002.

77. H. Crabtree: *Spinning Tops and Gyroscopic Motion*. Chelsea, 1909.

78. M. Crampin: Tangent bundle geometry for Lagrangian dynamics. *J. Phys. A: Math. Gen.* **16** (1983), 3755-3772.

79. P.E. Crouch: Geometric structures in systems theory. *Proc. IEE. D. Control Theory and Applications* **128** (5) (1981), 242-252.

80. R. Cushman, D. Kemppainen, J. Śniatycki, L. Bates: Geometry of nonholonomic constraints. *Rep. Math. Phys.* **36** (2/3) (1995), 275-286.

81. P. Dazord: Mécanique Hamiltonienne en présence de constraintes. *Illinois J. Math.* **38** (1) (1994), 148-175.

82. P.A.M. Dirac: *Lectures on Quantum Mechanics*. Belfer Graduate School of Science, Yeshiva University, New York, 1964.

83. V. Dragović, B. Gajić, B. Jovanović: Generalizations of classical integrable nonholonomic rigid body systems. *J. Phys. A: Math. Gen.* **31** (1998), 9861-9869.

84. M. Favretti: Equivalence of dynamics for nonholonomic systems with transverse constraints. *J. Dynam. Diff. Equations* **10** (4) (1998), 511-536.

85. M. Fliess: Fonctionelles causales non linéares et indeterminées non commutatives. *Bull. Soc. Math.* **109** (1981), 3-40.

86. Z. Ge, J.E. Marsden: Lie-Poisson Hamilton-Jacobi theory and Lie-Poisson integrators. *Phys. Lett. A* **133** (1988), 134-139.

87. F. Génot, B. Brogliato: New results on Painlevé paradoxes. *Eur. J. Mech. A/Solids* **18** (1999), 653-677.

88. G. Giachetta: Jet methods in nonholonomic mechanics. *J. Math. Phys.* **33** (1992), 1652-1665.

89. O. Gonzalez: Time integration and discrete Hamiltonian systems. *J. Nonlinear Sci.* **6** (1996), 449-467.

90. O. Gonzalez: Mechanical systems subject to holonomic constraints: Differential-algebraic formulations and conservative integration. *Physica D* **132** (1999), 165-174.

91. B. Goodwine: *Control of Stratified Systems with Robotic Applications*. PhD. Thesis, California Institute of Technology, 1998.

92. B. Goodwine, J.W. Burdick: Controllability of kinematic control systems on stratified configuration spaces. *IEEE Trans. Automat. Control* **46** (3) (2001), 358-368.

93. M.J. Gotay: *Presymplectic Manifolds, Geometric Constraint Theory and the Dirac-Bergmann Theory of Constraints.* PhD. Thesis, Center for Theoretical Physics, University of Maryland, 1979.

94. M.J. Gotay, J.M. Nester: Presymplectic Lagrangian systems I: the constraint algorithm and the equivalence theorem. *Ann. Inst. H. Poincaré* **A 30** (1979), 129-142.

95. M.J. Gotay, J.M. Nester: Presymplectic Lagrangian systems II: the second-order differential equation problem. *Ann. Inst. H. Poincaré* **A 32** (1980), 1-13.

96. X. Gràcia, J. Marín-Solano, M. Muñoz-Lecanda: Variational principles in mechanics: geometric aspects. Proc. of the VII Fall Workshop on Geometry and Physics, Valencia, Spain, 1998, Publicaciones de la RSME, vol. **1** (2000), pp. 81-94.

97. J. Grifone, M. Mehdi: On the geometry of Lagrangian mechanics with nonholonomic constraints. *J. Geom. Phys.* **30** (1999), 187-203.

98. R.W. Hamming: *Numerical Methods for Scientists and Engineers.* 2nd ed., Dover, New York, 1986, pp. 73.

99. H. Hermes: Control systems which generate decomposable Lie algebras. *J. Diff. Equations* **44** (1982), 166-187.

100. A. Ibort, M. de León, E.A. Lacomba, D. Martín de Diego, P. Pitanga: Mechanical systems subjected to impulsive constraints. *J. Phys. A: Math. Gen.* **30** (1997), 5835-5854.

101. A. Ibort, M. de León, E.A. Lacomba, J.C. Marrero, D. Martín de Diego, P. Pitanga: Geometric formulation of mechanical systems subjected to time-dependent one-sided constraints. *J. Phys. A: Math. Gen.* **31** (1998), 2655-2674.

102. A. Ibort, M. de León, E.A. Lacomba, J.C. Marrero, D. Martín de Diego, P. Pitanga: Geometric formulation of Carnot Theorem. *J. Phys. A: Math. Gen.* **34** (2001), 1691-1712.

103. A. Ibort, M. de León, G. Marmo, D. Martín de Diego: Nonholonomic constrained systems as implicit differential equations. *Rend. Sem. Mat. Univ. Pol. Torino* **54** 3 (1996), 295-317.

104. A. Ibort, M. de León, J.C. Marrero, D. Martín de Diego: Dirac brackets in constrained dynamics. *Fortschr. Phys.* **47** (5) (1999), 459-492.

105. A. Isidori: *Nonlinear Control Systems.* 3nd ed., Communications and Control Engineering Series, Springer-Verlag, Berlin, 1995.

106. L. Jay: Symplectic partitioned Runge-Kutta methods for constrained Hamiltonian systems. *SIAM J. Numer. Anal.* **33** (1996), 368-387.

107. B. Jovanović: Geometry and integrability of Euler-Poincaré-Suslov equations. *Nonlinearity* **14** (2001), 1555-1567.

108. T.R. Kane: *Dynamics.* Holt, Rinehart and Winston Inc., New York, 1968.

109. C. Kane, J.E. Marsden, M. Ortiz: Symplectic-energy-momentum preserving variational integrators. *J. Math. Phys.* **40** (1999), 3353-3371.

110. C. Kane, J.E. Marsden, M. Ortiz, M. West: Variational integrators and the Newmark algorithm for conservative and dissipative mechanical systems. *Int. J. Num. Math. Eng.* **49** (2000), 1295-1325.

111. M. Kawski: The complexity of deciding controllability. *Systems Control Lett.* **15** (1) (1990), 9-14.

112. M. Kawski: High-order small-time local controllability. In *Nonlinear Controllability and Optimal Control*, ed. H.J. Sussmann, Dekker, 1990, pp. 441-477.

113. M. Kawski: Geometric homogeneity and applications to stabilization. *Nonlinear Control Systems Design Symposium (NOLCOS)*, Tahoe City, 1995, pp. 251-256.

114. M. Kawski, H.J. Sussmann: Noncommutative power series and formal Lie-algebraic techniques in nonlinear control theory. In *Operators, Systems and Linear Algebra*, eds. U. Helmke, D. Pratzel-Wolters, E. Zerz, Teubner, 1997, pp. 111-128.

115. D. Kazhdan, B. Konstant, S. Sternberg: Hamiltonian group actions and dynamical systems of Calogero type. *Comm. Pure Appl. Math.* **31** (1978), 481-508.

116. J.B. Keller: Impact with friction. *ASME J. Appl. Mech.* **53** (1986), 1-4.

117. S.D. Kelly, R.M. Murray: Geometric phases and robotic locomotion. *J. Robotic Systems* **12** (6) (1995), 417-431.

118. H. Khennouf, C. Canudas de Wit, A.J. van der Schaft: Preliminary results on asymptotic stabilization of Hamiltonian systems with nonholonomic constraints. *Proc. IEEE Conf. Decision & Control*, New Orleans, USA, 1995, 4305-4310.

119. S. Kobayashi, K. Nomizu: *Foundations of Differential Geometry*. Interscience Tracts in Pure and Applied Mathematics. Interscience Publishers, Wiley, New-York, 1963.

120. J. Koiller: Reduction of some classical nonholonomic systems with symmetry. *Arch. Rational Mech. Anal.* **118** (1992), 113-148.

121. W.S. Koon, M.W. Lo, J.E. Marsden, S.D. Ross: Heteroclinic connections between periodic orbits and resonance transitions in celestial mechanics. *Chaos* **10** (2000), 427-469.

122. W.S. Koon, J.E. Marsden: The Hamiltonian and Lagrangian approaches to the dynamics of nonholonomic systems. *Rep. Math. Phys.* **40** (1997), 21-62.

123. W.S. Koon, J.E. Marsden: Poisson reduction of nonholonomic mechanical systems with symmetry. *Rep. Math. Phys.* **42** (1/2) (1998), 101-134.

124. V.V. Kozlov: Realization of nonintegrable constraints in classical mechanics. *Dokl. Akad. Nauk SSSR* **272** (3) (1983), 550-554; English transl.: *Sov. Phys. Dokl.* **28** (9) (1983), 735-737.

125. V.V. Kozlov: Invariant measures of the Euler-Poincaré equations on Lie algebras. *Funkt. Anal. Prilozh.* **22** (1988), 69-70; English transl.: *Funct. Anal. Appl.* **22** (1988), 58-59.

126. O. Krupková: Mechanical systems with nonholonomic constraints. *J. Math. Phys.* **38** (1997), 5098-5126.

127. O. Krupková: Higher-order mechanical systems with constraints *J. Math. Phys.* **41** (2000), 5304-5324.

128. M. Kummer: On the construction of the reduced phase space of a Hamiltonian system with symmetry. *Indiana Univ. Math. J.* **30** (1981), 281-291.

129. I. Kupka, W.M. Oliva: The nonholonomic mechanics. *J. Diff. Equations* **169** (2001), 169-189.

130. E.A. Lacomba, W.A. Tulczyjew: Geometric formulation of mechanical systems with one-sided constraints. *J. Phys. A: Math. Gen.* **23** (1990), 2801-2813.

131. G. Lafferriere, H.J. Sussmann: A differential geometric approach to motion planning. In *Nonholonomic Motion Planning*, eds. Z.X. Li, J.F. Canny, Kluwer, 1993, pp. 235-270.

132. B. Langerock: A connection theoretic approach to sub-Riemannian geometry. To appear in *J. Geom. Phys.*, 2002.

133. B.J. Leimkuhler, S. Reich: Symplectic integration of constrained Hamiltonian systems. *Math. Comp.* **63** (1994), 589-605.

134. B.J. Leimkuhler, R.D. Skeel: Symplectic numerical integrators in constrained Hamiltonian systems. *J. Comput. Phys.* **112** (1994), 117-125.

135. M. de León, J.C. Marrero, D. Martín de Diego: Nonholonomic Lagrangian systems in jet manifolds. *J. Phys. A: Math. Gen.* **30** (1997), 1167-1190.

136. M. de León, J.C. Marrero, D. Martín de Diego: Mechanical systems with nonlinear constraints. *Int. J. Theor. Phys.* **36** (4) (1997), 973-989.

137. M. de León, D. Martín de Diego: On the geometry of nonholonomic Lagrangian systems. *J. Math. Phys.* **37** (1996), 3389-3414.

138. M. de León, P.R. Rodrigues: *Methods of Differential Geometry in Analytical Mechanics*. North-Holland Math. Ser. **152**, Amsterdam, 1989.

139. N.E. Leonard: Mechanics and nonlinear control: Making underwater vehicles ride and glide. In *Nonlinear Control Systems Design (NOLCOS)*, Enschede, The Netherlands, 1998, pp. 1-6.

140. N.E. Leonard, P.S. Krishnaprasad: Motion control of drift-free, left-invariant systems on Lie groups. *IEEE Trans. Automat. Control* **40** (9) (1995), 1539-1554.

141. A.D. Lewis: *Aspects of Geometric Mechanics and Control of Mechanical Systems*. PhD. Thesis, California Institute of Technology, 1995.

142. A.D. Lewis: Local configuration controllability for a class of mechanical systems with a single input. *Proc. European Control Conference*, Brussels, Belgium, 1997.

143. A.D. Lewis: Affine connections and distributions with applications to nonholonomic mechanics. *Rep. Math. Phys.* **42** (1/2) (1998), 135-164.

144. A.D. Lewis: Simple mechanical control systems with constraints. *IEEE Trans. Automat. Control* **45** (8) (2000), 1420-1436.

145. A.D. Lewis, R.M. Murray: Variational principles for constrained systems: theory and experiments. *Int. J. Nonlinear Mech.* **30** (6) (1995), 793-815.

146. A.D. Lewis, R.M. Murray: Configuration controllability of simple mechanical control systems. *SIAM J. Control Optim.* **35** (3) (1997), 766-790.

147. A.D. Lewis, R.M. Murray: Configuration controllability of simple mechanical control systems. *SIAM Rev.* **41** (3) (1999), 555-574.

148. A.D. Lewis, J.P. Ostrowski, R.M. Murray, J.W. Burdick: Nonholonomic mechanics and locomotion: the Snakeboard example. *Proc. IEEE Conf. Robotics&Automation*, San Diego, California, 1994, 2391-2397.

149. P. Libermann, C.-M. Marle: *Symplectic Geometry and Analytical Mechanics*. D. Reidel Publ. Comp., Dordrecht, 1987.

150. K.M. Lynch: Nonprehensile robotic manipulation: Controllability and planning. PhD. Thesis, Carnegie Mellon University, 1996.
151. K.M. Lynch, N. Shiroma, H. Arai, K. Tanie: Collision-free trajectory planning for a 3-DOF robot with a passive joint. *Int. J. Robotics Research* **19** (12) (2000), 1171-1184.
152. W. Liu: Averaging theorems for highly oscillatory differential equations and iterated Lie brackets. *SIAM J. Control Optim* **35** (1997), 1989-2020.
153. W.S. Liu, H. Sussmann: Abnormal sub-Riemannian minimizers. In *Differential equations, dynamical systems, and control science*, Lecture Notes in Pure and Appl. Math. **152**, Dekker, New York, 1994, pp. 705-716.
154. M.J.P. Magill: *On a General Economic Theory of Motion*. Springer-Verlag, Berlin, 1970.
155. W. Magnus: On the exponential solution of differential equations for a linear operator. *Comm. Pure and Appl. Math.* **VII** (1954), 649-673.
156. C.-M. Marle: Reduction of constrained mechanical systems and stability of relative equilibria. *Commun. Math. Phys.* **174** (1995), 295-318.
157. C.-M. Marle: Various approaches to conservative and nonconservative nonholonomic systems. *Rep. Math. Phys.* **42** (1/2) (1998), 211-229.
158. J.E. Marsden: *Lectures on Mechanics*. London Mathematical Society Lecture Note Series **174**, Cambridge University Press, Cambridge, 1992.
159. J.E. Marsden, R. Montgomery, T.S. Ratiu: Reduction, symmetry and phases in mechanics. *Mem. Amer. Math. Soc.* **436**, 1990.
160. J.E. Marsden, G.W. Patrick, S. Shkoller (editors): *Integration Algorithms and Classical Mechanics*. Field Institute Communications **10**, AMS, 1996.
161. J.E. Marsden, S. Pekarsky, S. Shkoller: Discrete Euler-Poincaré and Lie-Poisson equations. *Nonlinearity* **12** (1999), 1647-1662.
162. J.E. Marsden, T.S. Ratiu: Reduction of Poisson manifolds. *Lett. Math. Phys.* **11** (1986), 161-169.
163. J.E. Marsden, T.S. Ratiu: *Introduction to Mechanics and Symmetry*. 2nd ed., Texts in Applied Mathematics **17**, Springer-Verlag, New York, 1999.
164. J.E. Marsden, J. Scheurle: Lagrangian reduction and the double spherical pendulum. *Z. Agnew. Math. Phys.* **44** (1993), 17-43.
165. J.E. Marsden, J. Scheurle: The reduced Euler-Lagrange equations. In *Dynamics and Control of Mechanical Systems, The Falling Cat and Related Problems*, ed. M.J. Enos, Fields Institute Communications **1**, AMS, 1993, pp. 139-164.
166. J.E. Marsden, A. Weinstein: Reduction of symplectic manifolds with symmetry. *Rep. Math. Phys.* **5** (1974), 121-130.
167. J.E. Marsden, M. West: Discrete mechanics and variational integrators. *Acta Numerica* **10** (2001), 357-514.
168. S. Martínez: *Geometric Methods in Nonlinear Control Theory with applications to Dynamic Robotic Systems*. PhD. Thesis, Universidad Carlos III de Madrid, 2002.
169. S. Martínez, J. Cortés: Motion control algorithms for mechanical systems with symmetry. Submitted to *Acta Appl. Math.*, 2001.

170. S. Martínez, J. Cortés, F. Bullo: On analysis and design of oscillatory control systems. Submitted to *IEEE Trans. Automat. Control*, 2001.

171. S. Martínez, J. Cortés, M. de León: The geometrical theory of constraints applied to the dynamics of vakonomic mechanical systems. The vakonomic bracket. *J. Math. Phys* **41** (2000), 2090-2120.

172. S. Martínez, J. Cortés, M. de León: Symmetries in vakonomic dynamics. Applications to optimal control. *J. Geom. Phys.* **38** (3-4) (2001), 343-365.

173. B.M. Maschke, A.J. van der Schaft: A Hamiltonian approach to stabilization of nonholonomic mechanical systems. *Proc. IEEE Conf. Decision&Control*, Orlando, USA, 1994, 2950-2954.

174. E. Massa, E. Pagani: Classical dynamics of nonholonomic systems: a geometric approach. *Ann. Inst. H. Poincaré (Phys. Théor.)* **55** (1991), 511-544.

175. E. Massa, E. Pagani: A new look at classical mechanics of constrained systems. *Ann. Inst. H. Poincaré (Phys. Théor.)* **66** (1997), 1-36.

176. R.I. McLachlan, C. Scovel: A survey of open problems in symplectic integration. In *Integration Algorithms and Classical Mechanics*, eds. J.E. Marsden, G.W. Patrick, S. Shkoller, Field Institute Communications **10**, AMS, 1996, pp. 151-180.

177. R.T. M'Closkey, R.M. Murray: Exponential stabilization of driftless nonlinear control systems using homogeneous feedback. *IEEE Trans. Automat. Control* **42** (5) (1997), 614-628.

178. K.R. Meyer: Symmetries and integrals in mechanics. In: *Dynamical Systems; Proceedings of Salvador Symposium on Dynamical Systems (1971, University of Bahia)*, ed. M.M. Peixoto, Academic Press, New York, 1973.

179. R. Montgomery: Abnormal minimizers. *SIAM J. Control Optim.* **32** (1994), 1605-1620.

180. P. Morando, S. Vignolo: A geometric approach to constrained mechanical systems, symmetries and inverse problems. *J. Phys. A: Math. Gen.* **31** (1998), 8233-8245.

181. J.J. Moreau: *Mécanique classique*. Tome II, Masson, Paris, 1971.

182. P. Morin, J.B. Pomet, C. Samson: Design of homogeneous time-varying stabilizing control laws for driftless controllable systems via oscillatory approximation of Lie brackets in closed loop. *SIAM J. Control Optim.* **38** (1999), 22-49.

183. J. Moser, A.P. Veselov: Discrete versions of some classical integrable systems and factorization of matrix polynomials. *Comm. Math. Phys.* **139** (1991), 217-243.

184. M. Muñoz-Lecanda, F. Yániz: Dissipative control of mechanical systems: a geometric approach. *SIAM J. Control Optim.* **40** (5) (2002), 1505-1516.

185. R.M. Murray: Nonlinear control of mechanical systems: a Lagrangian perspective. In *Nonlinear Control Systems Design (NOLCOS)*, Lake Tahoe, California, 1995, pp. 378-389. Also Annual Reviews in Control **21** (1997), 31-45.

186. R.M. Murray, Z.X. Li, S.S. Sastry: *A Mathematical Introduction to Robotic Manipulation*. CRC Press, Boca Ratón, 1994.

187. R.M. Murray, S.S. Sastry: Nonholonomic motion planning: steering using sinusoids. *IEEE Trans. Automat. Control* **38** (5) (1993), 700-716.

188. J. Neimark, N. Fufaev: *Dynamics of Nonholonomic Systems.* Transactions of Mathematical Monographs **33**, AMS, Providence, RI, 1972.

189. H. Nijmeijer, A.J. van der Schaft: *Nonlinear Dynamical Control Systems.* Springer-Verlag, New York, 1990.

190. P.J. Olver: *Applications of Lie groups to Differential Equations.* 2nd ed., Springer Verlag, Berlin, 1993.

191. R. Ortega, A. Loria, P.J. Nicklasson, H. Sira-Ramirez: *Passivity-Based Control of Euler-Lagrange Systems: Mechanical, Electrical and Electromechanical Applications.* Springer Verlag, New York, 1998.

192. R. Ortega, M. Spong, F. Gomez, G. Blankenstein: Stabilization of underactuated mechanical systems via interconnection and damping assignment. To appear in *IEEE Trans. Automat. Control*, 2002.

193. R. Ortega, A. van der Schaft, I. Mareels, B. Maschke: Putting energy back in control. *Control Systems Magazine* **21** (2001), 18-33.

194. M. Ortiz, L. Stainier: The variational formulation of viscoplastic constitutive updates. *Comp. Meth. Appl. Mech. Eng.* **171** (1999), 419-444.

195. J.P. Ostrowski: *Geometric Perspectives on the Mechanics and Control of Undulatory Locomotion.* PhD. Thesis, California Institute of Technology, 1995.

196. J.P. Ostrowski, J.W. Burdick: Controllability tests for mechanical systems with symmetries and constraints. *J. Appl. Math. Comp. Sci.* **7** (1997), 101-127.

197. J.P. Ostrowski, J.W. Burdick: The geometric mechanics of undulatory robotic locomotion. *Int. J. Robotics Research* **17** (1998), 683-702.

198. P. Painlevé: *Cours de Mécanique.* Tome I, Paris, Gauthier-Villars, 1930.

199. Y.H. Pao, L.S. Wang: The principle of virtual power and extended Appell's equations for dynamical systems with constraints. Preprint, 2001.

200. G. Pappas, J. Lygeros, D. Tilbury, S.S. Sastry: Exterior differential systems in control and robotics. In *Essays on Mathematical Robotics*, eds. J. Baillieul, S.S. Sastry, H.J. Sussmann, IMA Volumes in Mathematics and its Applications **104**, Springer-Verlag, 1998, pp. 271-372.

201. L.A. Pars: *A Treatise on Analytical Dynamics.* London, Heinemann, 1965.

202. Y. Pironneau: Sur les liaisons non holonomes non linéaires, déplacements virtuels à travail nul, conditions de Chetaev. *Proc. IUTAM-ISIMM Symp. Modern Developments Analytical Mech.*, Torino, Italy, 1982, Accademia delle Scienze di Torino **2** (1983), pp. 671-686.

203. J.B. Pomet: Explicit design of time-varying stabilizing control laws for a class of controllable systems without drift. *Systems Control Lett.* **18** (1992), 147-158.

204. M. Rathinam, R.M. Murray: Configuration flatness for Lagrangian systems underactuated by one control. *SIAM J. Control Optim.* **36** (1) (1998), 164-179.

205. S. Reich: Symplectic integration of constrained Hamiltonian systems by Runge-Kutta methods. Technical Report 93-13, University of British Columbia, 1993.

206. S. Reich: Symplectic integration of constrained Hamiltonian systems by composition methods. *SIAM J. Numer. Anal.* **33** (1996), 475-491.

207. R. Rosenberg: *Analytical Dynamics.* Plenum Press, New York, 1977.

208. E.J. Routh: *Treatise on the Dynamics of a System of Rigid Bodies.* MacMillan, London, 1860.

209. J. Ryckaert, G. Ciccotti, H. Berendsen: Numerical integration of the Cartesian equations of motion of a system with constraints: molecular dynamics of n-alkanes. *J. Comput. Phys.* **23** (1977), 327-341.

210. J.M. Sanz-Serna, M. Calvo: *Numerical Hamiltonian problems.* Chapman and Hall, London, 1994.

211. S. Sastry: *Nonlinear systems. Analysis, stability, and control.* Interdisciplinary Applied Mathematics Series **10**, Springer-Verlag, New York, 1999.

212. R. Sato, R.V. Ramachandran (editors): *Conservation Laws and Symmetry: Applications to Economics and Finance.* Kluwer, Boston, 1990.

213. D.J. Saunders, F. Cantrijn, W. Sarlet: Regularity aspects and Hamiltonization of non-holonomic systems. *J. Phys. A: Math. Gen.* **32** (1999), 6869-6890.

214. D. Schneider: *Nonholonomic Euler-Poincaré Equations and Stability in Chaplygin's Sphere.* PhD. Thesis, Univ. of Washington, 2000.

215. J.A. Schouten: *Ricci-Calculus. An Introduction to Tensor Analysis and its Geometrical Applications.* Springer-Verlag, Berlin, 1954.

216. A. Shapere, F. Wilczek: Geometry of self-propulsion at low Reynolds number. *J. Fluid. Mech.* **198** (1989), 557-585.

217. J.C. Simo, N. Tarnow, K.K. Wong: Exact energy-momentum conserving algorithms and symplectic schemes for nonlinear dynamics. *Comp. Meth. Appl. Mech. Eng.* **100** (1992), 63-116.

218. S. Smale: Topology and Mechanics. *Inv. Math.* **10** (1970), 305-331.

219. S. Smale: Topology and Mechanics. *Inv. Math.* **11** (1970), 45-64.

220. J. Śniatycki: Nonholonomic Noether theorem and reduction of symmetries. *Rep. Math. Phys.* **42** (1/2) (1998), 5-23.

221. J. Śniatycki: Almost Poisson spaces and nonholonomic singular reduction. *Rep. Math. Phys.* **48** (1/2) (2001), 235-248.

222. J. Śniatycki: The momentum equation and the second order differential equation condition. *Rep. Math. Phys.* **49** (2/3) (2002), 371-394.

223. E.D. Sontag: Controllability is harder to decide than accessibility. *SIAM J. Control Optim.* **26** (5) (1988), 1106-1118.

224. E.D. Sontag: *Mathematical Control Theory. Deterministic finite-dimensional systems.* 2nd ed., Texts in Applied Mathematics **6**, Springer-Verlag, New York, 1998.

225. M. Spivak: Calculus on Manifolds. Benjamin-Cummings, Reading, Ma, 1965.

226. S.V. Stanchenko: Nonholonomic Chaplygin systems (in Russian). *Prikl. Mat. Mehk.* **53** (1) (1989), 16-23.

227. D.E. Stewart: Rigid-body dynamics with friction and impact. *SIAM Review* **42** (2000), 3-39.

228. S. Stramigioli: *Modelling and IPC control of interactive mechanical systems. A coordinate-free approach.* Lecture Notes in Control and Information Sciences **266**, Springer-Verlag, London, 2001.

229. W.J. Stronge: Rigid body collisions with friction. *Proc. R. Soc. Lond.* A **431** (1990), 169-181.

230. W.J. Stronge: Friction in collisions: resolution of a paradox. *J. Appl. Phys.* **69** (2) (1991), 610-612.

231. H.J. Sussmann: Lie brackets and local controllability: a sufficient condition for scalar-input systems. *SIAM J. Control Optim.* **21** (5) (1983), 686-713.

232. H.J. Sussmann: A general theorem on local controllability. *SIAM J. Control Optim.* **25** (1) (1987), 158-194.

233. H.J. Sussmann: A product expansion of the Chen series. In *Theory and Applications of Nonlinear Control Systems*, eds. C.I. Byrnes, A. Lindquist, Elsevier, 1986, pp. 323-335.

234. H.J. Sussmann: New differential geometric methods in nonholonomic path finding. In *Systems, Models, and Feedback: Theory and Applications*, eds. A. Isidori, T.J. Tarn, Birkhäuser, Boston, Massachusetts, 1992, pp. 365-384.

235. J.Z. Synge: Geodesics in nonholonomic geometry. *Math. Annalen* **99** (1928), 738-751.

236. D. Tilbury: *Exterior differential systems and nonholonomic motion planning.* PhD. Thesis, University of California, Berkeley, 1994.

237. I. Vaisman: *Lectures on the Geometry of Poisson Manifolds.* Progress in Math. **118**, Birkhäuser, Basel, 1994.

238. A.J. van der Schaft: Linearization of Hamiltonian and gradient systems. *IMA J. Math. Contr. & Inf.* **1** (1984), 185-198.

239. A.J. van der Schaft: Implicit Hamiltonian systems with symmetry. *Rep. Math. Phys.* **41** (2) (1998), 203-221.

240. A.J. van der Schaft: *L2-Gain and Passivity Techniques in Nonlinear Control.* 2nd ed., Communications and Control Engineering Series, Springer Verlag, London, 2000.

241. A.J. van der Schaft, B.M. Maschke: On the Hamiltonian formulation of nonholonomic mechanical systems. *Rep. Math. Phys.* **34** (1994), 225-233.

242. A.J. van der Schaft, J.M. Schumacher: *An Introduction to Hybrid Dynamical Systems.* Lecture Notes in Control and Information Sciences **251**, Springer Verlag, New York, 1999.

243. A.M. Vershik: Classical and non-classical dynamics with constraints. In *Global Analysis-Studies and Applications I*, Lect. Notes Math. **1108**, Springer-Verlag, Berlin, 1984, pp. 278-301.

244. A.M. Vershik, L.D. Faddeev: Differential geometry and Lagrangian mechanics with constraints. *Sov. Phys. Dokl.* **17** (1) (1972), 34-36.

245. A.M. Vershik, V.Y. Gershkovich: Nonholonomic problems and the theory of distributions. *Acta Appl. Math.* **12** (2) (1988), 181-209.

246. A.P. Veselov: Integrable discrete-time systems and difference operators. *Funkts. Anal. Prilozhen.* **22** (1988), 1-13.

247. A.P. Veselov: Integrable Lagrangian correspondences and the factorization of matrix polynomials. *Funkts. Anal. Prilozhen.* **25** (1991), 38-49.

248. A.P. Veselov, L.E. Veselova: Flows on Lie groups with nonholonomic constraint and integrable nonhamiltonian systems. *Funkt. Anal. Prilozh.* **20** (1986), 65-66; English transl.: *Funct. Anal. Appl.* **20** (1986), 308-309.

249. A. Vierkandt: Über gleitende und rollende Bewegung. *Monats. der Math. u. Phys.* **3** (1892), 31-54.

250. G.T. Walker: On a dynamical top. *Quart. J. Pure Appl. Math.* **28** (1896), 175-184.

251. L.S. Wang, W.T. Chou: The analysis of constrained impulsive motion. Preprint, 2001.

252. L.S. Wang, P.S. Krishnaprasad: Gyroscopic control and stabilization. *J. Nonlinear Sci.* **2** (1992), 367-415.

253. F. Warner: *Foundations of differentiable manifolds and Lie groups.* Scott, Foresman, Glenview, Ill, 1973.

254. R.W. Weber: Hamiltonian systems with constraints and their meaning in mechanics. *Arch. Rational Mech. Anal.* **81** (1986), 309-355.

255. A. Weinstein: The local structure of Poisson manifolds. *J. Diff. Geom.* **18** (1983), 523-557.

256. J.M. Wendlandt, J.E. Marsden: Mechanical integrators derived from a discrete variational principle. *Physica D* **106** (1997), 223-246.

257. E.T. Whittaker : *A Treatise on the Analytical Dynamics of Particles and Rigid Bodies.* 4th ed., Cambridge University Press, Cambridge, 1959.

258. R. Yang: *Nonholonomic Geometry, Mechanics and Control.* PhD. Thesis, Systems Research Institute, Univ. of Maryland, 1992.

259. H. Yoshida: Construction of higher order symplectic integrators. *Phys. Lett. A* **150** (1990), 262-268.

260. G. Zampieri: Nonholonomic versus vakonomic dynamics. *J. Diff. Equations* **163** (2) (2000), 335-347.

261. D.V. Zenkov, A.M. Bloch, J.E. Marsden: The energy momentum method for the stability of nonholonomic systems. *Dyn. Stab. of Systems* **13** (1998), 123-166.

262. D.V. Zenkov, A.M. Bloch, J.E. Marsden: The Lyapunov-Malkin theorem and stabilization of the unicycle with rider. *Systems Control Lett.* **45** (4) (2002), 293-302.

Index

(almost-)Poisson
- bivector, 32
- bracket, 32
- manifold, 32
(almost-)symplectic manifold, 29

action, 21
- adjoint, 20
- coadjoint, 21
- free, 21
- Hamiltonian, 31
- lifted, 22
- proper, 21
- simple, 21
action sum, 143
affine connection, 25
- curvature, 26
- Levi-Civita, 26
- torsion, 25
algebra of exterior forms, 15
algorithm
- DEL, 143
- DLA, 146
- RDLA, 159
almost tangent structure, 35
alternation map, 15

Carnot's theorem, 126
- for generalized constraints, 132
case
- general, 84
- horizontal, 80
- purely kinematic or vertical, 67
Chaplygin system, 69, 103
Chetaev
- bundle, 55
- rule, 54
Christoffel symbols, 25, 115, 177

closure
- involutive, 176
- symmetric, 176
codistribution
- generalized, 17
- integrable
-- completely, 18
-- partially, 18
- rank, 17
- regular, 17
condition
- admissibility, 55
- compatibility, 55
- Sussmann's, 176
constraint, 43
- geometric, 43
- holonomic, 44
- homogeneous, 45
- ideal, 52
- impulsive, 125
- kinematic, 43
- nonholonomic, 44
contorsion, 28
contraction by a vector field, 16
coordinate system
- adapted bundle, 72
cotangent bundle, 14
covariant derivative, 25

dimension assumption, 156
discrete Legendre transformation, 144
distribution
- characteristic, 33
- generalized, 17
-- leaf, 18
- geodesically invariant, 60
- rank, 17
- regular, 17

Euler-Lagrange equations, 41
- forced, 42
examples
- ball
-- on a rotating table, 47
-- on a special surface, 135
- Benenti's example, 50
- counter example to Koiller's question, 119
- mobile robot with fixed orientation, 107
-- with a potential, 168
- nonholonomic free particle, 75, 96
-- modified, 100
- nonholonomic particle with a potential, 166
- particle with generalized constraint, 138
- planar rigid body, 190
- plate on an inclined plane, 78
- rolling disk, 45
-- vertical, 74
- Snakeboard, 49
- wheeled planar mobile robot, 109
exponential mapping, 20
exterior derivative, 15
exterior product, 15

force
- impulsive, 123
Formulation
- Constrained Coordinate, 149
- Generalized Coordinate, 148

geodesic
- curve, 25
- spray, 25
geometric homogeneity, 173
gradient, 25

Hamilton's principle, 41
holonomy, see phase, geometric
horizontal subspace, 23
horizontal symmetry, 65

impulse, 123
infinitesimal generator, 22
integrator
- energy, 142

- mechanical, 142
- momentum, 142
- nonholonomic, 146
- symplectic, 142
- variational, 143
-- generalized, 158

kinematic motion, 198
Koiller's question, 114

Lagrange-d'Alembert principle, 53
- discrete, 146
Lagrangian
- hyperregular, 36
- mechanical, 35
- natural, 35
- regular, 36
- singular, 36
Legendre transformation, 36
Lie algebra, 20
Lie bracket, 20
- bad, 196
- component, 196
-- length, 196
- degree, 196
- good, 196
Lie derivative, 16
Lie group, 19
lift
- complete, 37
- horizontal, 24
- vertical, 37
locally accessible at zero velocity (LA), 175
locally configuration accessible (LCA), 175
locally kinematically controllable, 199
locked inertia tensor, 83

manifold, 13
mechanical control system
- simple, 172
metric connection, 26
metric connection tensor, 106
momentum equation, 86
momentum mapping, 31
- CoAd-equivariant, 31
- constrained, 84
- discrete, 144

- nonholonomic, 85
-- discrete, 154

Nijenhuis bracket, 32, 68, 112
Noether's theorem, 64
- nonholonomic, 58
nonholonomic bracket, 57

one-form, 14
- Liouville, 30
- Poincaré-Cartan, 36

phase
- dynamic, 77
- geometric, 77
- total, 77
point
- regular, 17
- singular, 17
Poisson structure
- rank, 33
principal bundle, 23
principal connection, 23
- curvature, 24
- mechanical, 83
Principle of Virtual Work, 52
pullback, 15

Riemannian manifold, 24
Riemannian metric, 24

small-time locally configuration
    controllable (STLCC), 175
small-time locally controllable at zero
    velocity (STLC), 175
solution
- abnormal, 53
- controlled, 198

- singular, see solution, abnormal
space
- fiber, 23
- pose, 23
- shape, 23
structure constants, 20
symmetric product, 60
- bad, 176
- degree, 176
- good, 176
symplectic foliation, see distribution,
    characteristic

tangent bundle, 14
tensor field, 14
tensor product, 14
time scaling, 198
two-form
- Poincaré-Cartan, 36
- symplectic, 29
-- canonical on the cotangent bundle,
    30
-- globally conformal, 116

variation of constants formula, 200
vector field, 14
- decoupling, 198
- flow, 14
- fundamental, 22
- Hamiltonian, 29, 32
-- distributional, 57
- Killing, 25
- Liouville, 35, 173
vertical endomorphism, 35
virtual displacements, 52
virtual variations, 52

wedge product, see exterior product

Lecture Notes in Mathematics

For information about Vols. 1–1619
please contact your bookseller or Springer-Verlag

Vol. 1620: R. Donagi, B. Dubrovin, E. Frenkel, E. Previato, Integrable Systems and Quantum Groups. Montecatini Terme, 1993. Editors:M. Francaviglia, S. Greco. VIII, 488 pages. 1996.

Vol. 1621: H. Bass, M. V. Otero-Espinar, D. N. Rockmore, C. P. L. Tresser, Cyclic Renormalization and Automorphism Groups of Rooted Trees. XXI, 136 pages. 1996.

Vol. 1622: E. D. Farjoun, Cellular Spaces, Null Spaces and Homotopy Localization. XIV, 199 pages. 1996.

Vol. 1623: H.P. Yap, Total Colourings of Graphs. VIII, 131 pages. 1996.

Vol. 1624: V. Brýnzanescu, Holomorphic Vector Bundles over Compact Complex Surfaces. X, 170 pages. 1996.

Vol. 1625: S. Lang, Topics in Cohomology of Groups. VII, 226 pages. 1996.

Vol. 1626: J. Azéma, M. Emery, M. Yor (Eds.), Séminaire de Probabilités XXX. VIII, 382 pages. 1996.

Vol. 1627: C. Graham, Th. G. Kurtz, S. Méléard, Ph. E. Protter, M. Pulvirenti, D. Talay, Probabilistic Models for Nonlinear Partial Differential Equations. Montecatini Terme, 1995. Editors: D. Talay, L. Tubaro. X, 301 pages. 1996.

Vol. 1628: P.-H. Zieschang, An Algebraic Approach to Association Schemes. XII, 189 pages. 1996.

Vol. 1629: J. D. Moore, Lectures on Seiberg-Witten Invariants. VII, 105 pages. 1996.

Vol. 1630: D. Neuenschwander, Probabilities on the Heisenberg Group: Limit Theorems and Brownian Motion. VIII, 139 pages. 1996.

Vol. 1631: K. Nishioka, Mahler Functions and Transcendence. VIII, 185 pages. 1996.

Vol. 1632: A. Kushkuley, Z. Balanov, Geometric Methods in Degree Theory for Equivariant Maps. VII, 136 pages. 1996.

Vol. 1633: H. Aikawa, M. Essén, Potential Theory – Selected Topics. IX, 200 pages. 1996.

Vol. 1634: J. Xu, Flat Covers of Modules. IX, 161 pages. 1996.

Vol. 1635: E. Hebey, Sobolev Spaces on Riemannian Manifolds. X, 116 pages. 1996.

Vol. 1636: M. A. Marshall, Spaces of Orderings and Abstract Real Spectra. VI, 190 pages. 1996.

Vol. 1637: B. Hunt, The Geometry of some special Arithmetic Quotients. XIII, 332 pages. 1996.

Vol. 1638: P. Vanhaecke, Integrable Systems in the realm of Algebraic Geometry. VIII, 218 pages. 1996.

Vol. 1639: K. Dekimpe, Almost-Bieberbach Groups: Affine and Polynomial Structures. X, 259 pages. 1996.

Vol. 1640: G. Boillat, C. M. Dafermos, P. D. Lax, T. P. Liu, Recent Mathematical Methods in Nonlinear Wave Propagation. Montecatini Terme, 1994. Editor: T. Ruggeri. VII, 142 pages. 1996.

Vol. 1641: P. Abramenko, Twin Buildings and Applications to S-Arithmetic Groups. IX, 123 pages. 1996.

Vol. 1642: M. Puschnigg, Asymptotic Cyclic Cohomology. XXII, 138 pages. 1996.

Vol. 1643: J. Richter-Gebert, Realization Spaces of Polytopes. XI, 187 pages. 1996.

Vol. 1644: A. Adler, S. Ramanan, Moduli of Abelian Varieties. VI, 196 pages. 1996.

Vol. 1645: H. W. Broer, G. B. Huitema, M. B. Sevryuk, Quasi-Periodic Motions in Families of Dynamical Systems. XI, 195 pages. 1996.

Vol. 1646: J.-P. Demailly, T. Peternell, G. Tian, A. N. Tyurin, Transcendental Methods in Algebraic Geometry. Cetraro, 1994. Editors: F. Catanese, C. Ciliberto. VII, 257 pages. 1996.

Vol. 1647: D. Dias, P. Le Barz, Configuration Spaces over Hilbert Schemes and Applications. VII. 143 pages. 1996.

Vol. 1648: R. Dobrushin, P. Groeneboom, M. Ledoux, Lectures on Probability Theory and Statistics. Editor: P. Bernard. VIII, 300 pages. 1996.

Vol. 1649: S. Kumar, G. Laumon, U. Stuhler, Vector Bundles on Curves – New Directions. Cetraro, 1995. Editor: M. S. Narasimhan. VII, 193 pages. 1997.

Vol. 1650: J. Wildeshaus, Realizations of Polylogarithms. XI, 343 pages. 1997.

Vol. 1651: M. Drmota, R. F. Tichy, Sequences, Discrepancies and Applications. XIII, 503 pages. 1997.

Vol. 1652: S. Todorcevic, Topics in Topology. VIII, 153 pages. 1997.

Vol. 1653: R. Benedetti, C. Petronio, Branched Standard Spines of 3-manifolds. VIII, 132 pages. 1997.

Vol. 1654: R. W. Ghrist, P. J. Holmes, M. C. Sullivan, Knots and Links in Three-Dimensional Flows. X, 208 pages. 1997.

Vol. 1655: J. Azéma, M. Emery, M. Yor (Eds.), Séminaire de Probabilités XXXI. VIII, 329 pages. 1997.

Vol. 1656: B. Biais, T. Björk, J. Cvitanic, N. El Karoui, E. Jouini, J. C. Rochet, Financial Mathematics. Bressanone, 1996. Editor: W. J. Runggaldier. VII, 316 pages. 1997.

Vol. 1657: H. Reimann, The semi-simple zeta function of quaternionic Shimura varieties. IX, 143 pages. 1997.

Vol. 1658: A. Pumarino, J. A. Rodríguez, Coexistence and Persistence of Strange Attractors. VIII, 195 pages. 1997.

Vol. 1659: V, Kozlov, V. Maz'ya, Theory of a Higher-Order Sturm-Liouville Equation. XI, 140 pages. 1997.

Vol. 1660: M. Bardi, M. G. Crandall, L. C. Evans, H. M. Soner, P. E. Souganidis, Viscosity Solutions and Applications. Montecatini Terme, 1995. Editors: I. Capuzzo Dolcetta, P. L. Lions. IX, 259 pages. 1997.

Vol. 1661: A. Tralle, J. Oprea, Symplectic Manifolds with no Kähler Structure. VIII, 207 pages. 1997.

Vol. 1662: J. W. Rutter, Spaces of Homotopy Self-Equivalences – A Survey. IX, 170 pages. 1997.

Vol. 1663: Y. E. Karpeshina; Perturbation Theory for the Schrödinger Operator with a Periodic Potential. VII, 352 pages. 1997.

Vol. 1664: M. Väth, Ideal Spaces. V, 146 pages. 1997.

Vol. 1665: E. Giné, G. R. Grimmett, L. Saloff-Coste, Lectures on Probability Theory and Statistics 1996. Editor: P. Bernard. X, 424 pages, 1997.

Vol. 1666: M. van der Put, M. F. Singer, Galois Theory of Difference Equations. VII, 179 pages. 1997.

Vol. 1667: J. M. F. Castillo, M. González, Three-space Problems in Banach Space Theory. XII, 267 pages. 1997.

Vol. 1668: D. B. Dix, Large-Time Behavior of Solutions of Linear Dispersive Equations. XIV, 203 pages. 1997.

Vol. 1669: U. Kaiser, Link Theory in Manifolds. XIV, 167 pages. 1997.

Vol. 1670: J. W. Neuberger, Sobolev Gradients and Differential Equations. VIII, 150 pages. 1997.

Vol. 1671: S. Bouc, Green Functors and G-sets. VII, 342 pages. 1997.

Vol. 1672: S. Mandal, Projective Modules and Complete Intersections. VIII, 114 pages. 1997.

Vol. 1673: F. D. Grosshans, Algebraic Homogeneous Spaces and Invariant Theory. VI, 148 pages. 1997.

Vol. 1674: G. Klaas, C. R. Leedham-Green, W. Plesken, Linear Pro-p-Groups of Finite Width. VIII, 115 pages. 1997.

Vol. 1675: J. E. Yukich, Probability Theory of Classical Euclidean Optimization Problems. X, 152 pages. 1998.

Vol. 1676: P. Cembranos, J. Mendoza, Banach Spaces of Vector-Valued Functions. VIII, 118 pages. 1997.

Vol. 1677: N. Proskurin, Cubic Metaplectic Forms and Theta Functions. VIII, 196 pages. 1998.

Vol. 1678: O. Krupková, The Geometry of Ordinary Variational Equations. X, 251 pages. 1997.

Vol. 1679: K.-G. Grosse-Erdmann, The Blocking Technique. Weighted Mean Operators and Hardy's Inequality. IX, 114 pages. 1998.

Vol. 1680: K.-Z. Li, F. Oort, Moduli of Supersingular Abelian Varieties. V, 116 pages. 1998.

Vol. 1681: G. J. Wirsching, The Dynamical System Generated by the 3n+1 Function. VII, 158 pages. 1998.

Vol. 1682: H.-D. Alber, Materials with Memory. X, 166 pages. 1998.

Vol. 1683: A. Pomp, The Boundary-Domain Integral Method for Elliptic Systems. XVI, 163 pages. 1998.

Vol. 1684: C. A. Berenstein, P. F. Ebenfelt, S. G. Gindikin, S. Helgason, A. E. Tumanov, Integral Geometry, Radon Transforms and Complex Analysis. Firenze, 1996. Editors: E. Casadio Tarabusi, M. A. Picardello, G. Zampieri. VII, 160 pages. 1998.

Vol. 1685: S. König, A. Zimmermann, Derived Equivalences for Group Rings. X, 146 pages. 1998.

Vol. 1686: J. Azéma, M. Émery, M. Ledoux, M. Yor (Eds.), Séminaire de Probabilités XXXII. VI, 440 pages. 1998.

Vol. 1687: F. Bornemann, Homogenization in Time of Singularly Perturbed Mechanical Systems. XII, 156 pages. 1998.

Vol. 1688: S. Assing, W. Schmidt, Continuous Strong Markov Processes in Dimension One. XII, 137 page. 1998.

Vol. 1689: W. Fulton, P. Pragacz, Schubert Varieties and Degeneracy Loci. XI, 148 pages. 1998.

Vol. 1690: M. T. Barlow, D. Nualart, Lectures on Probability Theory and Statistics. Editor: P. Bernard. VIII, 237 pages. 1998.

Vol. 1691: R. Bezrukavnikov, M. Finkelberg, V. Schechtman, Factorizable Sheaves and Quantum Groups. X, 282 pages. 1998.

Vol. 1692: T. M. W. Eyre, Quantum Stochastic Calculus and Representations of Lie Superalgebras. IX, 138 pages. 1998.

Vol. 1694: A. Braides, Approximation of Free-Discontinuity Problems. XI, 149 pages. 1998.

Vol. 1695: D. J. Hartfiel, Markov Set-Chains. VIII, 131 pages. 1998.

Vol. 1696: E. Bouscaren (Ed.): Model Theory and Algebraic Geometry. XV, 211 pages. 1998.

Vol. 1697: B. Cockburn, C. Johnson, C.-W. Shu, E. Tadmor, Advanced Numerical Approximation of Nonlinear Hyperbolic Equations. Cetraro, Italy, 1997. Editor: A. Quarteroni. VII, 390 pages. 1998.

Vol. 1698: M. Bhattacharjee, D. Macpherson, R. G. Möller, P. Neumann, Notes on Infinite Permutation Groups. XI, 202 pages. 1998.

Vol. 1699: A. Inoue, Tomita-Takesaki Theory in Algebras of Unbounded Operators. VIII, 241 pages. 1998.

Vol. 1700: W. A. Woyczyński, Burgers-KPZ Turbulence, XI, 318 pages. 1998.

Vol. 1701: Ti-Jun Xiao, J. Liang, The Cauchy Problem of Higher Order Abstract Differential Equations, XII, 302 pages. 1998.

Vol. 1702: J. Ma, J. Yong, Forward-Backward Stochastic Differential Equations and Their Applications. XIII, 270 pages. 1999.

Vol. 1703: R. M. Dudley, R. Norvaiša, Differentiability of Six Operators on Nonsmooth Functions and p-Variation. VIII, 272 pages. 1999.

Vol. 1704: H. Tamanoi, Elliptic Genera and Vertex Operator Super-Algebras. VI, 390 pages. 1999.

Vol. 1705: I. Nikolaev, E. Zhuzhoma, Flows in 2-dimensional Manifolds. XIX, 294 pages. 1999.

Vol. 1706: S. Yu. Pilyugin, Shadowing in Dynamical Systems. XVII, 271 pages. 1999.

Vol. 1707: R. Pytlak, Numerical Methods for Optimal Control Problems with State Constraints. XV, 215 pages. 1999.

Vol. 1708: K. Zuo, Representations of Fundamental Groups of Algebraic Varieties. VII, 139 pages. 1999.

Vol. 1709: J. Azéma, M. Émery, M. Ledoux, M. Yor (Eds), Séminaire de Probabilités XXXIII. VIII, 418 pages. 1999.

Vol. 1710: M. Koecher, The Minnesota Notes on Jordan Algebras and Their Applications. IX, 173 pages. 1999.

Vol. 1711: W. Ricker, Operator Algebras Generated by Commuting Projéctions: A Vector Measure Approach. XVII, 159 pages. 1999.

Vol. 1712: N. Schwartz, J. J. Madden, Semi-algebraic Function Rings and Reflectors of Partially Ordered Rings. XI, 279 pages. 1999.

Vol. 1713: F. Bethuel, G. Huisken, S. Müller, K. Steffen, Calculus of Variations and Geometric Evolution Problems. Cetraro, 1996. Editors: S. Hildebrandt, M. Struwe. VII, 293 pages. 1999.

Vol. 1714: O. Diekmann, R. Durrett, K. P. Hadeler, P. K. Maini, H. L. Smith, Mathematics Inspired by Biology. Martina Franca, 1997. Editors: V. Capasso, O. Diekmann. VII, 268 pages. 1999.

Vol. 1715: N. V. Krylov, M. Röckner, J. Zabczyk, Stochastic PDE's and Kolmogorov Equations in Infinite Dimensions. Cetraro, 1998. Editor: G. Da Prato. VIII, 239 pages. 1999.

Vol. 1716: J. Coates, R. Greenberg, K. A. Ribet, K. Rubin, Arithmetic Theory of Elliptic Curves. Cetraro, 1997. Editor: C. Viola. VIII, 260 pages. 1999.

Vol. 1717: J. Bertoin, F. Martinelli, Y. Peres, Lectures on Probability Theory and Statistics. Saint-Flour, 1997. Editor: P. Bernard. IX, 291 pages. 1999.

Vol. 1718: A. Eberle, Uniqueness and Non-Uniqueness of Semigroups Generated by Singular Diffusion Operators. VIII, 262 pages. 1999.

Vol. 1719: K. R. Meyer, Periodic Solutions of the N-Body Problem. IX, 144 pages. 1999.

Vol. 1720: D. Elworthy, Y. Le Jan, X-M. Li, On the Geometry of Diffusion Operators and Stochastic Flows. IV, 118 pages. 1999.

Vol. 1721: A. Iarrobino, V. Kanev, Power Sums, Gorenstein Algebras, and Determinantal Loci. XXVII, 345 pages. 1999.

Vol. 1722: R. McCutcheon, Elemental Methods in Ergodic Ramsey Theory. VI, 160 pages. 1999.

Vol. 1723: J. P. Croisille, C. Lebeau, Diffraction by an Immersed Elastic Wedge. VI, 134 pages. 1999.

Vol. 1724: V. N. Kolokoltsov, Semiclassical Analysis for Diffusions and Stochastic Processes. VIII, 347 pages. 2000.

Vol. 1725: D. A. Wolf-Gladrow, Lattice-Gas Cellular Automata and Lattice Boltzmann Models. IX, 308 pages. 2000.

Vol. 1726: V. Marić, Regular Variation and Differential Equations. X, 127 pages. 2000.

Vol. 1727: P. Kravanja M. Van Barel, Computing the Zeros of Analytic Functions. VII, 111 pages. 2000.

Vol. 1728: K. Gatermann Computer Algebra Methods for Equivariant Dynamical Systems. XV, 153 pages. 2000.

Vol. 1729: J. Azéma, M. Émery, M. Ledoux, M. Yor Séminaire de Probabilités XXXIV. VI, 431 pages. 2000.

Vol. 1730: S. Graf, H. Luschgy, Foundations of Quantization for Probability Distributions. X, 230 pages. 2000.

Vol. 1731: T. Hsu, Quilts: Central Extensions, Braid Actions, and Finite Groups. XII, 185 pages. 2000.

Vol. 1732: K. Keller, Invariant Factors, Julia Equivalences and the (Abstract) Mandelbrot Set. X, 206 pages. 2000.

Vol. 1733: K. Ritter, Average-Case Analysis of Numerical Problems. IX, 254 pages. 2000.

Vol. 1734: M. Espedal, A. Fasano, A. Mikelić, Filtration in Porous Media and Industrial Applications. Cetraro 1998. Editor: A. Fasano. 2000.

Vol. 1735: D. Yafaev, Scattering Theory: Some Old and New Problems. XVI, 169 pages. 2000.

Vol. 1736: B. O. Turesson, Nonlinear Potential Theory and Weighted Sobolev Spaces. XIV, 173 pages. 2000.

Vol. 1737: S. Wakabayashi, Classical Microlocal Analysis in the Space of Hyperfunctions. VIII, 367 pages. 2000.

Vol. 1738: M. Émery, A. Nemirovski, D. Voiculescu, Lectures on Probability Theory and Statistics. XI, 356 pages. 2000.

Vol. 1739: R. Burkard, P. Deuflhard, A. Jameson, J.-L. Lions, G. Strang, Computational Mathematics Driven by Industrial Problems. Martina Franca, 1999. Editors: V. Capasso, H. Engl, J. Periaux. VII, 418 pages. 2000.

Vol. 1740: B. Kawohl, O. Pironneau, L. Tartar, J.-P. Zolesio, Optimal Shape Design. Tróia, Portugal 1999. Editors: A. Cellina, A. Ornelas. IX, 388 pages. 2000.

Vol. 1741: E. Lombardi, Oscillatory Integrals and Phenomena Beyond all Algebraic Orders. XV, 413 pages. 2000.

Vol. 1742: A. Unterberger, Quantization and Non-holomorphic Modular Forms. VIII, 253 pages. 2000.

Vol. 1743: L. Habermann, Riemannian Metrics of Constant Mass and Moduli Spaces of Conformal Structures. XII, 116 pages. 2000.

Vol. 1744: M. Kunze, Non-Smooth Dynamical Systems. X, 228 pages. 2000.

Vol. 1745: V. D. Milman, G. Schechtman, Geometric Aspects of Functional Analysis. VIII, 289 pages. 2000.

Vol. 1746: A. Degtyarev, I. Itenberg, V. Kharlamov, Real Enriques Surfaces. XVI, 259 pages. 2000.

Vol. 1747: L. W. Christensen, Gorenstein Dimensions. VIII, 204 pages. 2000.

Vol. 1748: M. Ruzicka, Electrorheological Fluids: Modeling and Mathematical Theory. XV, 176 pages. 2001.

Vol. 1749: M. Fuchs, G. Seregin, Variational Methods for Problems from Plasticity Theory and for Generalized Newtonian Fluids. VI, 269 pages. 2001.

Vol. 1750: B. Conrad, Grothendieck Duality and Base Change. X, 296 pages. 2001.

Vol. 1751: N. J. Cutland, Loeb Measures in Practice: Recent Advances. XI, 111 pages. 2001.

Vol. 1752: Y. V. Nesterenko, P. Philippon, Introduction to Algebraic Independence Theory. XIII, 256 pages. 2001.

Vol. 1753: A. I. Bobenko, U. Eitner, Painlevé Equations in the Differential Geometry of Surfaces. VI, 120 pages. 2001.

Vol. 1754: W. Bertram, The Geometry of Jordan and Lie Structures. XVI, 269 pages. 2001.

Vol. 1755: J. Azéma, M. Émery, M. Ledoux, M. Yor, Séminaire de Probabilités XXXV. VI, 427 pages. 2001.

Vol. 1756: P. E. Zhidkov, Korteweg de Vries and Nonlinear Schrödinger Equations: Qualitative Theory. VII, 147 pages. 2001.

Vol. 1757: R. R. Phelps, Lectures on Choquet's Theorem. VII, 124 pages. 2001.

Vol. 1758: N. Monod, Continuous Bounded Cohomology of Locally Compact Groups. X, 214 pages. 2001.

Vol. 1759: Y. Abe, K. Kopfermann, Toroidal Groups. VIII, 133 pages. 2001.

Vol. 1760: D. Pilipović, Consistency Problems for Heath-Jarrow-Morton Interest Rate Models. VIII, 134 pages. 2001.

Vol. 1761: C. Adelmann, The Decomposition of Primes in Torsion Point Fields. VI, 142 pages. 2001.

Vol. 1762: S. Cerrai, Second Order PDE's in Finite and Infinite Dimension. IX, 330 pages. 2001.

Vol. 1763: J.-L. Loday, A. Frabetti, F. Chapoton, F. Goichot, Dialgebras and Related Operads. IV, 132 pages. 2001.

Vol. 1764: A. Cannas da Silva, Lectures on Symplectic Geometry. XII, 217 pages. 2001.

Vol. 1765: T. Kerler, V. V. Lyubashenko, Non-Semisimple Topological Quantum Field Theories for 3-Manifolds with Corners. VI, 379 pages. 2001.

Vol. 1766: H. Hennion, L. Hervé, Limit Theorems for Markov Chains and Stochastic Properties of Dynamical Systems by Quasi-Compactness. VIII, 145 pages. 2001.

Vol. 1767: J. Xiao, Holomorphic Q Classes. VIII, 112 pages. 2001.

Vol. 1768: M.J. Pflaum, Analytic and Geometric Study of Stratified Spaces. VIII, 230 pages. 2001.

Vol. 1769: M. Alberich-Carramiñana, Geometry of the Plane Cremona Maps. XVI, 257 pages. 2002.

Vol. 1770: H. Gluesing-Luerssen, Linear Delay-Differential Systems with Commensurate Delays: An Algebraic Approach. VIII, 176 pages. 2002.

Vol. 1771: M. Émery, M. Yor, Séminaire de Probabilités 1967-1980. A Selection in Martingale Theory. IX, 553 pages. 2002.

Vol. 1772: F. Burstall, D. Ferus, K. Leschke, F. Pedit, U. Pinkall, Conformal Geometry of Surfaces in S^4. VII, 89 pages. 2002.

Vol. 1773: Z. Arad, M. Muzychuk, Standard Integral Table Algebras Generated by a Non-real Element of Small Degree. X, 126 pages. 2002.

Vol. 1774: V. Runde, Lectures on Amenability. XIV, 296 pages. 2002.

Vol. 1775: W. H. Meeks, A. Ros, H. Rosenberg, The Global Theory of Minimal Surfaces in Flat Spaces. Martina Franca 1999. Editor: G. P. Pirola. X, 117 pages. 2002.

Vol. 1776: K. Behrend, C. Gomez, V. Tarasov, G. Tian, Quantum Comohology. Cetraro 1997. Editors: P. de Bartolomeis, B. Dubrovin, C. Reina. VIII, 319 pages. 2002.

Vol. 1777: E. García-Río, D. N. Kupeli, R. Vázquez-Lorenzo, Osserman Manifolds in Semi-Riemannian Geometry. XII, 166 pages. 2002.

Vol. 1778: H. Kiechle, Theory of K-Loops. X, 186 pages. 2002.

Vol. 1779: I. Chueshov, Monotone Random Systems. VIII, 234 pages. 2002.

Vol. 1780: J. H. Bruinier, Borcherds Products on O(2,1) and Chern Classes of Heegner Divisors. VIII, 152 pages. 2002.

Vol. 1781: E. Bolthausen, E. Perkins, A. van der Vaart, Lectures on Probability Theory and Statistics. Ecole d' Eté de Probabilités de Saint-Flour XXIX-1999. Editor: P. Bernard. VIII, 466 pages. 2002.

Vol. 1782: C.-H. Chu, A. T.-M. Lau, Harmonic Functions on Groups and Fourier Algebras. VII, 100 pages. 2002.

Vol. 1783: L. Grüne, Asymptotic Behavior of Dynamical and Control Systems under Perturbation and Discretization. IX, 231 pages. 2002.

Vol. 1784: L.H. Eliasson, S. B. Kuksin, S. Marmi, J.-C. Yoccoz, Dynamical Systems and Small Divisors. Cetraro, Italy 1998. Editors: S. Marmi, J.-C. Yoccoz. VIII, 199 pages. 2002.

Vol. 1785: J. Arias de Reyna, Pointwise Convergence of Fourier Series. XVIII, 175 pages. 2002.

Vol. 1786: S. D. Cutkosky, Monomialization of Morphisms from 3-Folds to Surfaces. V, 235 pages. 2002.

Vol. 1787: S. Caenepeel, G. Militaru, S. Zhu, Frobenius and Separable Functors for Generalized Module Categories and Nonlinear Equations. XIV, 354 pages. 2002.

Vol. 1788: A. Vasil'ev, Moduli of Families of Curves for Conformal and Quasiconformal Mappings.IX, 211 pages. 2002.

Vol. 1789: Y. Sommerhäuser, Yetter-Drinfel'd Hopf algebras over groups of prime order. V, 157 pages. 2002.

Vol. 1790: X. Zhan, Matrix Inequalities. VII, 116 pages. 2002.

Vol. 1791: M. Knebusch, D. Zhang, Manis Valuations and Prüfer Extensions I: A new Chapter in Commutative Algebra. VI, 267 pages. 2002.

Vol. 1792: D. D. Ang, R. Gorenflo, V. K. Le, D. D. Trong, Moment Theory and Some Inverse Problems in Potential Theory and Heat Conduction. VIII, 183 pages. 2002.

Vol. 1793: J. Cortés Monforte, Geometric, Control and Numerical Aspects of Nonholonomic Systems. XV, 219 pages. 2002.

Vol. 1794: N. Pytheas Fogg, Substitution in Dynamics, Arithmetics and Combinatorics. Editors: V. Berthé, S. Ferenczi, C. Mauduit, A. Siegel. XVII, 402 pages. 2002.

Vol. 1795: H. Li, Filtered-Graded Transfer in Using Noncommutative Gröbner Bases. X, 197 pages. 2002.

Vol. 1796: J.M. Melenk, hp-Finite Element Methods for Singular Perturbations. XIV, 318 pages. 2002.

Vol. 1797: B. Schmidt, Characters and Cyclotomic Fields in Finite Geometry. VIII, 100 pages. 2002.

Vol. 1798: W.M. Oliva, Geometric Mechanics. XI, 270 pages. 2002.

Recent Reprints and New Editions

Vol. 1200: V. D. Milman, G. Schechtman, Asymptotic Theory of Finite Dimensional Normed Spaces. 1986. – Corrected Second Printing. X, 156 pages. 2001.

Vol. 1618: G. Pisier, Similarity Problems and Completely Bounded Maps. 1995 – Second, Expanded Edition VII, 198 pages. 2001.

Vol. 1629: J. D. Moore, Lectures on Seiberg-Witten Invariants. 1997 – Second Edition. VIII, 121 pages. 2001.

Vol. 1638: P. Vanhaecke, Integrable Systems in the realm of Algebraic Geometry. 1996 – Second Edition. X, 256 pages. 2001.

Vol. 1702: J. Ma, J. Yong, Forward-Backward Stochastic Differential Equations and Their Applications. 1999. – Corrected Second Printing. XIII, 270 pages. 2000.